FORTRAN FOR ENGINEERING

FORTRAN FOR ENGINEERING

WILLIAM SHICK

Professor of Electrical Engineering
Assistant Dean, College of Science and Engineering
Fairleigh Dickinson University

CHARLES J. MERZ, JR.

Professor and Chairman of the Department
of Mechanical Engineering
Fairleigh Dickinson University

McGRAW-HILL BOOK COMPANY

New York St. Louis San Francisco
Düsseldorf Johannesburg Kuala Lumpur
London Mexico Montreal
New Delhi Panama Rio de Janeiro
Singapore Sydney Toronto

FORTRAN FOR ENGINEERING

Library of Congress Catalog Card Number 79-172665

07-055276-2

567890BABA798765

This book was set in Press Roman, and was printed and bound by George Banta Company, Inc. The drawings were done by Oxford Illustrated Limited. The editors were B. J. Clark and M. E. Margolies. John A. Sabella supervised production.

To Jessie and Ann

Contents

Preface

Much has been written concerning the profound influence of the computer on industry, medicine, research, and education. Probably, many of the projections that have been made will fall short. In particular, the authors believe that the uses of the computer in the educational process have not yet been fully explored. In the university at which the authors teach, the growing importance of the computer has been recognized in two ways. In the first place, all students of science and engineering are required to take a course in the FORTRAN language. Secondly, the computer is utilized in junior and senior courses to facilitate the learning of various topics, to carry out calculations which would otherwise be difficult or unfeasible, and as a tool of engineering analysis and design.

This text is an outgrowth of the authors' participation in these educational activities, and is primarily intended to aid engineering students in learning how to write FORTRAN programs effectively. In the opinion of the authors, the treatment is sufficiently detailed to enable practicing engineers to master the text by self-study. The use of topics from electrical, mechanical, chemical, and civil engineering is a recognition of the interdisciplinary nature of modern engineering and the trend in engineering education.

Learning proceeds most rapidly when the subject of programming is logically divided into a number of relatively restricted topics treated in reasonable detail. The organization of the book is based on this philosophy. For example, Chapter 3 is limited to a discussion of the computer as a calculator, and Chapter 4 involves only transfer statements, such as the IF statement and the computed GO TO statement.

Many of the techniques discussed in the book are applied to engineering problems in the spirit of the COSINE Committee's admonition that "... motivation is an important factor in deciding what to teach and how to teach it; the student must be convinced that he is involved in meaningful studies leading to real-world application of the computer to engineering."

An attempt has been made to implement this recommendation of the COSINE Committee. For this reason, a majority of the computer programs presented in this book concern topics of current importance in science and engineering. In some cases, problems are solved in more than one way. In other cases, smaller programs are extended to form more general programs, or are incorporated in larger programs, because there is as much educational value in showing the development of a program as there is in displaying it "full-blown."

Each program is accompanied by a brief discussion of the relevant engineering theory, explanation of the variable names, specifications of the information required by the program, and a statement of the results provided by the program. All the programs are accompanied by at least one trial run. Flow charts are presented wherever their use significantly aids understanding. The range of subject matter used to illustrate programming techniques combined with the diversity of the problems at the end of each chapter is broad enough to allow instructors considerable flexibility in the selection of topics.

In conclusion, we would like to make a few comments about "software." We have not included a list of computer types, nor have we included tables of the capabilities of various compilers. Such tables are inadequate substitutes for a user's manual. The programmer should familiarize himself with the distinctive features of the computer he is using, and this knowledge is best acquired by reading the user's manual.

The authors would like to express their thanks to Mrs. Lee Bugiada, Mrs. Margaret Gorman, and Mrs. Helen Stanek for their assistance in the preparation of the manuscript. The authors also wish to express their appreciation for the painstaking and constructive criticism of Mr. Eric A. Weiss, Corporate Computer Sciences Consultant for the Sun Oil Company.

WILLIAM SCHICK
CHARLES J. MERZ, JR.

A Suggestion About The Use of This Book

In planning this book the authors hoped to organize the material to serve the needs of the reader who is a newcomer to FORTRAN. The beginner needs simple illustrations at the outset, followed by material of gradually increasing complexity as more and more of the vocabulary and grammar of the language are introduced. Later, he needs practice in the application of his skill as well as continual reinforcement of his knowledge by frequent reference to fundamentals.

To serve as a guide to the relative difficulty of the programming techniques discussed here, the following general grouping of the chapters is suggested:

Fundamentals	Chapters 1 and 2
Elementary Programming	Chapters 3–7
Intermediate Programming	Chapters 8–10
Advanced Programming	Chapters 11 and 12

The rules of grammar of the FORTRAN language have been presented in this book at the places where their discussion seemed natural or logical; however, the reader may benefit from an overview of the language, arranged according to types of statements or categories of operations. Accordingly, in the following outline, articles concerned with the language, *per se*, have been grouped under topical headings. This summary can be used by the reader to locate articles that touch on related material, and thereby expand his knowledge in a particular area.

PROGRAMMING

DATA REPRESENTATION

FORTRAN ARITHMETIC

FORTRAN LOGIC

PROGRAM CONTROLS

EVALUATION OF VARIABLES

UNCONDITIONAL TRANSFERS

CONDITIONAL TRANSFERS

ITERATION

ARRAYS

DECLARATIONS

DATA INPUT AND OUTPUT

SPECIFICATIONS

SUBPROGRAMS

1
Fundamentals of the FORTRAN Language

In this chapter we will discuss the essential nomenclature of the FORTRAN language and some of the principles of programming that make it possible to perform arithmetic calculations using that language. Simple illustrations will be used to clarify the rules. The major emphasis will not be on programming techniques but on definitions.

1.1 COMPUTER PROGRAMS

All calculations are an ordered sequence of arithmetic and logic operations. To carry out a calculation, either a man or a machine must be instructed to perform the appropriate operations at the proper time. The set of instructions is called a *program* and can be prepared as a list of instructions on a sheet of paper, a properly arranged deck of cards with one instruction on each card, a punched paper tape, etc.

If the program is to be executed by a machine (a computer), the instructions must be written in *machine language.* Machine language is a set of numerical codes, each of which is interpreted by the computer as a *command* to carry out a specific operation the machine has been designed to perform. A machine-language command also designates the *locations* or *storage addresses* (within the computer) of the numbers to be operated on.

The *object program* is a program written in machine language and stored in the *memory* or *storage* of the computer. After each operation has been completed, the computer automatically refers to the stored object program to determine the next operation.

A *general-purpose computer* is a machine that can store, on separate occasions, different object programs and thereby carry out different sets of instructions.

1.2 FORTRAN LANGUAGE

The FORTRAN language is one of a group of *procedure-oriented* computer languages that are based on the structure of the original language in which a problem is expressed and on the particular procedure employed in solving that problem.

The procedure-oriented languages are intended to relieve the human programmer of the chore of reducing all computation and decision making to the elementary steps required by the limited repertoire of operations available at the machine-language level. In addition, for procedure-oriented languages, the computer assumes the clerical task of keeping track of the contents of all locations in the computer storage.

FORTRAN is an acronym for *formula translation* and is the name of a procedure-oriented language designed primarily for handling scientific formulas and applying numerical methods to the solution of problems.

The FORTRAN language is a formal set of rules that makes it possible to prepare a set of computer instructions called a *source program.* The language consists of *code words* and *symbols* that can be interpreted by the computer and transformed into machine language.

At the present writing, the FORTRAN language has been standardized in two versions:

1. Standard FORTRAN—the most powerful version of the language; follows the rules stated in American National Standard Institute (ANSI) X3.9-1966 (commonly called FORTRAN IV).
2. Basic FORTRAN—a subset of standard FORTRAN for use in smaller computers; follows ANSI X3.10-1966.

The majority of the programs in this book are written in standard FORTRAN and were processed on a time-sharing system[1] using as a remote access terminal a standard teletype unit fitted with a tape punch and tape reader. A simplified means of communicating with the computer (i.e., *reading-in* data and *reading-out* results) was employed in most cases.

1.3 FORTRAN COMPILER

The FORTRAN compiler is a computer program that translates the instructions of the source program into machine language and produces an object program.

During *compilation*, the compiler program is stored in the computer. The computer than *reads* the statements of the source program, *interprets* the code words and symbols, and *generates* a set of machine-language commands that will produce the desired action in the computer. The object program produced in this manner is stored in the computer and used to perform the desired calculations.

Figure 1.3.1 shows the steps involved in running a FORTRAN program. For simplicity, the sketches show the source program, object program, and program data as being prepared on punched cards. The reader should realize that this is only symbolic and that actually *any* external storage medium (e.g., paper tape, magnetic tape, magnetic disk, etc.) can be used.

In addition to its translating function, the compiler is programmed to *detect* and *diagnose errors* in the syntax of the source program. When an error has been found, the computer produces an *error message* which usually classifies the error and gives its

[1] J. K. Rice and J. R. Rice, "Introduction to Computer Science," Holt, New York, 1969, p. 144.

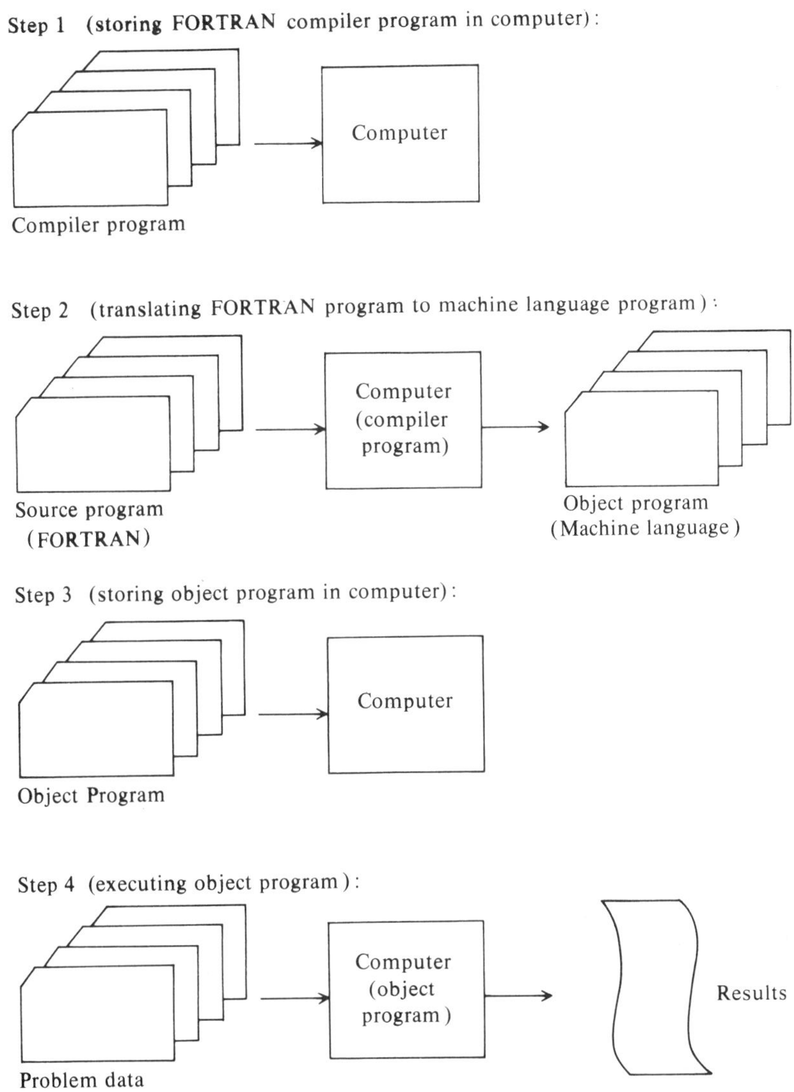

Fig. 1.3.1 Compilation and execution of a FORTRAN program.

location in the source program. The reader should note that this error detection is limited to the "grammar" of the language and does *not* locate errors in the *logic* of the programmed procedure, which is the responsibility of the programmer.

The FORTRAN compiler must necessarily be programmed in machine language, and therefore each computer will have a unique compiler. This individuality of compilers is furthered by the competitive efforts of the various computer manufacturers to provide better diagnostic routines and special relaxations of the FORTRAN syntax in their compiler. Because of these differences in operation of compilers, FORTRAN is no longer the machine-independent language it was originally intended to be and actually exists in many "dialects." A FORTRAN source program written for a particular computer and successfully compiled on that computer may not work on another machine. The discussion in this book is intended to be as general as possible. However, the programmer should check the FORTRAN manual for the particular computer to be used to determine if any special restrictions must be applied or if any advantageous features are available.

1.4 NUMBERING SYSTEMS

A *numbering system* is a means of recording the *total* obtained during a *counting process.* Basically, a *number* is described by a *string of symbols* in which each symbol represents the size of a subtotal obtained in the counting process.

The size of each *term* or *subtotal* is indicated by a combination of the *value* represented by a particular symbol and the *position* of the symbol in the string. This latter component of the value is called the *place value* of the term. The place value of each position in the string depends on the *base* or *radix* of the numbering system and is equal to some preassigned *power* of the base. The value of a *term* in a string is the *product* of the value of the symbol used and the place value of the position occupied by the symbol. The value of a *number* represented by a string of symbols is the *sum* of the values of all the terms.

For example, a number represented by the string of symbols UVW, in the base B, would have the value

$$\text{Value} = (U \times B^{PU}) + (V \times B^{PV}) + (W \times B^{PW})$$

where PU, PV, and PW (the exponents of the base) would have values assigned to them according to the positions of their respective symbols in the string.

In all present numbering systems the exponents applied to the base are decimal integers starting with *zero* for the position immediately to the left of the *system point* and *increasing* by one for each successive position to the left. Fractional numbers are represented in the usual manner, but the exponent is *minus one* for the position immediately to the right of the system point and *decreases* by one for successive positions to the right.

To illustrate how this procedure works, let us first consider the *decimal system.* Any of the 10 familiar digits of the decimal system can be used in a string of symbols to represent a number. Consider the decimal number 2608. To indicate that this number is represented in the base 10, or decimal system, we will write it as $2608._{(10)}$ and, following the rules outlined above, can expand it as

$$\begin{aligned} 2608._{(10)} &= (2 \times 10^3) + (6 \times 10^2) + (0 \times 10^1) + (8 \times 10^0) \\ &= \quad 2000 \quad + \quad 600 \quad + \quad 0 \quad + \quad 8 \end{aligned}$$

If the number were $26.08_{(10)}$, we would expand it as

$$\begin{aligned} 26.08_{(10)} &= (2 \times 10^1) + (6 \times 10^0) + (0 \times 10^{-1}) + (8 \times 10^{-2}) \\ &= \quad 20 \quad + \quad 6 \quad + \quad 0 \quad + \quad 0.08 \end{aligned}$$

In the *binary system* only two digits, 0 and 1, are used. Let us examine the binary number $1011._{(2)}$ and determine its value in the decimal system. (Again the subscript in parentheses indicates the base of the system.)

$$\begin{aligned} 1011._{(2)} &= (1 \times 2^3) + (0 \times 2^2) + (1 \times 2^1) + (1 \times 2^0) \\ &= \quad 8 \quad + \quad 0 \quad + \quad 2 \quad + \quad 1 \\ &= 11._{(10)} \end{aligned}$$

If the number were $10.11_{(2)}$, it would be evaluated as

$$\begin{aligned} 10.11_{(2)} &= (1 \times 2^1) + (0 \times 2^0) + (1 \times 2^{-1}) + (1 \times 2^{-2}) \\ &= \quad 2 \quad + \quad 0 \quad + \quad 0.5 \quad + \quad 0.25 \\ &= 2.75_{(10)} \end{aligned}$$

1.5 STORAGE OF DATA

Because the storage area in electronic digital computers employs devices which may exist in only two possible states, the binary number system (which requires only two digits, as noted above) is ideally suited for representing numbers. The *binary digits*, 0 and 1, are commonly referred to as *bits*.

If a computer is designed to store or operate on number systems other than the binary system discussed in Sec. 1.4, a method of representing each symbol with a special *binary code* is employed. The code consists of *unique groups* of binary digits, *one group* for *each symbol* in the number system. One such system for coding decimal digits is as follows.

Decimal digit	*Coding group*
0	0000
1	0001
2	0010
3	0011
4	0100
5	0101
6	0110
7	0111
8	1000
9	1001

To demonstrate the difference between using a coding scheme and using the binary number system, let us consider two examples. First, let us take the decimal number $149._{(10)}$:

1. Using the binary coding system tabulated above, we have

$$149._{(10)} = \underbrace{0001}_{1}\underbrace{0100}_{4}\underbrace{1001}_{9}$$

2. Using the binary number system directly, we have

$$\begin{aligned} 149._{(10)} &= 10010101_{(2)} \\ &= (1 \times 2^7) + (0 \times 2^6) + (0 \times 2^5) + (1 \times 2^4) + (0 \times 2^3) \\ &= \quad 128 \quad + \quad 0 \quad + \quad 0 \quad + \quad 16 \quad + \quad 0 \\ &\qquad + (1 \times 2^2) + (0 \times 2^1) + (1 \times 2^0) \\ &\qquad + \quad 4 \quad + \quad 0 \quad + \quad 1 \end{aligned}$$

Second, consider the decimal number $38._{(10)}$:

1. Using the binary coding system, we have

$$38._{(10)} = \underbrace{0011}_{3}\underbrace{1000}_{8}$$

2. Using the binary number system, we have

$$\begin{aligned} 38._{(10)} &= 100110_{(2)} \\ &= (1 \times 2^5) + (0 \times 2^4) + (0 \times 2^3) + (1 \times 2^2) + (1 \times 2^1) + (0 \times 2^0) \\ &= \quad 32 \quad + \quad 0 \quad + \quad 0 \quad + \quad 4 \quad + \quad 2 \quad + \quad 0 \end{aligned}$$

Note that in the coding system *one group* is needed to represent *one symbol.* (In our illustrations the symbols were the 10 decimal digits.) If we wanted to represent a number consisting of a string of 5 decimal digits, we would require a string of 5 code groups with 4 binary digits in each group, or a total of 20 binary digits.

It is interesting to note that in *this* coding system 12 bits were required to represent $149._{(10)}$, whereas only 8 bits were needed to represent the same number in the binary number system. Similarly, for $38._{(10)}$ the coding system needed 8 bits and the binary system required only 6 bits. In general, the binary number system can represent a *number* with fewer bits than are required by a coding system. Characters other than numbers must necessarily be encoded for storage.

In any binary coding system a group consisting of m bits can be used to encode 2^m pieces of information. The coding system described above employed 4 binary digits in each group, and therefore one group could be used to encode 2^4 or 16 pieces of information. (In our illustration of the possible 16 arrangements, we used only 10 to represent the decimal digits; the remaining 6 arrangements were not used, but *might have been* used to represent any other 6 symbols.) If the size of the code group is expanded from 4 to 6 bits, there are 64 unique arrangements possible and 64 pieces of information (that is, 64 *symbols*) can be encoded. With this increase in the number of arrangements available, the 10 decimal digits, the 26 alphabetic characters, and a maximum of 28 *special* characters (such as: . , () $ % + — = &) can be coded. The system using 6 bits is called the *binary coded decimal* (BCD) code. When the coding group is enlarged to 8 bits, it is possible to encode a total of 256 characters. The *extended binary coded decimal information code* (EBCDIC) employs 8 bits.

Ultimately, all information is stored in the computer using binary digits. Numerical data are stored either as pure binary numbers or as binary coded numbers. Alphabetic data and special characters are stored using some binary coding scheme. Most computers use the binary number system for representing numerical data, and therefore it is appropriate that we examine a potential source of error that is inherent in the system.

The reader is undoubtedly familiar with the fact that not all proper fractions can be exactly represented with a finite number of digits using the decimal number system. For example:

$$\tfrac{1}{7} \approx 0.14285711428571\ldots_{(10)}$$

$$\text{and} \quad \tfrac{1}{12} \approx 0.08333333333333\ldots_{(10)}$$

This same shortcoming exists in the binary number system and also applies to some fractions that are not affected in the decimal system. For example, in the decimal system the following fraction can be represented exactly:

$$\tfrac{1}{10} = 0.1000000\ldots_{(10)}$$

However, in the binary system the same fraction can only be approximated:

$$\tfrac{1}{10} \approx 0.0001100110011\ldots_{(2)}$$

$$0.0001100110011_{(2)} = 0.0999755859375_{(10)}$$

$$0.00011001100110011_{(2)} = 0.09999847412109375_{(10)}$$

The fact that *inexact* representation of numerical data can exist in the system may occasionally lead to the *accumulation of appreciable errors.* Since engineering computations are so firmly grounded in the decimal system, the programmer should be particularly alert to the possibility of introducing errors from normally unexpected sources through the use of factors such as 10 percent, or by making a great many incremental steps of 0.001, etc.

1.6 TYPES OF DATA

Within the structure of the FORTRAN language, provision has been made for representing data in any of several *modes.* These modes are:

1. Integer—whole numbers
2. Real—numbers containing a decimal point
3. Double precision—integer or real numbers stored with additional precision
4. Complex—numbers containing both real and imaginary parts
5. Logical—logical or "truth" values
6. Alphameric—literal information

The first four *types* or *modes* are used to represent numerical values during arithmetic operations. The logical mode is used to represent the logical values, TRUE and FALSE, for operations with Boolean algebra. The alphameric mode is used to represent literal data and can contain numeric, alphabetic, and special characters. This mode provides the means for transmitting table headings, units, etc.

In each of the first five modes listed above, the language makes provision for both *constants* and *variables.* The *value* of a *constant* remains unchanged while a program is being run; the *value* of a *variable* may change.

FORTRAN constants are always expressed literally: If the constant is an integer, a string of decimal digits is used to express its value; if the constant is a logical value, TRUE or FALSE is used to express its value; etc. (All modes are treated in detail below.)

As in algebra, FORTRAN variables are represented by *names.* In operation, the name is actually the *address* of a storage location in the computer in which some value is stored. Citing the variable name in a program is equivalent to calling forth the *current contents* of the "named" storage location and using this current value in carrying out the required calculations. It is important to note that using this value does *not* erase it in storage.

For both constants and variables, the FORTRAN language distinguishes between integer and real numbers by the manner in which they are represented in the program, that is, by *implication.* The symbols used to represent the number are interpreted as designating a particular mode of number. *Integer constants* are always written *without* a decimal point; *real constants* are always written *with* a decimal point. Names of *integer variables* always begin with a letter from the following group: I, J, K, L, M, N; names of *real variables* always start with any letter *except* those in the specified group.

An optional alternative, and overriding, means of specifying integer or real mode is by use of a type declaration statement (see Sec. 8.1). A declaration statement is *always* used to specify mode for double precision, complex, and logical variables.

1.7 INTEGER CONSTANTS

Integer (or *fixed point*) constants are expressed in literal form. They are always written *without* a decimal point.

The number of digits permitted in any constant, as well as the maximum value of the constant, depends on the particular computer being used. For example, a computer system may allow constants containing from one to seven decimal digits and still limit the maximum value to 8388607 (rather than the expected 9999999).

Constants may be either positive or negative. If no algebraic sign is shown, the number is assumed to be positive.

Examples of integer constants are

7 632 −40 +16 0

1.8 STATEMENT NUMBERS

Statement numbers are special reference numbers used to *label* particular *statements* or *instructions* in a FORTRAN source program. The statement number makes it possible to designate the beginning or end statement in a series of instructions. It is the programmer's only means of referring to a particular statement within a source program.

Statement numbers, which are always positive integer constants containing from one to five digits, may be selected at random by the programmer and do not have to be in sequence. Naturally, because they are reference numbers, no one statement number may be used to identify more than one FORTRAN statement.

Every statement in a program does not have to be numbered. To conserve storage space in the computer it is advisable to number only those statements that must be referred to by other statements in the program.

It should be noted that statement numbers bear no relationship to the integer constants used in making calculations as a program is executed. The use of a constant in a program does not prevent its use as a statement number.

1.9 INTEGER VARIABLES

All *integer* (or *fixed point*) variables are expressed as *named* quantities; that is, each variable is referred to by a *unique name.* The name actually designates a location in storage in which the *current value* of a particular variable is stored.

Rules for naming integer variables are as follows:

1. Any group of from *one* to *six* characters may be used.
2. Characters may be letters or decimal digits.
3. *No* special characters are permitted.
4. The group must start with an alphabetic character.
5. To imply integer mode the first character must be one of the following: I, J, K, L, M, N.

Some examples of integer variables are

N ITEM LINE3 M26IT KOUNT

1.10 REAL CONSTANTS

Real (or *floating point*) constants may be expressed in either of two ways: (1) *without* an exponent or (2) *with* an exponent.

Without an exponent the value of the constant is expressed literally, using from one to nine[1] decimal digits with a decimal point in front of the first digit, or in back of the last digit, or between any two of the digits. The value of the number may be either positive or negative. If no algebraic sign is present, the number is assumed to be positive.

Some examples of real constants expressed in this manner are

+3.14159265 −416.03 .003 25. 0.09

With an exponent the value of the constant is represented by a notation similar to scientific notation. Each number is represented by a real constant written without an exponent, as described above, followed by a positive or negative one- or two-digit exponent[2] of the decimal base 10 which acts as a multiplier for the real constant. The exponent *may* be zero but must *not* be a blank. The letter E is used to separate the number from the exponent. Some examples are shown below:

FORTRAN notation	*Scientific notation*	*Decimal notation*
−41.60E+01	-41.60×10^{1}	−416.0
30.E−4	$30. \times 10^{-4}$	0.003
.25E3	0.25×10^{3}	250.
3.14E0	3.14×10^{0}	3.14

1.11 REAL VARIABLES

All *real* (or *floating point*) variables are expressed as *names.* As in the case of an integer variable, the name identifies the contents of a storage location in the computer. Using the name in an instruction to the computer actually calls from memory the number stored in the location at that particular time.

Rules for naming real variables are as follows:

1. Any group containing from *one* to *six* characters may be used.
2. Characters may be letters or decimal digits.
3. *No* special characters are permitted.
4. Group must start with an alphabetic character.
5. To imply real mode the first character must be any letter *other than* one of the following: I, J, K, L, M, N.

Some examples of real variable names are

BASE A X3 XITEM Y24T SI AMT

[1] The limit on the number of digits permitted depends on the computer system being used.

[2] The maximum value of the exponent depends on the system being used.

1.12 ARITHMETIC OPERATORS

In FORTRAN the available *arithmetic operators* and their *symbols* are as follows:

Addition	+
Subtraction	−
Multiplication	*
Division	/
Exponentiation	**

All the above operators are *binary operators.* This nomenclature does not refer to binary numbers, but rather to the fact that each of these operators requires *two operands.* The operands may be constants or variables. For example:

A+B   N*L   Q**3   W/386.

The symbol used for the subtraction operator *can also* be used as a *unary operator*, with a single operand, to perform the arithmetic operation of *negation* (forming the negative of a number). For example:

−A   −B**3   −(C+D)

To avoid ambiguity, no two arithmetic operators may be placed side by side in a string of symbols describing a series of calculations (see Sec. 1.17). For this reason, if the *unary minus* is *not* applied as the *leading* operator in a computation, the unary minus *and* the operand to which it applies must be enclosed in parentheses:

```
B+(−A)                        P/(−(C+D))
  ↑↑                              ↑ ↑
  │└──Operand                     │ └─────Operand
  └───Unary minus                 └───────Unary minus
```

Except for the operation of exponentiation (see Sec. 1.15), *both* the operands in a computation must be in the same number mode. *Some* compilers will accept *mixed-mode* computations, that is, calculations using both integer and real number types. The programmer should check the manual for his particular computer system before attempting to use this relaxation of the rules.

1.13 INTEGER ARITHMETIC

Addition, subtraction, and multiplication give correct results if the answer is within the range allowed for fixed point numbers by the particular FORTRAN system being used. This allowable range lies between the positive and negative values of the largest integer number that can be represented by a particular computer. The programmer should consult the user's manual to determine these numbers.

Division of integer numbers, unless it is exact, will return the *greatest integer value* as the answer. That is, the true quotient will be *truncated* to the next smaller integer and the *remainder* will be discarded.

Some examples of integer division are

4/3 = 1   8/2 = 4   9/2 = 4   3/5 = 0

1.14 REAL ARITHMETIC

With real-mode numbers, addition, subtraction, multiplication, and division give correct results with an accuracy which is usually nine significant digits, providing the answer is within the range of real numbers allowed by the computer system. The exact number of significant digits, and the range of values that can be expressed, depends on the system being used. If necessary, greater significance can be obtained by employing *double precision* representation of numbers (see Secs. 8.2 to 8.5).

Rounding error in floating point arithmetic can be introduced by computing systems that round up. It is the responsibility of the programmer to guard against the accumulation of serious errors through rounding. The reader should note that many computers do *not* round up the ninth significant digit if the tenth digit is greater than 5; the tenth digit and following digits are simply dropped. This creates *truncation error.*

Some examples of computer-produced error are

(1./3.)*3. = .999999999

1000000. + .009 = 1000000.00

1.15 EXPONENTIATION

In FORTRAN, both integer *and* real numbers can be raised to a power. For *integers* the exponent must also be an integer. For *reals* the exponent can be either an integer *or* a real number.

If the exponent is a real number, the same result is obtained whether the base number is positive or negative; that is, (+A)**B is the same as (−A)**B because the computer calculates the result as | A | **B using logarithms. The actual evaluation is accomplished using a compiler-supplied function (see Sec. 1.22) that determines the value of A**B by the following procedure: EXP(B*ALOG(ABS(A))).

If the exponent is an integer number, the result is calculated by repeated multiplication and the rules of algebra apply; that is, (−2.)**3 = −8. because the computer calculates the result as (−2.)*(−2.)*(−2.) = −8. This procedure is usually faster than the method using logarithms. Note that

(−2.)**3 = −8.

and (−2.)**3. = +8.

Most systems give an error message with the second result.

The reader should further note that exponentiation is the only arithmetic operation for which operands of different mode are permitted by *all* FORTRAN compilers.

1.16 ARITHMETIC EXPRESSIONS

An *arithmetic expression* may be a single variable or a single constant, or it may consist of several numbers (constants and/or variables) and prescribe a series of operations to be performed. In most FORTRAN systems *mixed-mode* arithmetic expressions are not allowed; i.e., all elements in a multielement expression must be of one type (either integer or real). The reader should recall the exceptions noted in Sec. 1.12.

For *real* arithmetic expressions *all* compilers permit two exceptions to this one-mode rule:

1. Exponents *may* be integers (see Sec. 1.15).
2. Subscripts *must* be integers (see Sec. 7.1).

Some examples of arithmetic expressions are

A   −32.16   A*B+C   (M/N)*K

1.17 HIERARCHY OF ARITHMETIC OPERATIONS

When an arithmetic expression involves several arithmetic operations, the *normal order* in which they are performed (regardless of their position in the expression) is according to the following *precedence*:

1. Exponentiation
2. Arithmetic negation (forming the negative of a value)
3. Multiplication and/or division
4. Addition and/or subtraction

If there are two or more operations having the same precedence rating, these particular operations are performed in *sequence* from left to right. For example, the expression

A/B + C − D**3/9.

would be evaluated according to the following sequence:

1. D**3 will be evaluated and stored (*highest* precedence).
2. A will be divided by B and the result stored (*first* multiplication or division operation reading from left to right).
3. The stored value obtained in step 1 will be divided by 9. and the result stored (*next* multiplication or division operation).
4. The result of step 2 will be added to C and the result stored (*first* addition or subtraction operation reading from left to right).
5. The result of step 3 will be subtracted from the result obtained in step 4 and the value of the expression stored (*next* addition or subtraction operation).

Parentheses may be used to change the normal precedence order of operations. That part of the computation contained in the *innermost* pair of parentheses is completed first, then the next outermost, etc. For example, the expression

(A/(B+C) − D**3)/9.

would be evaluated according to the following sequence:

1. C will be added to B and the result stored (*innermost* parentheses).
2. D**3 will be evaluated and stored (*highest* precedence operation in the *next* outermost parentheses).
3. A will be divided by the result obtained in step 1 and the result stored (*first* multiplication or division operation within the controlling parentheses).
4. The result of step 2 will be subtracted from step 3 and the result stored (*first* addition or subtraction operation within the currently controlling parentheses).

5. The result of step 4 will be divided by 9. and the value of the expression will be stored (*highest* precedence, and in this case the *only*, operation outside the outermost parentheses).

There is no limit on the number of parentheses that may be used in an expression; however, they must be used in pairs (left and right hand).

Parentheses may also be used to separate what would normally be adjacent operation symbols. For example, "forty divided by minus eight" must be written as 40./(–8.) and *not* as 40./–8. The latter arrangement violates the requirement that no two arithmetic operators be written side by side (see Sec. 1.12).

1.18 FORTRAN STATEMENTS

FORTRAN statements are the means by which the programmer directs the operation of the computer. A set of FORTRAN statements, arranged in proper order, constitutes a *source program* (see Sec. 1.2).

FORTRAN statements can be divided into two main categories: *executable* statements and *nonexecutable* statements. There are several types of statements under each of these categories.

In general, *executable statements produce some action* by the computer during the program run. The *types* of executable statements are:

1. Assignment statements
2. Control statements
3. Unconditional transfer statements
4. Conditional transfer statements
5. Input and output statements

In a broad sense, the *nonexecutable statements provide information* that is needed for the proper execution of the program. The *types* of nonexecutable statements are:

1. Specification declarations
2. Type declarations
3. Definition declarations

1.19 ARITHMETIC ASSIGNMENT STATEMENT

The *arithmetic assignment* (or *replacement*) statement is an executable statement that has the following general form:

$$v = exp$$

where v is any *unsigned* integer, real, double precision, or complex *variable*, and *exp* is any *valid* numeric *expression* as defined in Sec. 1.16.

The arithmetic assignment or replacement statement is used for the following purposes:

1. To calculate the value of an expression and store the value
2. To store a particular value, i.e., to *set* the value of a variable
3. To change the mode of a number (where permitted)

In interpreting a replacement statement, the phrase *is to be replaced by* should be read where the algebraic *equals* symbol appears. For example:

X = A*B	is read as "the current value of X is to be replaced by the product of the current value of A and the current value of B."
K = K + 1	is read as "the current value of K is to be replaced by the current value of K increased by one."

The following are examples of arithmetic assignment statements:

```
A = B + C        X = A*B + C**Z
Y = 27.92        N = X + Y
K = 0            D = Z - P**(1./3.)
```

1.20 CHANGE OF NUMBER MODE

A replacement statement can be used to *change the mode* of a numeric value, that is, to change the manner in which it is represented in the computer storage.

Integer to real

The following arithmetic assignment statement is an example of the process of *floating* the value of an integer (changing an *integer* value to a *real* value, that is, changing from *fixed point* mode to *floating point* mode):

```
A = N
```

If 3 is stored in location N, the above statement will store 3. in location A. The 3 will remain *unchanged* in storage location N.

This transition can be used when a fixed point variable is needed in an expression. For example, Y = A*B*K is incorrect because the expression is in mixed mode (A and B are reals and K is an integer). The expression can be corrected by the following sequence of statements:

```
X = K
Y = A*B*X
```

Real to integer

The following arithmetic assignment statement illustrates the process of *fixing* or *truncating* a real value (changing a *real* value to an *integer* value, that is, from *floating point* to *fixed point* mode):

```
N = A
```

If 17.65 is stored in location A, the above statement will store 17 in location N, *and the contents of A will remain unchanged.*

Note that a real value can be truncated and returned to real mode by the following sequence of replacement statements:

If B has the value 6.81 stored in it

writing M = B will store 6 in location M

and then B = M will store 6. in location B (i.e., the 0.81 has been removed)

In a more general sense, the transition of number mode is applicable to the values of expressions:

$v = exp$

where v is a variable of the desired mode and *exp* is an expression evaluated in a different mode. Considered in this manner it is possible to change from single precision real to double precision real, from integer to double precision real, etc. The following list indicates the valid mode transitions:

Variable mode		*Expression mode*
Integer	=	Real (single precision)
Integer	=	Real (double precision)
Real (single precision)	=	Integer
Real (single precision)	=	Real (double precision)
Real (double precision)	=	Integer
Real (double precision)	=	Real (single precision)
Complex	=	Real (single precision)

Of course any of the variables can be used in an arithmetic assignment statement with an expression of the *same* mode.

It should be noted that in *fixing* the value of a real expression, if the expression consists of several terms, it will be evaluated in floating point mode and *then* truncated and stored. For example:

In M = B*C: if 6.6 is stored in location B and 2.0 is stored in location C, then 13 will be stored in location M

1.21 FORTRAN-SUPPLIED FUNCTIONS

In addition to the expressions that can be written using the arithmetic operations, the FORTRAN language permits the use of several *FORTRAN-supplied functions.* These functions are evaluated by *prewritten routines* that are automatically provided to the object program when required. The compiler handles this procedure, and the programmer is not concerned with the internal operation of the computer during the execution of these functions. However, as in all programming, certain rules must be observed and these are discussed below.

A FORTRAN-supplied function is *used* by writing its *assigned name* in the proper place in the source program, just as any other variable is used in an expression. When it has been evaluated, a *single value* is returned to the position occupied by the function in the expression and the arithmetic calculations are completed using this value.

It should be noted that the evaluation of these functions has a *higher operational precedence* than the arithmetic operators (see Sec. 1.17) and therefore all the functions "called" in an expression will be evaluated *before* the normal hierarchy of operations is brought into play. However, if the argument of the function is itself an expression, then normal precedence will be used in evaluating the *argument.*

There are two *types* of FORTRAN-supplied functions: *intrinsic* (also called in-line) and *external* (also called out-of-line).

1.22 EXTERNAL FUNCTIONS

External functions are machine-language subprograms (see Sec. 10.2) that are selected by the compiler from the FORTRAN library (which is part of the compiler) and *appended* to the main program. Each time the function is required by the main program, the compiler generates instructions to *branch* to the subprogram, *transmit* the value of the argument, *evaluate* the function according to the instructions found there, and *return* the computed value to the main program.

The external functions that are available in *all* FORTRAN systems are as follows.

Meaning	*Algebraic notation*	*FORTRAN symbol*
Sine	$\sin x$	SIN(X)
Cosine	$\cos x$	COS(X)
Exponential	e^x	EXP(X)
Natural logarithm	$\ln x$	ALOG(X)
Square root	$\sqrt{x}$	SQRT(X)
Arctangent	$\arctan x$	ATAN(X)

Most compilers have other functions available in the library. The programmer should consult the manual for the system he is using.

The following rules apply to external functions.

1. The argument must be enclosed in parentheses.
2. The argument may be any *real* variable, or expression, or constant.
3. The value returned by the function will be in *real mode.*[1]
4. The arguments for the sine and cosine functions, and the value returned by the arctangent function, are in *radians.*

Some examples of the use of external functions are

```
Y = SIN(A*3.14159*X/B)
P = W + SQRT(COS(X))
THETA = ATAN(Y/X)
```

1.23 INTRINSIC FUNCTIONS

Intrinsic functions are machine-language procedures that are *inserted* in the main program by the compiler at *every location* where the function is used. Most compilers have a variety of these functions available, and the programmer should obtain a list of the functions at his disposal. A typical intrinsic function included in *all* systems is the following:

Meaning	*Algebraic notation*	*FORTRAN symbol*
Absolute value	$\|x\|$	ABS(X)

[1] Some special functions return double precision or complex values.

The rules that apply to *external* functions (see Sec. 1.22) also apply to *intrinsic* functions.

1.24 PAUSE STATEMENT

The PAUSE *statement* is an executable statement having the form

PAUSE

This statement causes the computer to *halt temporarily* during the execution of the object program. The halt permits the computer operator to inspect the results obtained up to this point and possibly make a decision concerning the further execution of the program. Pressing the "Start" button of the computer console causes the computer to go to the next instruction in the program and resume execution.

If the program is to be run under control of a *monitor program,*[1] the PAUSE statement *should not* be used because it requires the intervention of the human operator to restart the machine and is therefore wasteful of computer time. This statement *should never* be used in a time-sharing system because the computer controls are not accessible to the (remote) user.

1.25 STOP STATEMENT

The STOP *statement* is an executable statement having the form

STOP

This statement causes the computer to *halt* and the word STOP to be printed as output. Pressing the "Start" button on the computer console causes a repetition of the above action. The program *does not* resume operation. The STOP statement is used when the programmer intends to *permanently halt execution* of the program and vacate the computer. If the program is running under control of a monitor program, the STOP statement returns control to the monitor. In a time-sharing system, the STOP statement produces the printing action described above and the system then proceeds to fulfill the requests of the next user in the time-sharing queue.

1.26 END STATEMENT

The END *statement* is a nonexecutable statement of the form

END

This statement *defines* the end of a FORTRAN source program and *must be* the *last* statement in a *source program or a subprogram* (see Sec. 10.2). It is a signal to the compiler that there are no more statements to be translated into machine language in the program currently being processed.

The reader should note that there may be several STOP statements in a program, but there can be *only one* END statement. For example, there could be two STOP statements in a program, one located at the point where the calculations would normally be concluded and one located where it might be desirable to halt the program prematurely because of some detectable failure in its execution.

[1] A monitor program is a program that controls the operation of a computer during the processing of several source programs. When one job is completed, the monitor automatically loads, compiles, and executes the next program in the input device.

1.27 INTRODUCTION TO FLOW CHARTING

Flow charting is a method of graphically outlining the steps of a program. A flow chart displays the *logic* of the program by indicating the *flow of control* (see Sec. 4.1) for the decisions included in the program.

To facilitate reading, specially shaped symbols (specified by ANSI X3.5) are used to represent the various types of operations in a program. The shapes used in this text for the elementary operations are shown below. Additional discussions of flow charting also appear in Secs. 4.8, 5.7, and 10.11.

PROGRAM TERMINUS

This symbol is used to indicate the beginning of a program and is also used with the STOP (Sec. 1.25) and PAUSE (Sec. 1.24) statements. When it is used to show the start of a program, the name assigned to the program by the authors is shown in quotation marks within the block.

PROGRAM PATH LINE

This symbol is used to indicate the flow of control from one program step to another.

PATH LINE CONNECTORS

On page

Off page

These symbols are used to show the *continuation* of a path line on another sheet or on the same sheet when it would be confusing to show the path line in its entirety. These symbols are always used in pairs with a distinctive code symbol in each member of the pair. For example:

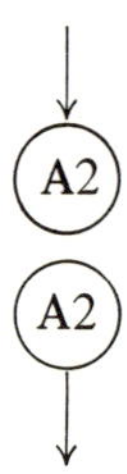

Indicates a *break* in the path line

Shows the *continuation* of the path line

INPUT OR OUTPUT BLOCK

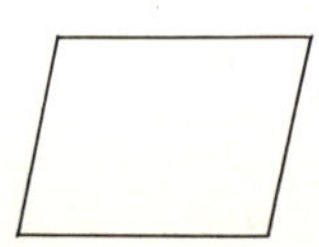

This symbol is used to indicate the *transmission* of information, either *to* the computer or *from* the computer. It serves a dual purpose and symbolizes *data being transmitted* to the computer as well as *results being obtained* from the computer.

The input or output block (I/O block) is used in connection with the following FORTRAN statements:

1. READ (Section 2.1)
2. WRITE (Section 2.2)
3. PRINT (Section 2.2)
4. PUNCH (Section 2.2)

In a flow chart, the symbol contains one of the FORTRAN commands named above together with a list of the variable names to be processed.

PROCESSING BLOCK

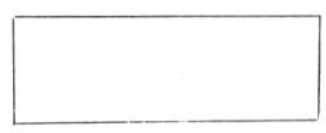

This symbol is used to show one or more *initialization or calculation* steps. To clarify the operations, the authors have included descriptive commands, such as "Set" or "Calculate," in the upper left-hand corner of the processing blocks when it was deemed appropriate.

STATEMENT NUMBER INDICATOR

To assist the student in correlating the source program and the flow chart, the FORTRAN *statement numbers* used in the program have been shown, in parentheses, outside the upper left-hand corner of the appropriate block. The following flow-chart segment illustrates this procedure.

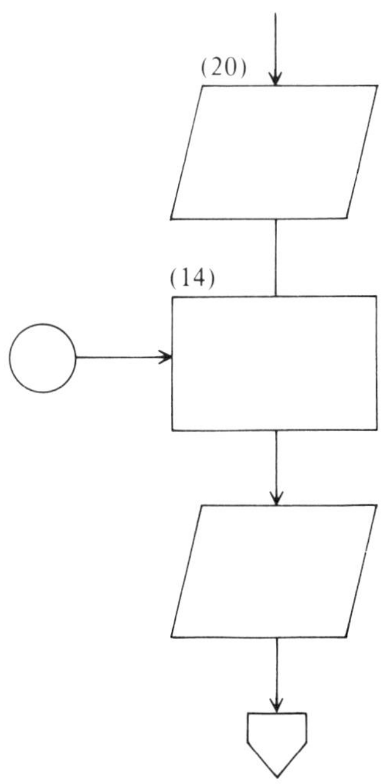

1.28 TWO ILLUSTRATIVE PROGRAMS

In order to enter data into the computer, and for the computer to print answers in the manner desired by the programmer, most computers require that FORMAT statements (see Sec. 2.3) be included in the source program. However, time-sharing computers permit the programmer to *omit* FORMAT statements. To enter data into time-sharing computers, an executable statement of the following form is required:

READ, *list*

where *list* is an ordered sequence of one or more variable names separated by commas. Similarly, data currently stored in the computer can be printed by use of an executable statement of the form

PRINT, *list*

where, once again, *list* is an ordered sequence of one or more variable names separated by commas. The features just described are called *unformatted* or *free-field* input and output, respectively. Standard FORMAT statements are discussed in Chap. 2.

At this point, we will use the option of unformatted input and output to present two simple programs which illustrate some of the concepts discussed in this chapter. Other programs of approximately the same complexity are discussed in Chap. 3.

The first program to be discussed, shown in Fig. 1.28.1, reads-in a temperature in degrees Fahrenheit and converts the temperature to its equivalents in degrees Celsius, Rankine, and Kelvin. In discussing the programs in this and in later sections, we will usually indicate the positions of various statements in the programs by naming the *line sequence numbers* of the lines on which these statements occur. (The line sequence numbers are the five-digit numbers in the column to the left of the program and should not be confused with the *statement numbers* discussed in Sec. 1.8.) The letter C following line sequence numbers 00001, 00002, and 00003 indicates that the statements on those lines are comments and are not regarded by the compiler as part of the program. The formulas for calculating degrees Celsius, Rankine, and Kelvin are on lines 00020, 00030, and 00040, respectively. Notice that the name DK, rather than K, is used for degrees Kelvin because K denotes an integer variable. If K had been used, the calculation would have been carried out in real mode (C + 273.), but the answer would have been in integer mode, and significant figures might have been lost.

The segment of the program from lines 00050 through 00080 consists of several PRINT statements. On line 00050, the statement PRINT produces a blank line above the printed answers. A trial run is shown in Fig. 1.28.2. In the particular computer system used, whenever an input command is encountered during execution, the line sequence number on which the READ statement appears is printed. Then, on the next

```
00001C    TEMPER
00002C    THIS PRØGRAM CØNVERTS DEGREES FAHRENHEIT TØ DEGREES
00003C    CELSIUS, DEGREES RANKINE, AND DEGREES KELVIN.
00010     READ,F
00020     C=5.*(F-32.)/9.
00030     R=F+460.
00040     DK=C+273.
00050     PRINT,
00060     PRINT,C
00070     PRINT,R
00080     PRINT,DK
00090     STØP
00100     END
```

Fig. 1.28.1 Program TEMPER.

```
INPUT:00010
 ?212.


       100.0000
       672.0000
       373.0000
STØP
```

Fig. 1.28.2 Trial run for TEMPER.

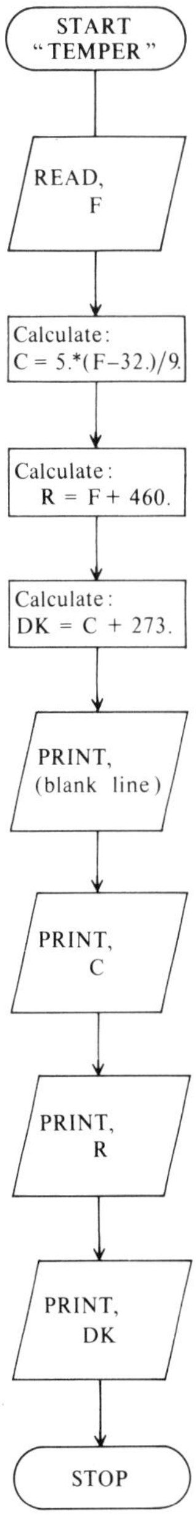

Fig. 1.28.3 Flow chart of program TEMPER.

```
00001C  | CØSINE
00002C  | THIS PRØGRAM CALCULATES THE LENGTH ØF THE SIDE ØF A
00003C  | TRIANGLE BY THE LAW ØF CØSINES.
00010   | READ,A,B,ANGLE
00020   | GAMMA=3.1415927*ANGLE/180.
00030   | C=SQRT(A**2+B**2-2.*A*B*CØS(GAMMA))
00040   | PRINT,
00050   | PRINT,C
00060   | STØP
00070   | END
```

Fig. 1.28.4 Program COSINE.

line, a question mark is printed. This is the signal for data to be entered. In the trial run shown, a value of 212°F was entered. The equivalent temperatures in degrees Celsius, Rankine, and Kelvin were then printed on succeeding lines.

A flow chart for the program is shown in Fig. 1.28.3 to illustrate the diagramming procedure.

As a second example, let us consider the program shown in Fig. 1.28.4. In this program, the length of side C of a triangle is calculated by the law of cosines in accordance with the formula:

$$C = (A^2 + B^2 - 2AB \cos \theta)^{\frac{1}{2}}$$

As the program shows, the lengths of sides A and B and the size of the angle between them, expressed in degrees, are entered using unformatted input (line 00010). The symbolic name ANGLE denotes the angle between sides A and B. On line 00020, ANGLE is converted to radians, and the number of radians is stored in GAMMA. Finally, C is calculated on line 00030. Five trial runs are shown in Fig. 1.28.5.

```
INPUT:00010
 ?10.,10.,60.


        10.0000

INPUT:00010
 ?5.,7.35,46.


         5.2882

INPUT:00010
 ?5.,7.35,162.


        12.2035

INPUT:00010
 ?3.,4.,90.


         5.0000

INPUT:00010
 ?5.,7.35,180.


        12.3500
```

Fig. 1.28.5 Trial runs for COSINE.

PROBLEMS

Write the following FORTRAN constants in ordinary decimal notation:

1-1. .36921E+02
1-2. .36921E–3
1-3. 3E7
1-4. .31416E+01
1-5. .0042E06
1-6. .35E0

Write the following real numbers as FORTRAN constants, using E notation:

1-7. 6.06×10^{23}
1-8. 32.16
1-9. 3.14159
1-10. 0.30103
1-11. 2.718
1-12. 186284.

Indicate which of the following labels are for fixed point variables, which are for floating point variables, and which are not permitted in FORTRAN:

1-13. NAME
1-14. SUM
1-15. ABC
1-16. ANAME
1-17. 2ND
1-18. ITEM
1-19. ANSWER1
1-20. MEAN
1-21. LABEL
1-22. X
1-23. J
1-24. JOB3
1-25. LENGTH2
1-26. ZMARK
1-27. KOUNT

Calculate the answer to each of the following expressions using integer arithmetic:

1-28. 3/7 + 4
1-29. 7/3 + 4
1-30. (6 + 3)/2
1-31. 29/(6 + 2)
1-32. 2 × 9/16
1-33. 2 × (9/16)

Write FORTRAN expressions for the following algebraic expressions:

1-34. $\dfrac{a^2 + b^2}{2}$
1-35. $mx + b$
1-36. $\dfrac{ac}{d} + \dfrac{b}{c}$
1-37. $p^3y - q^2x$
1-38. $\dfrac{PL^3}{3EI}$
1-39. $\dfrac{-b + \sqrt{b^2 - 4ac}}{2a}$

The following incorrect FORTRAN expressions are intended to represent the algebraic expression on the left. Make the necessary corrections.

1-40. $\dfrac{4x^3}{b}$ (4*X**3)/B
1-41. $\dfrac{2ab}{cd}$ 2.*A*B/C*D
1-42. $6(x^2 + y^3)$ 6.*X**2+Y**3
1-43. $\dfrac{a^2 + b^2}{2c^2}$ A**2+B**2/2.*C**2

Write the algebraic expressions that have been coded by the following FORTRAN expressions:

1-44. 2.*X**3+3.*Y**3/A*B
1-45. (2.*X**3+3.*Y**3)/A*B
1-46. (2.*X**3+3.*Y**3)/(A*B)
1-47. (2.*X**3+3.*Y**3)/A/B

Factor the following algebraic expressions and write FORTRAN expressions for them:

1-48. $a^2 + ax - 2ab - 2bx$
1-49. $x^4 + 3x^3 - 3x^2 + 3x + 2$

Outline the sequence of steps by which the following FORTRAN expressions will be evaluated:

1-50. (R*T – A)/(R – 1.)
1-51. S*R/((1. + R)**N – 1.)
1-52. 1. + ((A – B)/(A + B))**2/4.

1.53. 2.*SIN((A + B)/2.)*COS((A − B)/2.)

1-54. 0.5*A*(EXP(X/A) + EXP(−X/A))

Write FORTRAN statements to make the following calculations:

1-55. Temperature scale conversion:

$$C = \tfrac{5}{9}\,(F - 32)$$

where C = Celsius temperature, °C
F = Fahrenheit temperature, °F

1-56. Area of trapezoid:

$$A = \frac{h}{2}\,(b_1 + b_2)$$

where A = area, in^2
h = altitude, in
b_1 = base, in
b_2 = side parallel to base, in

1.57. Uniformly accelerated displacement:

$$s = v_0 t + \tfrac{1}{2} a t^2$$

where s = displacement, in
v_0 = initial velocity, in/s
a = acceleration, in/s^2
t = time, s

1-58. Compound interest formula:

$$A = P(1 + \frac{i}{m})^{nm}$$

where A = amount, \$
P = principal, \$
i = interest rate, % (as a decimal)
m = number of times interest is compounded each year
n = number of years

1-59. Heating value of a solid fuel:

$$HV = 145(\text{C}) + 620\left(\text{H} - \frac{\text{O}}{8}\right) + 40(\text{S})$$

where HV = Heating value, Btu/lb
C = carbon content, %
H = hydrogen content, %
O = oxygen content, %
S = sulfur content, %

1-60. Grashof number:

$$\text{GR} = \frac{\rho^2 \beta g L^3 (T_s - T_f)}{\mu^2}$$

where GR = Grashof number
ρ = density, lb-m/ft^3
β = coefficient of thermal expansion, 1/°F
L = characterizing length, ft
g = gravitational acceleration, ft/s^2
T_s = temperature of surface, °F
T_f = temperature of fluid, °F
μ = viscosity, lb-m/ft-s

1-61. Diesel cycle efficiency:

$$\text{Eff} = 1 - \left(\frac{1}{r_1}\right)^{k-1} \frac{r_2{}^k - 1}{k(r_2 - 1)}$$

where Eff = efficiency, % (as a decimal)
r_1 = compression ratio
r_2 = cutoff ratio
k = specific heat ratio (*not* integer)

1-62. Water flow through a horizontal venturi meter:

$$Q = C \frac{A_1 A_2}{\sqrt{A_1^2 - A_2^2}} \quad \sqrt{2g(h_1 - h_2)}$$

where Q = flow rate, ft^3/s
C = meter coefficient
A_1 = area of pipe cross section, ft^2
A_2 = area of throat cross section, ft^2
h_1 = pressure heat in main pipe, ft
h_2 = pressure heat in throat, ft
g = gravitational acceleration, ft/s^2

1-63. Oscillation of a simple pendulum:

$$t = \pi \sqrt{\frac{L}{g}}$$

where t = time of oscillation, s
L = length of pendulum, in
g = gravitational acceleration, in/s^2

1-64. Drive shaft size:

$$D = \sqrt[3]{\frac{16T}{\pi S}}$$

where D = shaft diameter, in
T = torsional moment, lb-in
S = allowable shear stress, lb/in^2

1-65. Depreciation by fixed percentage of balance:

$$D = 1 - \sqrt[Y]{\frac{S}{C}}$$

where D = depreciation per year, % (as a decimal)
Y = number of years
S = salvage value after Y years, \$
C = cost when new, \$

1-66. Maximum height of a projectile:

$$H = \frac{V^2 \sin^2 \alpha}{2g}$$

where H = maximum height of trajectory (neglecting air resistance), ft
V = projection velocity, ft/s
α = projection angle above horizontal, rad
g = gravitational acceleration, ft/s^2

1-67. Alternating current power:

$$P = EI \cos \phi$$

where P = power, W
E = effective voltage, V
I = effective current, A
ϕ = phase angle, degrees

1-68. Snell's law of refraction:

$$N = \frac{\sin \theta_i}{\sin \theta_r}$$

where N = index of refraction
θ_i = angle of incidence, degrees
θ_r = angle of refraction, degrees

1-69. Weibull probability density function:

$$Y = bx^{(b-1)} e^{-x^b}$$

where Y = function value
b = positive constant
x = positive random variable

1-70. Log mean temperature difference:

$$T_m = \frac{(T_1 - T_a) - (T_2 - T_b)}{\ln [(T_1 - T_a)/(T_2 - T_b)]}$$

where T_m = mean temperature difference, °F
T_1 = temperature of hot fluid entering, °F
T_2 = temperature of hot fluid leaving, °F
T_a = temperature of cold fluid entering, °F
T_b = temperature of cold fluid leaving, °F

2
Communicating with the Computer

In this chapter the basic methods of transmitting data *to* the computer and obtaining results *from* the computer are discussed. Some of the detailed discussion presented here can be omitted by the student during the first reading. To identify this material, most sections after the introduction of the topic have been divided into two parts, A and B. In all cases, Part A contains the material that is deemed essential for the student's understanding of the succeeding chapters; Part B contains examples of usage and discussion of some of the fine points of input/output operations. The authors suggest that the student bypass parts B the first time through the chapter, and then refer back to them for clarification as questions arise.

2.1 READ STATEMENT

The READ *statement* is an executable statement with the following general form:

READ (*m*,*n*) *list*

where *m* is an integer constant *or* an integer variable that refers to a particular input device, *n* is the statement number of a FORMAT statement (see Sec. 2.3), and *list* is an ordered sequence of one or more variable names separated by commas.

PART A

During execution, the READ statement causes the referenced input device to *scan* a data file and *transmit* the information found there to the computer storage locations that are designated by the names in the *list*. The *order* in which the scanned numbers are transmitted to storage is determined by the order in which the variables are named in the list. The *first number scanned* will be stored in the location reserved for the *first variable named* in the list, the second number scanned will be transmitted to the second variable's storage location, etc.

The numbering of the input devices[1] is usually determined by the computer system being used and the peripheral equipment that is available. The programmer must ascertain the numbers being used with the system on which his program is to be run. For example, 1 may refer to a card reader, 4 to a magnetic disk, etc.

The *form* of the data transmitted, that is, the *mode* and *manner of representation* (see Sec. 2.4), is determined by the referenced FORMAT statement. The mode of the data transmitted must correspond to the mode *implied* by the name of the variable or *specified* for the variable name in a declaration statement (see Sec. 8.1). That is, an integer value must be stored in a location reserved for an integer variable; a real number, in a location designated by a real variable name; a logical value, in a location addressed by a name that has been declared to be logical mode (see Sec. 9.2); etc.

On some small computer systems and on most time-sharing systems[2] only one input device is available and therefore it is not necessary to show a device reference number. The READ statement, for these systems, can be written as

READ *n, list*

where *n* and *list* have the same meaning as described above.

Some FORTRAN compilers make it possible to read data into the computer without specifying the form in which the data must be presented. Such input operations are called *unformatted input,* or *free-field input,* and the data are usually arranged as a string of values separated by commas. Where this feature is available the READ statement is written as

READ, *list*

where *list* has the same meaning as before and the *absence of a statement number n* indicates that unformatted input is being employed.

Some examples of READ statements are

```
READ (1,20) A,B,C
READ (1,16) ITEM,X,Y,S(10)
READ 52, XL,DIAM,F
READ 1, RPM
READ, U,F,A,T
```

[1] Some of the input devices that may be available are:

1. Electric typewriter
2. Card reader
3. Paper tape reader
4. Magnetic tape reader
5. Optical scanner
6. Cathode-ray tube

[2] The computer system used for solving the majority of the problems in this book was a time-sharing system for which only one input device, a teletypewriter, was available. The input command was READ, with the modifications indicated in the text. The most frequently used input statements were

	READ *n, list*	Formatted input from teletypewriter
and	READ, *list*	Unformatted input from teletypewriter

PART B

In Part A we discussed the use of constants for the value of *m,* that is, using the READ statement with a *particular* computer system for which the assigned device codes are known. Occasionally a program must be prepared *without knowing the computer system* that will be used and, therefore, *without knowing in advance the numbers assigned* to the input devices for that system. Since *m* may also be an integer variable, the programmer can prepare the READ statement with a variable name (in place of *m*) and *initialize* the value of the variable with a replacement statement that appears *before* the READ statement. This procedure still requires that the programmer, or program user, know the value assigned to the particular input device; however, when the actual computer has been chosen, it is relatively simple to insert the appropriate replacement statement just prior to the compilation of the program.

For example, assume a program has been prepared with the variable name IN1 in the position specified for *m.* At some subsequent time this program is to be run on a computer where the correct input device is designated by 7. Just before the source program is compiled, a statement reading

IN1 = 7

can be inserted in the program, thereby *initializing* the value of IN1 (and therefore *m*) at 7. The programmer would, of course, have to be certain that this eleventh-hour insertion were made at a point in the program *before* the READ statement were encountered.

Another form of the READ statement is available for use when data that were previously stored in binary form are being read from a file. Under these circumstances it is not necessary to perform any conversion between external representation and internal representation, and therefore the READ statement need only indicate the device where the data can be found. This is accomplished by the statement

READ (*m*) *list*

where *m* is the integer code for the device (or *file*) being read and *list* has the same meaning as before.[1]

Some additional examples of READ statements are

READ (K,100) X,Y,Z
READ (4) K,V,J

2.2 WRITE STATEMENT

The WRITE *statement* is an executable statement having the general form

WRITE (*m,n*) *list*

where *m* is an integer constant *or* an integer variable referring to a particular output device, *n* is a statement number referencing a FORMAT statement (see Sec. 2.3), and *list* is an ordered sequence of one or more variable names separated by commas.

[1] The time-sharing system used for solving most of the problems in this book had a teletypewriter as the only I/O device. Therefore *m* was used to reference data files or what might be considered as *simulated* input devices. The commands employed were

READ (*m,n*) *list*	Formatted input from file *m*
READ (*m,*) *list*	Unformatted input from file *m*
READ (*m*) *list*	Binary input from file *m*

The WRITE statement causes the data currently stored in the locations named in the *list* to be transmitted in the order in which they are listed to the output device[1] designated by *m* and transcribed according to the form specified in the referenced FORMAT statement.

Some examples of WRITE statements are

```
WRITE (3,66) ANSWER
WRITE (3,4000)X,Y,XY,XSQ
WRITE (5,101) HP, SPEED
```

PART A

As in the case of the READ statement (see Sec. 2.1), variants of the general form of the WRITE statement are also possible with computer systems having limited output facilities.[2] The PRINT *statement,* an example of one of these variants, has the form

PRINT *n, list*

where *n* and *list* have the meanings defined above. The PRINT statement causes the data to be printed on a typewriter or line printer according to the specified format.

There is also an *unformatted output* statement available with some compilers. This feature is usually associated with a PRINT statement and is *indicated by the absence of a statement number reference:*

PRINT, *list*

with *list* having the same meaning as before.

Some examples of PRINT statements are

```
PRINT 70,A,X,Y
PRINT, A,B,C
```

PART B

The technique of using an *integer variable* in place of *m* in a READ statement, described in Part B, Sec. 2.1, is also applicable to the WRITE statement. This procedure is only used when the programmer does not know in advance the output device code numbers for the computer system on which his program will be run.

Provision has been made in the FORTRAN language so that data may be written on external storage devices in the same manner in which they are stored internally in the computer, that is, in *binary form.* When this is done, no conversion is necessary

[1] The numbers assigned as reference numbers for the available output devices are determined by the computer system being employed. Some of the output devices that may be available are:

1. Electric typewriter
2. Line printer
3. Card punch
4. Paper tape punch
5. Magnetic tape
6. Magnetic disk
7. Cathode-ray tube

[2] The time-sharing system used for illustrating most of the problems in this book was accessed through a teletypewriter. Both formatted and unformatted output was possible. The most commonly used output statements for this system were

PRINT *n, list*	Formatted output to teletypewriter
PRINT, *list*	Unformatted output to teletypewriter

and therefore no FORMAT specification is required. However, the device code number for the piece of equipment to which the data are being transmitted is needed in the WRITE statement. The form of this type of statement is

WRITE (m) *list*

where m is the device code for the storage unit being used[1] and *list* has the same meaning as before.

Some additional examples of WRITE statements are

```
WRITE (IOUNIT,75) K,B,P
WRITE (5) P,Q
```

A special output command that is available on some computer systems is the PUNCH *statement* having the form

PUNCH n, *list*

where n and *list* have the meaning defined earlier in this section.

The PUNCH statement accomplishes essentially the same output action as the PRINT statement except the data are punched into cards rather than being printed out as hard copy. The same action can be accomplished using the WRITE statement if the output device code for the card punch is substituted for m.

2.3 FORMAT STATEMENT

The FORMAT *statement* is a nonexecutable statement of the form

n FORMAT ($s_1,s_2,s_3, \ldots ,s_m$)

where n is a statement number referenced by an input or output statement, and $s_1,s_2,s_3, \ldots ,s_m$ is an ordered list of format specifications (see Sec. 2.4).

The FORMAT statement *provides information* to the computer by specifying the *data mode* and the *form of representation* (or *conversion*) that is to be used in transmitting data into *or* out of the computer storage. Each specification in the ordered sequence within the parentheses refers to a variable in the corresponding location in the I/O *list* of a READ or WRITE statement.

The following example illustrates the general use of the FORMAT statement. If an *output* statement reads

PRINT 400, X,Y,XY,XSQ

and the referenced FORMAT statement is

400 FORMAT (s_1,s_2,s_3,s_4)

[1] On the time-sharing system used by the authors for running the majority of the problems in this text, the output commands that were available were PRINT and WRITE. The PRINT statement actuated the teletypewriter and the data were printed out; the WRITE statement transcribed the data onto a *data file* where it could be retained for future use. The WRITE commands available were

WRITE (m,n) *list*	Formatted output to file m
WRITE (m,) *list*	Unformatted output to file m
WRITE (m) *list*	Binary output to file m

the value stored in X will be printed according to specification s_1; the value stored in Y will be printed according to specification s_2; the value stored in XY will be printed according to specification s_3; and the value stored in XSQ will be printed according to specification s_4.

FORMAT statements are unique in that they may be located *anywhere in the* source program as long as they are properly referenced. The one exception to this rule is that a FORMAT statement *may not* be placed in a location which requires an *executable* statement (see Sec. 5.3). A convenient procedure followed by some programmers is to locate the FORMAT statement immediately after the READ or WRITE statement to which it applies.

The reader should note that the FORMAT statement only provides information to the computer, and, therefore, *one* FORMAT statement may be referenced by *several* input and/or output statements in a single program.

2.4 FORMAT SPECIFICATIONS

Format specifications are groups of symbols that provide information to the computer describing the *form* in which data are to be read-in or printed out. These specifications are denoted by $s_1, s_2, \ldots,$ in the general form of the FORMAT statement discussed in Sec. 2.3.

The following *types* of format specifications are available in the FORTRAN language:

1. Numeric specifications
2. Alphameric specifications
3. Control specifications

The first deals with the *transmission of numerical data;* the second deals with the *transmission of literal data;* the third is concerned with the *spacing* of data *during input and output operations.*

2.5 I SPECIFICATION

The *I type specification* is a *numeric* specification for *integer constants* and has the general form

where *w* is an unsigned integer constant indicating the *field width* (i.e., the number of *spaces reserved* for the numerical value).

PART A

When the I specification is used for *input operations,* the following rules apply:

1. The number must be *right-justified* in the specified field.
2. If no algebraic sign is shown, the number is assumed positive.
3. Positive numbers may completely fill the field.

When used for *output operations:*

1. The number will be right-justified in the field.
2. The field must provide one space for the algebraic sign.
3. *No* algebraic sign will be printed if the value is positive (but a space must be provided, nevertheless).

PART B

The following examples of I specifications show how the data would appear for both input and output operations. (The symbol "b" is used in this and succeeding sections to indicate a *blank space.*)

Input operations

EXAMPLE 1

Actual input field = bbb+26
FORMAT specification = I6
Value stored = +26

The + sign is optional. It would be equally valid to use the input field bbbb26. If the data are not right-justified in the field, the low-order positions in the field are "filled" with zeros during the READ operation and the number transmitted to storage is not the one that was intended. (For example, if the value were represented as bbb26b, the computer would store 260.)

EXAMPLE 2

Actual input field = b−40
FORMAT specification = I4
Value stored = −40

The data must be right-justified, and, because the value being represented is negative, the − sign must be shown. Obviously, a space in the field is required for the minus sign.

EXAMPLE 3

Actual input field = 52
FORMAT specification = I2
Value stored = +52

No space has been provided in the input field for an algebraic sign, and therefore the value being read-in is assumed to be positive. The value will be transmitted correctly.

Output operations

EXAMPLE 1

Value in storage = +26
FORMAT specification = I6
Output field = bbbb26

The value is automatically right-justified in the field during the printing operation. The + sign is suppressed by the output device, and its absence implies that the value is positive.

EXAMPLE 2

Value in storage = −40
FORMAT specification = I4
Output field = b−40

The value is right-justified, and the − character is printed in a space provided for it by the field width specification.

EXAMPLE 3

Value in storage = +52
FORMAT specification = I2
Output field = (*Error*)

An *error message* will be printed because the specification does not provide enough space for printing the algebraic sign even though it is not printed when the sign is positive. The data in this example cannot be printed if *w* is less than 3.

2.6 F SPECIFICATION

real

The *F type specification* is a *numeric* specification for *real and double precision*[1] *constants without exponents* and has the form

F*w* · *d*

where *w* is an unsigned integer constant specifying the field width, and *d* is an unsigned integer constant indicating the number of digits to the *right* of the decimal point.

PART A

When the F specification is used for *input operations,* the following rules apply:

1. If *no* algebraic sign is shown, the number is assumed to be positive.
2. If *no* decimal point is shown, the number is assumed to be right-justified in the field and the decimal point is inserted in the specified position by the computer.
3. If a decimal point *is* shown (note that the decimal point will occupy one space in the specified field), its location will override the *d* specification if they conflict.

When used for *output operations:*

1. One space must be reserved for the algebraic sign.
2. One space must be reserved for the decimal point.
3. *No* algebraic sign will be printed if the number is positive.
4. The printed number will be right-justified in the field.

[1] See Sec. 8.4.

PART B

The following examples of F specifications show how the data would appear for both input and output operations.

Input operations

EXAMPLE 1

Actual input field = bbb259
FORMAT specification = F6.1
Value stored = +25.9

The specification indicates that one digit in the field is to be located to the right of the decimal point; therefore, in transmitting the data to storage, the computer will *insert* a decimal point between the 5 and the 9 and the value will be stored as 25.9.

EXAMPLE 2

Actual input field = −25.93
FORMAT specification = F6.1
Value stored = −25.93

The presence of a decimal point in one of the positions in the field *overrides* the specification, and the value will be stored as −25.93. The negative sign must be shown in the input field to indicate that the value is negative.

EXAMPLE 3

Actual input field = 26387
FORMAT specification = F5.2
Value stored = +263.87

In this example the specification exactly fits the data being stored. Since the value is positive, no algebraic sign needs to be shown; therefore, no space is required for this purpose in the input field, and w was chosen equal to the number of *digits* being transmitted, i.e., *five*. The value of d was chosen as *two* to provide for the insertion of a decimal point between the 3 and the 8 during the storage operation.

EXAMPLE 4

Actual input fields = (*a*) bbb26387
= (*b*) 263.87bb
FORMAT specification = F8.2
Value stored = +263.87

Field a The data are right-justified in the field and since the specification requires the last two digits in the field to be located to the right of the decimal point, the value will be stored correctly.

Field b In this example the data are *not* right-justified, but the presence of a decimal point overrides the specification and the value will be stored exactly as printed in the field. The absence of an algebraic sign implies a positive value.

Output operations

EXAMPLE 1

Value in storage = +25.90
FORMAT specification = F6.1
Output field = bb25.9

The decimal point is printed and occupies one of the spaces in the field. Because the value is positive, the algebraic sign is not printed.

EXAMPLE 2

Value in storage = −25.93
FORMAT specification = F6.1
Output field = b−25.9

The specification has requested that only *one* digit be displayed after the decimal point; consequently, only the 9 is printed, and the 3 in the second decimal place is lost. Since the value in storage is negative, the algebraic sign *is* printed.

EXAMPLE 3

Value in storage = +263.87
FORMAT specification = F5.2
Output field = (*Error*)

An *error message* will be printed because the specification does not provide enough space for printing the algebraic sign (even though it would be suppressed for this *positive* value), the decimal point, and the five numeric characters. (Some compilers relax these requirements and automatically compensate for the erroneous format specification.)

EXAMPLE 4

Value in storage = +263.87
FORMAT specification = F8.2
Output field = bb263.87

The value is printed with two digits to the right of the decimal point, just as specified. The + sign is *not* printed.

2.7 E SPECIFICATION

The *E type specification* is a *numeric* specification for *real constants with exponents* and has the form

E$w \cdot d$

where w is an unsigned integer constant specifying the field width, and d is an unsigned integer constant indicating the number of digits to the right of the decimal point.

PART A

When the E specification is used for *input operations,* the following rules apply:

1. If *no* algebraic sign is shown, the number is assumed to be positive.
2. The exponent must be right-justified in the field or it will be "filled" with trailing zeroes.
3. The exponent does *not* have to occupy *four* spaces; for example, the following are all legitimate exponents if right-justified in the field:

   E+02 = E02 = E+2 = E2 = +02 = +2

   E−04 = E−4 = −04 = −4
4. If *no* decimal point is shown, the decimal portion of the number is assumed to be right-justified to the exponent, and the decimal point will be inserted in the specified position by the computer.
5. If a decimal point *is* shown, its position will override the specification if there is a conflict.

When used for *output operations:*

1. *One* space must be reserved for the *algebraic sign.*
2. If the number is positive, *no* algebraic sign will be printed.
3. *One* space must be provided for a *leading zero.*
4. *One* space must be reserved for the *decimal point.*
5. *Four* spaces must be reserved for the exponent.
6. The number will be printed in the *normalized form,* that is, the decimal portion will *follow* the decimal point, *unless* a scale factor is used (see Sec. 2.12).
7. The *maximum* number of digits in the decimal part of the number is the maximum number of significant digits retained by the particular computer being used.
8. The *actual* number of significant digits that will be printed is equal to *d.* If (*w*—*d*) is greater than *seven* (the number of spaces *reserved* for the E conversion; rules 1, 3, 4, and 5 above), *leading blanks* are used to fill the field.

PART B

The following examples of E specifications illustrate how the data would appear for both input and output operations.

Input operations

EXAMPLE 1

Actual input fields	= (*a*) bb30E+06
	= (*b*) bbb3E+07
	= (*c*) bbbbb3E7
	= (*d*) bbbbb3+7
	= (*e*) bbbb.3E8
FORMAT specification	= E8.0
Value stored	= + 30. × 10^6

Field a In this illustration the number is right-justified in the eight-space field and is written in a form closely resembling scientific notation. The fact that the specification calls for zero digits to the right of the decimal point will cause the computer to insert a decimal point between 0 and E, that is, *after* the string of digits representing the constant part of the number. No algebraic sign is present, and therefore the value is assumed to be positive and will be stored correctly.

Field b In this representation of the same value, only the significant 3 has been shown. The specification will cause a decimal point to be inserted between the 3 and the E. The exponent has been changed to +07 to compensate for the leftward displacement of the decimal point and therefore will represent the intended value.

Field c This illustration is the same as (*b*) except the exponent has been represented more succinctly. The absence of an algebraic sign for the exponent implies a positive exponent.

Field d Another method of showing the exponent in an abbreviated form is shown here. Because the symbol E has been omitted, the algebraic sign of the exponent *must* be shown.

Field e In this representation, the presence of the decimal point *overrides* the specification and the constant portion of the value will be transmitted as 0.3. The exponent has been adjusted, and the value will be stored correctly.

EXAMPLE 2

Actual input field = bb34E+06
FORMAT specification = E8.1
Value stored = + 3.4×10^6

The number is right-justified in the field, and the absence of an algebraic sign implies a positive value. The specification will cause a decimal point to be inserted between the 3 and the 4, and the value will be stored correctly.

EXAMPLE 3

Actual input fields = (*a*) bb265830E+00
= (*b*) bbb26583E+01
= (*c*) bb265.83E+00
= (*d*) bb.26583E+03
FORMAT specification = E12.3
Value stored = + 265.83

Field a As the specification calls for three digits after the decimal point, a zero was added to the string of digits representing the decimal part of the number. The computer will insert a decimal point between the 5 and the 8, and, as the exponent is zero, the value will be stored correctly.

Field b Unlike case (*a*), no trailing zero was added, and, therefore, the computer will insert a decimal point between the 6 and the 5. However, the exponent has been represented as +01 and will therefore *relocate* the decimal point and the value will be stored correctly.

Field c The presence of the decimal point overrides the format specification, and the value will be correctly transmitted to storage.

Field d In this illustration the number has been represented in the normalized form. The presence of the decimal point overrides the specification, and the adjustment of the exponent means the value will be correctly transmitted to storage.

EXAMPLE 4

Actual input field = b—2936741E—1
FORMAT specification = E12.4
Value stored = — 29.36741

The representation of the data is right-justified in the field, and the negative sign preceding the 2 indicates that the *value* is negative. The specification will cause the computer to insert a decimal point between the 3 and the 6. The exponent of —1 will *shift* the decimal point one space to the left, and the value will be stored correctly.

Output operations

EXAMPLE 1

Value in storage = + 30. × 10^6
FORMAT specification = E8.0
Output field = bb0.E+08

No significant digits are obtained because the computer always prints in the *normalized form* and the specification called for zero digits after the decimal point. The order of magnitude and algebraic sign of the value *are* indicated.

EXAMPLE 2

Value in storage = + 3.4 × 10^6
FORMAT specification = E8.1
Output field = b0.3E+07

As in Example 1, the value is represented in *normalized form* but with only one significant digit to the right of the decimal point. The exponent has been adjusted to represent the true order of magnitude.

EXAMPLE 3

Value in storage = 265.83
FORMAT specification = E12.3
Output field = bbb0.265E+03

The value is represented in the *normalized form.* As required by the specification, only three significant digits are shown after the decimal point. The exponent will relocate the decimal point, and the value is represented to the desired significance.

EXAMPLE 4

Value in storage = — 29.36741
FORMAT specification = E12.4
Output field = b—0.2936E+02

Once again, representation of the value is in *normalized form* with the four significant digits called for by the specification. The leading negative sign *has* been printed to indicate that the value in storage is a negative value.

2.8 MULTIPLE SPECIFICATIONS

A *multiple specification* in a FORMAT statement has the form

FORMAT $(n_1s_1, n_2s_2, \ldots, n_ms_m)$

where $n_1, n_2, \ldots, n_m$ are unsigned integer constants and $s_1, s_2, \ldots, s_m$ are format specifications. The constants act as *multipliers* or repetition counts for the specifications they *precede.*

PART A

When a specification must be repeated in a FORMAT statement (e.g., when several variables in the *list* are being transmitted under control of the *same* specification), repetition can be indicated by placing an unsigned integer constant in *front* of the specification to indicate the number of times it is to be used. For example, if the first four variables in a list are to be printed as F10.0 and the next two variables as E20.8, the FORMAT statement may be written

```
FORMAT(4F10.0,2E20.8)
```

Note that it is not necessary to use repetition counters with *every* specification in a FORMAT statement. When no counter is shown, it is understood to be 1. The following is a valid FORMAT statement:

```
FORMAT(I6,2F8.2,F10.0)
```

PART B

If the number of items in an input or output list *exceeds* the number of numeric specifications in the referenced FORMAT statement, the specifications are repeated from the *last open parenthesis.* For example, the statements

```
   PRINT 65,A,B,C,X
65 FORMAT(2F10.0,E12.8)
```

will print A and B as F10.0 and C as E12.8 *on one line,* and X will be printed as F10.0 *on the next line.* If the operation had been *input,* and the executable statement was

```
   READ 65,A,B,C,X
```

then A, B, and C would have been read (according to their respective formats) from the *first* data card, and X would have been read from the *next* data card as F10.0.

If the number of items in an input or output list is *less than* the number of numeric specifications in the referenced FORMAT statement, the input or output operation is ended when all the listed variables have been read or printed. For example, the statements

```
   READ 90,A,B,C,D
90 FORMAT(6F8.2,2I3)
```

will cause the computer to read *four* fields under the F8.2 specifications; the remaining *two* F8.2 and the *two* I3 specifications will be ignored.

2.9 H SPECIFICATION

The *H (or Hollerith) type specification* is an *alphameric* specification for alphabetic *and* numeric data and has the form

wH

where *w* is an unsigned integer constant indicating the field width.

Under control of the Hollerith specification, the computer reads or prints the alphameric characters (including special characters *and* blanks) that are contained in the designated field. As it is used in the usual FORMAT statement, the *designated field* consists of the *w* characters *immediately* following the symbol H in the statement. For example, in the specification 8HANSWERb= the *field* consists of the 8 characters ANSWERb=; and in the specification 16HPOUNDSbPERbSQbIN the *field* consists of the 16 characters POUNDSbPERbSQbIN.

PART A

When used for *output*, the *w* characters that immediately follow the H in the FORMAT statement (*or* the *w* characters that *replaced* those shown in the FORMAT statement; see Part B of this section) are printed.

Some examples of the use of the Hollerith specification follow.

EXAMPLE 1

If storage location X contains the value +812.39645, the statements

```
   PRINT 30, X
30 FORMAT(8HANSWERb=F10.3)
```

will cause the computer to print

```
ANSWERb=bbb812.396
```

EXAMPLE 2

If storage location S contains a value of +28563.714, the statements

```
   PRINT 76, S
76 FORMAT(8HSTRESSb=F10.1,4HbPSI)
```

will cause the computer to print

```
STRESSb=bbb28563.7bPSI
```

PART B

When used for *input* operations, the *w* characters in the field being read *will replace* (in the computer storage) the *w* characters that actually follow the H in the format specification. For example, if a data card being read by the computer contains the following symbols in the first 15 columns:

```
    EXCHANGEbSETbbb
```

the statements

```
    PRINT 10
 10 FORMAT(15HbSAMPLEbFORMATb)
    READ 10
    PRINT 10
```

will cause the computer to print the two lines

```
    bSAMPLEbFORMATb
    EXCHANGEbSETbbb
```

The READ 10 statement *replaced* bSAMPLEbFORMATb with the *new* legend. Unless another similar revision of FORMAT statement 10 is made, all subsequent PRINT 10 commands will cause EXCHANGEbSETbbb to be printed.

Some FORTRAN compilers permit *literal alphameric data* (that is, strings of alphabetic, numeric, and special characters) to be transmitted from a FORMAT statement if the data are included between either *apostrophes* or *quotation marks.*[1] The effect produced with both input *and* output operations is the same as that produced with the Hollerith specification. For output operations, the "quoted literal" has an advantage over the Hollerith specification in that it is not necessary to obtain an exact count of the characters being printed. The computer will print exactly what is contained between the quotation marks. For this reason, the feature is quite attractive to the programmer.

2.10 X SPECIFICATION

The *X type* or *blank field specification* is a control specification having the form

$$wX$$

where w is an unsigned integer constant. This specification is used to indicate the width of a blank field.

PART A

When used for *output* operations, a blank field w spaces wide is inserted in the output data. For example, if 16.28 is stored in location A and 6 is stored in location N, the statements

```
    PRINT 40, A,N
 40 FORMAT(F6.1,5X,I3,6HbSTEPS)
```

will print the line

```
    bb16.2bbbbbbb6bSTEPS
      6     5   3   6
```

[1] The time-sharing system used for preparing most of the programs in this book employed quotation marks.

PART B

When used for input operations, the *w* characters in the *record* being scanned by the reading device are *replaced* with blanks in the computer storage. For example, if a card with

```
bbb6 83215bbbb
 4    4    5
```

punched in the first 13 columns is read under control of

```
   READ 10,N,D
10 FORMAT(I4,4X,F5.2)
```

the computer will store 6 in location N and 500.00 in location D. The four-digit group 8321 was skipped during the scanning process because of the blank field specification 4X.

2.11 CONTROL SPECIFICATIONS

There are two types of specification that are used as part of the FORMAT statement to *control line spacing* during *output* operations: the *solidus* (or *slash*) and the *printer control character.* The solidus may also be used to control *input* operations.

PART A

In any FORMAT statement, the solidus (/) has the effect of returning the *printer* (or *reader*) to the beginning of a new *output line* (or *input record*) before continuing with the rest of the specifications in the statement. For example, the *output* operations controlled by the statements

```
   PRINT 22,A,B,C,D
22 FORMAT(2F8.0/2F10.3)
```

would print the contents of A and B on one line according to specification F8.0, and the contents of C and D on the *next* line according to specification F10.3.

Several slashes grouped together in a FORMAT statement have the effect of producing *blank lines* during *output,* or of *ignoring* data cards (i.e., data records) during *input.* For example, the *output* operation indicated by the statements

```
   PRINT 83,X,Y,Z
83 FORMAT(E20.8///F8.3/F8.3///)
```

would print the contents of location X on the first line (specification E20.8), skip *two* lines (one for *each space* between the slashes), print the contents of Y in the first eight spaces on the *fourth* line (specification F8.3), return to the left-hand margin on the next line (the *fifth*) and print the contents of Z in the first eight spaces (specification F8.3), skip *three* lines (one for *each space* between slashes and parenthesis), and return to the left-hand margin of line nine to await further printing commands.

Skipping a print line, as controlled by the space between a slash and a parenthesis, may be effected at either the beginning or the end of the FORMAT statement specification list. For example, the statements

```
   PRINT 15,A
15 FORMAT(/F8.3)
```

would produce a blank line *followed* by the contents of location A on the second line. The statements

```
    PRINT 16,A
16 FORMAT(F8.3/)
```

would produce the contents of location A on the first line, and the *second* line would be blank.

PART B

As noted above, the *solidus* may also be used to control *input* operations. For example, on *input,* the statements

```
    READ(1,20)J,K,L,A
20 FORMAT(2I4/I3,10XF8.2)
```

would cause the card reader to read values for J and K from the first 8 columns of a card (each value according to specification I4), return to the *first* column of the *next* card and read a value for L from the first 3 columns (specification I3), *skip* over the next 10 columns (specification 10X), and read a value for A from the next 8 columns (specification F8.2). Note that a comma is *not* required after an X specification.

A *group of several slashes* will cause the card reader to ignore one or more data cards. For example, during *input*, the statements

```
    READ(1,40)A,B,C
40 FORMAT(F8.0///2F10.5)
```

would cause the card reader to read a value for A from the first 8 columns of the first card (specification F8.0), *skip* the next *two* cards (one for *each space* between the slashes), and read values for B and C from the first 20 columns of the *fourth* card (specification 2F10.5).

The skipping of an input card, as controlled by the space between a slash and a parenthesis, operates in a similar manner to the process described in Part A of this section for printer control.

If the computer *output is to be printed by a high-speed printer,* the *first* character of each printed line must be a *control character.* The control character is suppressed during the printing operation; that is, it is not actually printed, but it must be present in the output format specification.

The most commonly available control characters, and the action they produce are:

1. b–causes the printer to *single-space*
2. 0–causes the printer to *double-space*
3. 1–causes the printer to jump to the next page

The control character can be placed at the beginning of an output line by using a Hollerith specification in the FORMAT statement. For example, the statements

```
    WRITE(3,19)A,B
19 FORMAT(1H0,2F8.1)
```

would cause the printer to double-space *before* printing the contents of A and B according to specification 2F8.1 on the *second* line. That is, *one* blank line would be produced, with the values of A and B on the *next* line.

2.12 SCALE FACTORS[1]

The *scale factor* is a modifying specification that may be applied to *numeric* specifications for *real* and *double precision* values (see Sec. 8.4), and has the form:

*n*P

where *n* is a positive or negative integer constant. The scale factor has the effect of *relocating* the decimal point in a number undergoing conversion. When it is used, the scale factor is part of the FORMAT statement and is written *in front of* the *first* numeric specification to which it applies. *All* the real and double precision specifications *following* the scale factor in a FORMAT statement *are affected by it.* A zero scale factor has *no* effect and can be used to remove the effect of a leading scale factor in a FORMAT statement.

During either input *or* output operations, with F specification numbers, the decimal point is moved *n* places to the *right* if the scale factor is positive and *n* places to the *left* if the scale factor is *negative.*

Some examples of scaling F specifications during *input* operations follow.

EXAMPLE 1

Actual input field = −25.693
FORMAT specification = 1PF7.0
Value stored = − 256.93

EXAMPLE 2

Actual input field = bb391264
FORMAT specification = −3PF8.2
Value stored = 3.91264

EXAMPLE 3

Actual input field = 125
FORMAT specification = 4PF3.0
Value stored = 1,250,000.

Some examples of scaling F specifications during *output* operations follow.

EXAMPLE 1

Value in storage = + 267.854
FORMAT specification = −2PF6.3
Output field = b2.678

[1] This entire section may be omitted on first reading.

EXAMPLE 2

Value in storage = + 0.03875
FORMAT specification = 3PF4.0
Output field = b38.

The reader should note that the scale factor, when applied to F specifications, *changes the value* of the number undergoing conversion.

The scale factor is applicable to E and D (see Sec. 8.4) type conversions *only* during *output* operations. You will recall that for normal output under E (and D) specification control, the decimal point leads the string of numeric characters representing the significant digits of the value. When a scale factor is applied, the decimal point in front of the string is *moved,* in the same manner as described above for F specification conversion. However, the exponent part of the representation is simultaneously *adjusted* to compensate for the relocation of the decimal point, and the number being represented *remains unchanged in value.*

Some examples of scaling E specifications during *output* operations follow.

EXAMPLE 1

Value in storage = + 262.831
FORMAT specification = 0PE 10.3
Output field = bb.262E+03

EXAMPLE 2

Value in storage = + 262.831
FORMAT specification = 1PE10.3
Output field = b2.628E+02

Note that while a scale factor *changes the value* of F specification numbers on output, it merely *changes the appearance* of E specification numbers.

PROBLEMS

For all the problems in this chapter, assume the system you are working on has the following device code numbers for the various pieces of hardware connected to the computer:

1 = card reader
2 = card punch
3 = line printer
4 = magnetic disk
5 = magnetic tape
6 = paper tape reader
7 = paper tape punch
66 = electric typewriter or teletype

In the following problems, describe the input or output action produced by each of the statements.

2-1. READ 50,A,B,C

2-2. READ, A,B,C

2-3. READ(66,60)X,Y,Z

2-4. READ(66,)X,Y,Z

2-5. READ(1,75)A,B,N

2-6. READ(4,10)ITEM,VALUE

2-7. PRINT,A,X

2-8. PRINT 65,A,X

2-9. READ(5,100)T,F

2-10. WRITE(66,250)A,B,Z

2-11. WRITE(7,250)A,B,Z

2-12. READ(6,65)A,X

2-13. WRITE(4,250)A,B,Z

2-14. WRITE(5,100)T,F

2-15. WRITE(66,)A,X

In the following group of problems, show the form of the output produced by each of the given FORMAT statements. Show numerical data by writing an X for each digit in the number. In the case of real numbers, show the location of the decimal point. (For example, to represent a number printed under conversion specification F8.3, write xxxx.xxx.) Blanks are to be indicated by b, and literal alphameric data by the appropriate characters.

2-16. FORMAT(F7.3,I6)

2-17. FORMAT(F7.3,5XI6)

2-18. FORMAT(F7.3/I6)

2-19. FORMAT(2F7.3,I4)

2-20. FORMAT(2F7.3/I4)

2-21. FORMAT(F7.3,10HCUBICbFEET)

2-22. FORMAT(5HbbAb=F7.3,5HbFEET)

2-23. FORMAT(22H**bTABLEbOFbRESULTSb**//)

2-24. FORMAT(5HbDEPT6X14HUNITSbPRODUCED5X5HHOURS)

2-25. FORMAT(E18.8)

2-26. FORMAT(E12.4,E20.6)

2-27. FORMAT(E18.6/E16.4)

2-28. FORMAT(2PE16.4)

2-29. FORMAT(1PE16.4,E18.6)

2-30. FORMAT(1PE16.4,0PE18.6)

2-31. FORMAT(1P2E16.4)

2-32. FORMAT(1PE16.4,–1PE16.4)

For the next set of problems, show the decimal representation of the values stored in the named locations when the following set of data cards are read:

Column number:	1	2	3	4	5	6	7	8	9	10	11	12	13	14	15	16	17	18
First card:			2	4						7	6	9	8	5				
Second card:				8	4	3								4	2	1		
Third card:	1	2	.	9							7	1	.	5				
Fourth card:				6	3					5	6	.	3	9	5		4	7

```
2-33.      READ(1,100)A,B,C
       100 FORMAT(3F6.2)
2-34.      READ(1,101)A,B,C
       101 FORMAT(F6.2/F6.2,F12.7)
2-35.      READ(1,102)A,B,C,N
       102 FORMAT(//2F10.0/F6.3,I4)
2-36.      READ(1,103)A,B,C,N
       103 FORMAT(2F3.0,F6.2,I2)
2-37.      READ(1,104)A,B,C,N,K
       104 FORMAT(10XF4.2/2F6.2//I4,2XI4)
```

For the next group of problems, assume the following set of data cards are being read. Show the representation of what is stored in the computer.

Column number:	1	2	3	4	5	6	7	8	9	10	11	12	13	14	15	16
First card:	8	6	5	9	E	+	8		3	1	6	4	9	8		
Second card:		4	0	6				−	2	2	0	8	0	3	7	
Third card:	B	E	S	T		V	A	L	U	E						

```
2-38.       READ(1,105)A,B,N
        105 FORMAT(E7.2,F7.2/I3)
2-39.       READ(1,106)A,B,C,N
        106 FORMAT(10X3F1.0,I1)
2-40.       READ(1,107)AMT,MO,NDA,NYR
        107 FORMAT(/F5.0,5X3I2)
2-41.       READ(1,108)A,B
            READ(1,109)
        108 FORMAT(10XF6.3/F6.1)
        109 FORMAT(10HSUBSTITUTE)
2-42.       READ(1,110)A
        110 FORMAT(F4.2//3X3H***)
```

For the next group of problems, assume the decimal representation of the values stored in the named computer locations are

A =	+263.8472	I = +42
B =	+69431.3779	J = −113
C =	+0.0000062812	K = +9
X = +160000000.		
Y =	−6.8319	
Z =	+91.274	

Assume *unformatted output* is governed by the following standard:

Real values are converted in the form 1PE13.5.

Integer values are converted in the form I9.

Show the computer output produced by the following statements:

```
2-43.       PRINT,A,B,J
2-44.       PRINT 36,A,B,J
         36 FORMAT(2F10.2,I5)
2-45.       PRINT 37,A,B,J
         37 FORMAT(2F10.2//I8)
2-46.       PRINT 38,A,I
         38 FORMAT(6HROOTb=F10.3,I10,11HbITERATIONS)
2-47.       WRITE(66,37)A,B,J
         37 FORMAT(2F10.2//I8)
2-48.       WRITE(3,39)A,C
         39 FORMAT(1X2E15.4)
2-49.       WRITE(3,40)A,C
         40 FORMAT(1H0F10.2,E12.3)
2-50.       WRITE(3,41)A,C
         41 FORMAT(1XF10.2,E12.3)
2-51.       WRITE(3,42)X,Y,Z
         42 FORMAT(1H13E12.3)
2-52.       WRITE(3,43)X,Y,Z
         43 FORMAT(1H11P3E12.3)
2-53.       WRITE(3,44)I,J,K
         44 FORMAT(6X1HI5X1HJ5X1HK/1X3I6)
```

2-54.
```
   WRITE(3,45)K,Z
45 FORMAT(1H112HTRIALbNUMBERI3/6X10HVELOCITYb=F10.2, 7HbFT/SEC)
```
2-55.
```
   WRITE(2,46)A,B,X
46 FORMAT(3E11.4)
```
2-56.
```
   WRITE(66,b)A,B,J
```
2-57.
```
   WRITE(5)I
```
2-58.
```
   WRITE(4)A,B
```
2-59.
```
   WRITE(3,47)A,X
47 FORMAT(1X–3PF6.3,F15.3)
```
2-60.
```
   WRITE(3,48)
48 FORMAT(23Hb**bTABLEbOFbRESULTSb**/)
```
2-61.
```
   WRITE(3,49)K,Y
49 FORMAT(17H1COMPUTEDbRESULTS/12H0CASEbNUMBERI5/6X3HYb=F8.3)
```
2-62.
```
   PRINT,A
   PRINT,
   PRINT,B
```
2-63.
```
   WRITE(3,50)Z,I,A
50 FORMAT(1X1PF8.2,I6,F10.2)
```
2-64.
```
   WRITE(3,51)Z,I,A
51 FORMAT(1X1PF8.2,I6,E12.4)
```
2-65.
```
   WRITE(3,52)Z,I,A
52 FORMAT(1X1PF8.2,I6,0PE12.4)
```

3
The Computer as a Calculator

In this chapter, we will examine a variety of short problems in which the computer is used as a calculator. The experience gained will provide a useful background for understanding the programs of the subsequent chapters. As previously mentioned, most of the programs in this and in later chapters have been written for and tested on a time-shared computer system. This computer system permits the programmer to violate some of the rules of standard FORTRAN in order to simplify the writing of programs. However, the authors have avoided the use of most of these simplifications so that the programs discussed in this book can be used on any computer. One important exception to this policy is the use of unformatted input and output, which is discussed below. As the reader will see, the use of unformatted input and output presents no conceptual difficulty to the beginner. On the contrary, it enables the programmer to concentrate his attention on the logic of the program.

3.1 THE EQUATION OF A PARABOLA

The equation of a parabola is

$$y = x^2$$

Figure 3.1.1 is a program for evaluating the function x^2 and storing the numerical value in location Y. The five-digit numbers at the left, 00001, 00002, 00003, . . . , are called *line sequence numbers.* They represent a convenient and necessary method of identifying the sequential position of the statements which constitute the program. Usually, the order in which statements appear is critical. The line sequence numbers increase in steps of 10 as a matter of convenience, and if the

```
00001C |    PARAB1
00002C |    THIS PRØGRAM EVALUATES A FUNCTIØN AND ALSØ ILLUSTRATES
00003C |    VARIØUS PRINT-ØUT METHØDS.
00010  |    READ, X
00020  |    Y=X**2
00030  |    PRINT 20, Y
00040  | 20 FØRMAT(F6.3)
00050  |    PRINT 30, Y
00060  | 30 FØRMAT(2HY=F6.3)
00070  |    PRINT, Y
00080  |    PRINT, "Y=", Y
01000  |    STØP
01010  |    END


INPUT:00010
 ?2.5

 6.250
Y= 6.250
        6.2500
Y=          6.2500
STØP
```

Fig. 3.1.1 Program PARAB1 and trial run.

programmer wishes to add additional statements *between* those that presently exist, he can easily do so. The zeros before the digits were added to maintain uniformity of appearance as the line sequence numbers increase from one to two or more digits (for example, 00001 and 00030).

Immediately to the right of line sequence numbers 00001, 00002, and 00003 there is a letter C. This denotes that the information appearing on these lines represents *comments* for user convenience. Comments are not part of the executed program. In the computer used in these programs, the first column to the right of the line sequence numbers is always blank *unless a comment is to be indicated.* The statement READ, X on line 00010 signifies that when the program is run, a numerical value is to be supplied for the real variable X.

```
00001C |    PARAB2
00002C |    THIS PRØGRAM ILLUSTRATES A PRØGRAMMING ERRØR.
00010  |    Y=X**2
00020  |    READ, X
00030  |    PRINT 20, Y
00040  | 20 FØRMAT(F6.3)
00050  |    PRINT 30, Y
00060  | 30 FØRMAT(2HY=F6.3)
00070  |    PRINT, Y
00080  |    PRINT, "Y=", Y
01000  |    STØP
01010  |    END


INPUT:00020
 ?2.5

 0.000
Y= 0.000
        0.0000
Y=          0.0000
STØP
```

Fig. 3.1.2 Program PARAB2 and trial run.

The computer then calculated the square of X, or 6.25, and this number was stored in location Y. The answer was printed four times because four format statements were written into the program. Lines 00030 and 00040 show a conventional format for printing the answer. The same format appears on line 00050 and on line 00060, combined with the use of Hollerith conversion, so that Y = is printed in *front* of the answer. On lines 00070 and 00080 two examples of unformatted output are shown. This feature is not available on all computers.

In Fig. 3.1.2, the program of Fig. 3.1.1 is repeated. However, one significant change was made. Lines 00010 and 00020 were interchanged, and, as a result, a value of Y = 0 was calculated. The reason is that the computer always executes the statements of the program sequentially, unless programmed to do otherwise. Thus, line 00010 (the calculation statement) was executed before line 00020 (the READ statement). Therefore, Y was evaluated as the square of X. No data had yet been supplied for X, so X was taken as zero. Therefore, the calculation yielded Y = 0.

3.2 ILLUSTRATION OF THE E AND THE F FORMATS

In Fig. 3.2.1 we show a short program for calculating sin x and storing the value in location Y. The value of Y has been printed twice, once using F specification, and once using E specification (see lines 00040 and 00060). The program was run three times: the first time with X = .1; the second with X = .01; and finally with X = .001.

The F specification is 11.8, and the E specification is 15.8. Thus, in each case the program specifies eight digits to the right of the decimal point. In the F specification, the eight digits include zeros to the right of the decimal point. In the E specification, the first significant digit *immediately* follows the decimal point, and any

```
00001C |   LIBRA1
00002C |   IN THIS PRØGRAM A CØMPUTER SUPPLIED PRØGRAM IS USED TØ
00003C |   EVALUATE THE SINE FUNCTIØN.ALSØ,DIFFERENCES BETWEEN THE
00004C |   E AND F FØRMATS ARE ILLUSTRATED.
00010  |   READ,X
00020  |   Y=SIN(X)
00030  |   PRINT 5,Y
00040  | 5 FØRMAT(2HY=F11.8)
00050  |   PRINT 7,Y
00060  | 7 FØRMAT(2HY=E15.8)
00070  |   STØP
00080  |   END


INPUT:00010
 ?.1

Y= 0.09983342
Y= 0.99833417E-01

INPUT:00010
 ?.01

Y= 0.00999983
Y= 0.99998333E-02

INPUT:00010
 ?.001

Y= 0.00100000
Y= 0.99999983E-03
```

Fig. 3.2.1 Program LIBRA1 and trial runs.

leading zeros are incorporated in the exponential part of the printout. Therefore, the E specification supplies more significant figures. In fact, in the third run, with X = .001, roundoff occurs, and the answer using F specification contains only one significant digit, namely unity.

3.3 NUMBER OF TURNS REQUIRED FOR A SPECIFIED INDUCTANCE[1]

The inductance L of a coil with square cross section is approximately

$$L = 0.000008\left[N^2 S \ln \frac{D}{B} + 0.3794 + 0.4714\frac{B}{D} - 0.0143\left(\frac{B}{D}\right)^2 - 0.029\left(\frac{B}{D}\right)^4\right] \tag{3.3.1}$$

The symbols are defined as follows.

Formula symbol	*Computer name*	*Meaning*	*Units*
L	XL	Inductance	mH
N	XN	Turns	
B	B	Winding length	cm
S	S	Length of side of coil cross section	cm
D	DIAM	Effective diameter	cm

For convenience, let the expression in brackets be represented by the symbol F. Equation (3.3.1) can now be represented compactly as

$$L = 0.000008 N^2 SF \tag{3.3.2}$$

The usual engineering problem is: Given a specified inductance, find the required number of turns. Therefore, Eq. (3.3.2) is solved for N, the required number of turns, as follows:

$$N = \sqrt{\frac{L}{0.000008 SF}} \tag{3.3.3}$$

In Fig. 3.3.1, a program called INDUCT is given for the solution of this problem. The data supplied for the problem at hand are

Inductance (XL) = 3 mH
Side of square winding form = 32 cm
Available winding length = 2.54 cm

Equation (3.3.3) appears on line 00110 of the program. Notice that in the program the inductance L is given the real name XL. This is done because inductance is usually not an integer; furthermore, the argument of the square root function must not be in integer mode. Therefore, all constants and variables on the right side of the equation are in real mode, and the number of turns N is calculated in real mode. In general, a fractional number of turns has no physical significance. By equating the expression on the right side of the arithmetic statement to an integer N, the digits to the right of the decimal point are truncated and the answer is converted to integer mode.

[1] F. W. Grover, "Inductance Calculations," Dover, New York, 1962, p. 175.

```
C     INDUCT
C     THIS PRØGRAM CALCULATES THE INDUCTANCE ØF SHØRT SQUARE
C     CØILS.  (SEE GRØVER, PAGE 175)
      PRINT,"INDUCTANCE IN MHY."
      READ,XL
      PRINT,"LENGTH ØF SIDE ØF CØIL IN CM."
      READ,S
      PRINT,"WINDING LENGTH IN CM."
      READ,B
      DIAM=1.414*S
      RAT=DIAM/B
      REC=1./RAT
      F=ALØG(RAT)+.3794+.4714*REC-.0143*(REC**2)-.029*(REC**4)
      N=SQRT(XL/(S*F*8.E-6))
      PRINT 20,N
   20 FØRMAT(15HREQUIRED TURNS=I6)
      STØP
      END
```

```
INDUCTANCE IN MHY.
INPUT:00020
 ?3.

LENGTH ØF SIDE ØF CØIL IN CM.
INPUT:00040
 ?32.

WINDING LENGTH IN CM.
INPUT:00060
 ?2.54

REQUIRED TURNS=     59
STØP
```

Fig. 3.3.1 Program INDUCT and trial run.

Another feature of interest in this program is that captions are printed before the question marks when data is requested. In this way, the user of the program is assisted in entering the data in the proper sequence.

3.4 BODY RESISTANCE OF A SEMICONDUCTOR DIODE[1]

The equation for the current through a biased germanium junction diode is

$$I = I_o e^{qV/KT} - 1 \tag{3.4.1}$$

The symbols are defined as follows.

Formula symbol	*Computer name*	*Meaning*	*Units*
I	XI	Diode current	A
I_o	XIO	Reverse saturation current	A
q	Q	Electron charge, 1.6×10^{-19}	C
V	VJ	Voltage across the junction of the diode	V
K	XK	Boltzmann constant, 1.38×10^{-23}	J/degree
T	T	Junction temperature	°K

[1] J. Millman and C. C. Halkias, "Electronic Devices and Circuits," McGraw-Hill, New York, 1968, p. 127.

This equation does not take into account the ohmic resistance of the semiconductor material of which the diode is made. Therefore, if a battery of voltage *VB* is connected in series with the diode so as to forward bias it, the current which flows is *less* than would be expected from Eq. (3.4.1).

If the current I for a given value of *VB* is known, the ohmic resistance of the diode can be found. First, solve for V from Eq. (3.4.1):

$$V = \frac{KT}{q} \ln \left(\frac{I}{I_o} + 1 \right) \tag{3.4.2}$$

In Eq. (3.4.2), I is the current found when forward bias *VB* is applied. I_o is found by laboratory measurement with reverse bias. Next, use Ohm's law to find the ohmic resistance *RB*:

$$RB = (VB - V)/I \tag{3.4.3}$$

The program is shown in Fig. 3.4.1. Equations (3.4.2) and (3.4.3) appear on lines 00040 and 00050, respectively. The symbol V was replaced by *VJ* to emphasize that it represents junction voltage. The data used in running the program were (in proper order)

$$\begin{aligned} I &= 0.0217 \text{ A} \\ I_o &= 10^{-5} \text{ A} \\ VB &= 0.74 \text{ V} \\ T &= 300^\circ\text{K} \end{aligned}$$

This is the sequence required by the input statement on line 00010.

If one wishes to use integer names in floating point mode, the predefined convention of implicitly indicating data mode by the initial letter of the variable name can be overridden by means of a REAL declaration statement (see Sec. 8.1). In Fig. 3.4.2, the program of Fig. 3.4.1 has been modified by inclusion of a REAL declaration.

```
00001C    | DIØDE1
00002C    | THIS PRØGRAM CALCULATES SEMICØNDUCTØR DIØDE BØDY
00003C    | RESISTANCE.
00010     | READ,XI,XIØ,VB,T
00020     | XK=1.38E-23
00030     | Q=1.6E-19
00040     | VJ=((XK*T)/Q)*ALØG(XI/XIØ+1.)
00050     | RB=(VB-VJ)/XI
00060     | PRINT,"VJ=",VJ
00070     | PRINT,"RB=",RB
00080     | STØP
00090     | END


INPUT:00010
 ?.0217,1.E-5,.74,300.

VJ=        0.1988
RB=       24.9403
STØP
```

Fig. 3.4.1 Program DIODE1 and trial run.

```
C  DIØDE2
C  THIS PRØGRAM CALCULATES SEMICØNDUCTØR DIØDE BØDY
C  RESISTANCE.
   REAL I,IØ,K
   READ,I,IØ,VB,T
   K=1.38E-23
   Q=1.6E-19
   VJ=((K*T)/Q)*ALØG(I/IØ+1.)
   RB=(VB-VJ)/I
   PRINT,"VJ=",VJ
   PRINT,"RB=",RB
   STØP
   END


INPUT:00020
 ?.0217,1.E-5,.74,300.

VJ=        0.1988
RB=       24.9403
STØP
```

Fig. 3.4.2 Program DIODE2 and trial run.

3.5 PERFORMANCE OF A ONE-STAGE TRANSISTOR AMPLIFIER[1]

If the H parameters of the transistor used in an amplifier stage are known, the current gain, input resistance, voltage gain, and output resistance of the amplifier can be calculated as a function of the source resistance and load resistance. A program which performs these calculations for a transistor amplifier operating in the midband frequency range is shown in Fig. 3.5.1.[2]

```
C    AMPLI1
C    THIS PRØGRAM CALCULATES THE VØLTAGE AND CURRENT GAIN
C    AND THE INPUT AND ØUTPUT RESISTANCE ØF A TRANSISTØR
C    AMPLIFIER STAGE GIVEN THE H PARAMETERS ØF THE TRANSISTØR,
C    THE LØAD RESISTANCE,AND THE SØURCE RESISTANCE.
     READ,HI,HØ,HF,HR,RL,RS
     DEL=HI*HØ-HF*HR
     AI=HF/(1.+HØ*RL)
     RI=(HI+DEL*RL)/(1.+HØ*RL)
     AV=(-HF*RL)/(HI+DEL*RL)
     RØ=(HI+RS)/(DEL+HØ*RS)
     PRINT 5,HI,HØ,HF,HR
   5 FØRMAT(3HHI=I6,3X3HHØ=E8.2,3X3HHF=I5,3X3HHR=E8.2)
     DECVØL=8.686*ALØG(ABS(AV))
     PRINT 6,AI,AV,RI,RØ
   6 FØRMAT(3HAI=F7.2,3X3HAV=F7.2,3X3HRI=F8.1,3X3HRØ=F8.1)
     PRINT,"DECIBEL VØLTAGE GAIN"
     PRINT,DECVØL
     STØP
     END

INPUT:00010
 ?2500.,5.E-5,49.,2.E-3,2000.,1000.

HI=  2500   HØ=0.50E-04   HF=   49   HR=0.20E-02
AI=  44.55   AV= -38.37   RI=  2321.8   RØ= 45454.5
DECIBEL VØLTAGE GAIN
      31.6805
STØP
```

Fig. 3.5.1 Program AMPLI1 and trial run.

[1] Millman and Halkias, *op. cit.*, p. 311.

[2] In the midband frequency range, the voltage and current are relatively constant.

These calculations are shown in Fig. 3.5.1. The equations involved are

$$A_I = \frac{-h_f}{1 + h_o R_L}$$

$$R_i = \frac{h_i + (\text{del})R_L}{1 + h_o R_L}$$

$$A_v = \frac{-h_f R_L}{h_i + (\text{del})R_L}$$

$$R_o = \frac{h_i + R_s}{(\text{del}) + h_o R_s}$$

$$\text{del} = h_i h_o - h_f h_r$$

The symbols are defined as follows.

Formula symbol	*Computer name*	*Meaning*	*Units*
h_f	HF	Short-circuit forward current gain	
h_r	HR	Reverse open-circuit voltage amplification	
h_i	HI	Input resistance with output short-circuited	ohms
h_o	HO	Output conductance with input open-circuited	ohms
A_I	AI	Current gain	
A_V	AV	Voltage gain	
R_i	RI	Input resistance	ohms
R_o	RO	Output resistance	ohms
R_L	RL	Load resistance	ohms
R_s	RS	Source resistance	ohms

The equations for current gain, input resistance, voltage gain, and output resistance are on lines 00030, 00040, 00050, and 00060, respectively. Because the quantity $(h_i)(h_o) - (h_f)(h_r)$ appears in three of these equations, it is convenient, when programming, to give this expression a name DEL and then use this name in the remainder of the program.

Voltage gain (dB) = $20 \log_{10} A_V$.

This formula appears on line 00090. The out-of-line function ALOG(X) (see Sec. 1.22) returns a value of the natural logarithm of X to the program, and so it is necessary to replace the constant 20 by 8.686 (that is, $20 \log_{10} e \approx 8.686$). Also notice that the argument of the log function is specified as ABS(AV) because voltage gain often is a negative quantity, and the logarithm of a negative quantity is undefined.

3.6 ECCENTRICALLY LOADED COLUMN

The maximum stress produced in the eccentrically loaded column shown in Fig. 3.6.1 is given by the secant formula[1] as

$$\sigma_{max} = \frac{P}{A}\left[1 + \frac{ec}{r^2}\sec\left(\frac{L}{2r}\sqrt{\frac{P}{AE}}\right)\right] \tag{3.6.1}$$

Formula symbol	*Computer name*	*Meaning*	*Units*
σ_{max}	S	Maximum stress	lb/in^2
P	P	Applied load	lb
A	A	Cross-sectional area of column	in^2
e	ECCEN	Eccentricity of loading	in
c	C	Distance from neutral axis to edge of section	in
r	R	Radius of gyration of the cross section	in
L	XL	Length of column	in
E	E	Modulus of elasticity	lb/in^2

Recalling that sec $\theta = 1/(\cos\theta)$, we can rewrite the expression for computer evaluation as

$$\sigma_{max} = \frac{P}{A}\left[1 + \frac{ec}{r^2\cos\left(\frac{L}{2r}\sqrt{\frac{P}{AE}}\right)}\right] \tag{3.6.2}$$

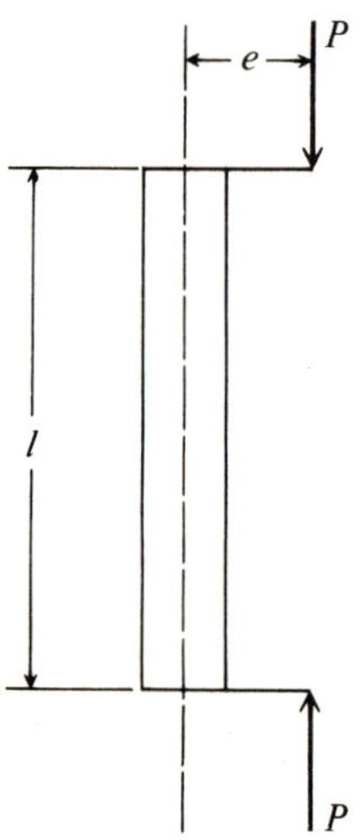

Fig. 3.6.1 Eccentrically loaded column.

[1] S. Timoshenko and D. H. Young, "Elements of Strength of Materials," 4th ed., Van Nostrand, Princeton, N.J., 1962, p. 283.

```
C    CØLUM1
C    THIS PRØGRAM EVALUATES THE MAXIMUM STRESS IN AN ECCENTRIC-
C    ALLY LØADED CØLUMN BY MEANS ØF THE SECANT FØRMULA.
     READ,P,A,ECCEN,C,R,XL,E
     S=P*(1.+ECCEN*C/(R*R*CØS(XL*SQRT(P/A/E)/2./R)))/A
     PRINT 20,S
20   FØRMAT(16HMAXIMUM STRESS =E12.4,2X3HPSI/)
     STØP
     END


INPUT:00010
?4.E4,29.43,15.,5.56,4.61,120.,3.E7

MAXIMUM STRESS =  0.6713E+04  PSI

STØP
```

Fig. 3.6.2 Program COLUM1 and trial run.

The change from secant to cosine was necessary to make it possible to use the available FORTRAN-supplied function for the cosine. Note that there is a square root within the argument for the cosine function. This square root must *itself* be evaluated by the appropriate FORTRAN-supplied function before the cosine function can be evaluated.

The program for evaluating the expression in Eq. (3.6.2) is shown in Fig. 3.6.2 with a sample run. This program illustrates the concept that the argument of a function can be *any* expression in the *proper mode.*

3.7 CRITICAL GAP FOR ENTRY INTO A TRAFFIC STREAM[1]

One of the factors used in evaluating the effectiveness of a ramp leading to a highway is the critical gap T. It is assumed that a driver waits on a ramp until he estimates a time gap between successive vehicles which he regards as sufficiently long for him to safely enter the traffic stream. This critical gap T is a function of the angle between the ramp and the highway, and the length of the acceleration lane (see Fig. 3.7.1). The critical gap can be calculated from the following equation:

$$T = 5.547 + 0.828\theta - 1.043D + 0.045D^2 - 0.042\theta^2 - 0.847S \qquad (3.7.1)$$

where T = critical gap, s
θ = angle between ramp and highway, degrees
D = length of acceleration lane/100 feet (normalized length)
S = shape factor (1 for a tapered ramp; 0 for a parallel ramp)

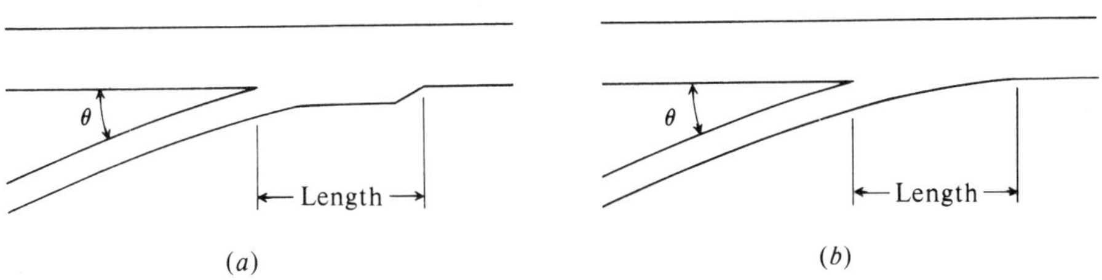

Fig. 3.7.1 (*a*) Parallel and (*b*) tapered lane ramps.

[1] D. R. Drew, "Traffic Flow Theory and Control," McGraw-Hill, New York, 1968, p. 218.

```
C     RAMP1
C     THIS PRØGRAM CALCULATES THE LENGTH ØF THE CRITICAL GAP,T,
C     IN SECØNDS. T IS A FUNCTIØN ØF:TH,THE ANGLE BETWEEN THE
C     RAMP AND HIGHWAY;D,THE NØRMALIZED LENGTH ØF THE ACCELERA-
C     TIØN LANE;AND S,THE RAMP FACTØR. S IS ZERØ FØR A
C     TAPERED RAMP,AND ØNE FØR A PARALLEL RAMP.
      READ,TH,D,S
      T=5.547+.828*TH-1.043*D+.045*D*D-.042*TH*TH-.847*S
      PRINT 10,T
   10 FØRMAT("CRITICAL GAP=",F6.2," SECØNDS"//)
      STØP
      END
```

Fig. 3.7.2 Program RAMP1.

```
INPUT:00010
 ?10.,2.,0.

CRITICAL GAP=  7.72  SECØNDS


STØP
```

Fig. 3.7.3 Trial run for RAMP1.

A program for calculating the critical gap in accordance with Eq. (3.7.1) is shown in Fig. 3.7.2. The same symbols are used in the program as in Eq. (3.7.1), except that θ is denoted by TH. A trial run is also shown in Fig. 3.7.3. In the trial run, the critical gap T has been calculated for the case of a 10° ramp angle, an acceleration lane 200 feet long, and a shape factor of $S = 0$.

3.8 OPTIMUM PIPE SIZE[1]

In the chemical process industry, an important pipe-line design problem is that of choosing the optimum pipe size for greatest economy of operation. Briefly, the problem is this: As the pipe size increases, the *installation and maintenance costs increase*; however, for a given flow rate, as the pipe diameter increases, the friction losses, and therefore the *pumping costs, decrease.* The optimum pipe size obviously depends on a trade-off between *pipe costs* and *pumping costs.* Figure 3.8.1 summarizes the problem in graphical form.

The analytical solution of the problem can be obtained by finding an expression for the *total annual costs* (maintenance costs *plus* pumping costs) as a function of the pipe diameter and minimizing this expression with respect to the pipe diameter. Solving the minimized function for diameter gives

$$D_{i(\text{optimum})} = \frac{\left(\dfrac{w}{1000}\right)^{0.448} \mu_c^{0.025}}{\gamma^{0.315}} \left[\frac{0.0232yk(1+j)}{(a+b)(f+1)\,xe}\right]^{0.158} \tag{3.8.1}$$

[1] M. S. Peters, "Elementary Chemical Engineering," McGraw-Hill, New York, 1954, pp. 283-285.

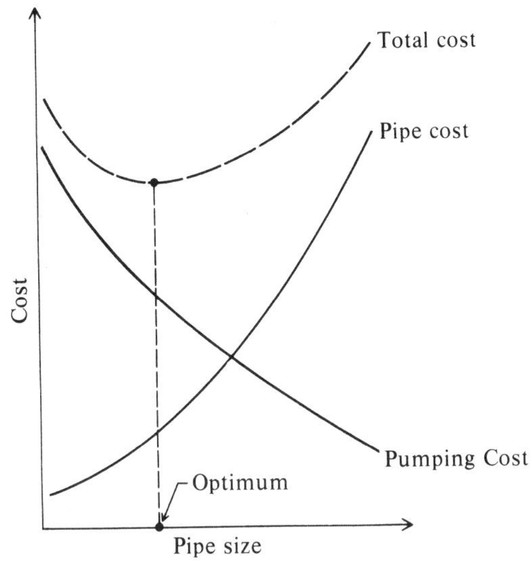

Fig. 3.8.1 Selection of optimum pipe size.

The symbols in the equation are defined as follows.

Symbol	*Computer name*	*Meaning*	*Units*
D_i	DIAM	Inside diameter of pipe	in
w	W	Mass flow rate	lb/hr
μ_c	XMU	Viscosity of fluid	cP
γ	GAMMA	Specific weight of fluid	lb/ft^3
y	Y	Yearly operation period	h
k	XK	Cost of electrical energy	\$/kWh
j	XJ	Frictional loss due to fittings expressed as a percentage of the loss in a straight pipe	% (as a decimal)
a	A	Annual fixed costs expressed as a percentage of installation costs	% (as a decimal)
b	B	Annual maintenance costs expressed as a percentage of installation costs	% (as a decimal)
f	F	Ratio of cost of fittings plus cost of erection to cost of pipe	
x	X	Cost of 1-in standard steel pipe	\$/ft
e	E	Efficiency of motor-pump combination	% (as a decimal)

Figure 3.8.2 shows a computer program for evaluating the optimum diameter. The arithmetic assignment statement on line 00020 illustrates the use of fractional exponents. The student should note that the FORTRAN expression for the right-hand

```
00001C   PIPE1
00002C   PRØGRAM TØ CALCULATE THE ØPTIMUM PIPE SIZE FØR
00003C   FLUID FLØW BASED ØN MINIMUM ØPERATING CØST
00010    READ,W,XMU,GAMMA,Y,XK,XJ,A,B,F,X,E
00020    DIAM=(W/1000.)**0.448*XMU**0.025/GAMMA**0.315
00025&    *(0.0232*Y*XK*(1.+XJ)/((A+B)*(F+1.)*X*E))**0.158
00030    PRINT,DIAM
00040    STØP
00050    END
```

Fig. 3.8.2 Program PIPE1.

side of Eq. (3.8.1) was *too long to fit on one line.* On line 00025, the *ampersand*[1] immediately following the line sequence number indicates that the line contains a *continuation* of the FORTRAN statement appearing on the preceding line. Unformatted input and output operations, lines 00010 and 00030, respectively, were employed for simplicity.

Figure 3.8.3 shows a trial run of the program. The answers given by this method of calculation are *conservative*, and, therefore, the program user would pick the *largest* standard steel pipe whose inside diameter is *smaller than* the value found for DIAM. For the values used in the illustration (19,000 lb/h of glycol at 140°F) DIAM was found to be equal to 2.2885 in. The engineer would use a 2-in standard steel pipe with an actual inside diameter of 2.067 in.

Figure 3.8.4 shows an alternative method of making the same calculation. The identical expression is again evaluated on lines 00020, 00025, and 00026, but *without* using a continuation statement. In the arithmetic assignment statement on line 00020, a portion of the calculation is completed and the result stored in DIAM. This result is then used as a factor of the expression shown on line 00025 and the calculation is completed. The same procedure is used on line 00026, and the *final* result is stored in the location designated as DIAM. Obviously, this method is *not* as efficient as that used in Fig. 3.8.2 and is mentioned here only to demonstrate to the student that it is not always necessary to create a separate storage location to *temporarily* hold a value during a series of calculations. A storage location intended to hold one of the final results being sought can frequently be used at other places in the program to store interim results. Ordinarily, storage space in the computer is more than adequate, but, with very large programs, it is sometimes necessary to conserve storage. Judicious application of the principle illustrated in Fig. 3.8.4, that is, using a single storage location for holding several values (at different times) during program execution, can reduce the storage requirements of a program.

```
INPUT:00010
?19000.,98.,78.8,8500.,0.011,0.3,0.1,0.02,2.,0.15,0.4

      2.2885
STØP
```

Fig. 3.8.3 Trial run for PIPE1.

[1] Used in this way, this symbol is peculiar to the time-sharing system employed by the authors. The standard method of indicating a continuation line is to place *any* character, *except* a blank or a zero, in the *sixth space* on a line. The first five spaces are reserved for statement numbers, the sixth space is *ordinarily* blank, and *all* FORTRAN statements begin in the seventh space or to the right of the seventh space. (As mentioned earlier in the book, if a C appears in the *first* space of a line, the line is understood by the compiler to be merely a *comment* inserted for the programmer's convenience.)

```
C     PIPE2
C     REVISIØN ØF PRØGRAM TØ CALCULATE ØPTIMUM PIPE SIZE
C     FØR FLUID FLØW BASED ØN MINIMUM ØPERATING CØST
      READ,W,XMU,GAMMA,Y,XK,XJ,A,B,F,X,E
      DIAM=(W/1000.)**0.448*XMU**0.025/GAMMA**0.315
      DIAM=DIAM*(0.0232*Y*XK*(1.+XJ))**0.158
      DIAM=DIAM/((A+B)*(F+1.)*X*E)**0.158
      PRINT,DIAM
      STØP
      END
```

Fig. 3.8.4 Program PIPE2.

```
INPUT:00010
?19000.,98.,78.8,8500.,0.011,0.3,0.1,0.02,2.,0.15,0.4

        2.2885
STØP
```

Fig. 3.8.5 Trial run for PIPE2.

In Fig. 3.8.5 we have the results of a run made on the *revised* program using the same data as were used with the *original* version. As we expected, the result is the same.

PROBLEMS

3-1. Write a FORTRAN program to calculate the heating value of a solid fuel, given the composition as percentages of sulphur, hydrogen, oxygen, and carbon (see Prob. 1-59).

A typical Pennsylvania bituminous coal shows the following weight analysis:

Sulphur = 0.87%
Hydrogen = 5.16%
Oxygen = 9.19%
Carbon = 74.51%

3-2. The Stefan-Boltzmann law is

$$E_b = \sigma T^4$$

where E_b = energy radiated by a blackbody, Btu/h-ft^2
T = absolute temperature, °R
$\sigma = 0.1714 \times 10^{-8}$

Write a FORTRAN program that will accept a temperature in degrees Fahrenheit and printout the energy radiated (see Sec. 1.28 for temperature conversion formula).

Sample data

Set	°*F*
1	200
2	60
3	600

3-3. Stokes' law describes the drag on a sphere moving through a fluid in the laminar flow region:

$$D = 3\pi V d\mu$$

where D = drag, lb-ft
V = velocity, ft/s
d = diameter of sphere, ft
μ = viscosity, lb-s/ft^2

Write a FORTRAN program that will read values of $V, d,$ and μ, and will print the drag D.

Sample data

Set	V	d	μ	
1	0.1	0.1	0.02	(Glycerin @ 65°F)
2	0.1	0.5	2.2×10^{-5}	(Water @ 65°F)
3	0.1	0.25	3.2×10^{-5}	(Mercury @ 65°F)

3-4. The uniformly accelerated motion equation of Prob. 1-57 can be modified as follows to show the effect of an initial displacement:

$$s = s_0 + v_0 t + \tfrac{1}{2}at^2$$

where s = displacement at time t, in
s_0 = initial displacement, in
v_0 = initial velocity, in/s
a = acceleration, in/s^2
t = time, s

Write a computer program that will read values of s_0, v_0, a, and t, and will printout the value of s.

Sample data

Set	s_0	v_0	a	t
1	+390	+8	−386	2
2	+500	+120	−386	10
3	+120	−150	+40	6

3-5. The time required for one oscillation of a simple pendulum is expressed by the formula given in Prob. 1-63. Write a program that will accept the length of the pendulum as input data and will printout the oscillation time.

Sample data

$L = 22$

3-6. In a manufacturing plant where parts are made in large lots, placed in storage, and used as required in assembling the product, a common problem is the determination of the most economic lot size. The following expression is used to calculate the number of pieces that should be made in a run:

$$X = \sqrt{\frac{2SDU}{(T+R)C + 2A(1-U/P)}}$$

where X = lot size, number of pieces
S = setup and takedown cost, \$/run
D = working days per year
U = usage rate, pieces/day
P = production rate, pieces/day
T = taxes and insurance, % (as a decimal)
R = desired rate of return on investment, % (as a decimal)
C = cost of material and labor, \$/piece
A = annual storage charges, \$/piece

Write a complete FORTRAN program that will read-in values of the parameters and will print the economic lot size.

Sample data

$S = 400.$	$T = 0.05$
$D = 243.$	$R = 0.15$
$U = 300.$	$C = 0.72$
$P = 1500.$	$A = 0.25$

3-7.* The parallel resonant frequency f_p of a quartz crystal is about 0.3 percent higher than its series resonant frequency f_s. The formulas for f_p and f_s are

$$f_s = \frac{P}{\sqrt{LC}}$$

$$f_p = P\sqrt{(1/L)(1/C + 1/C')}$$

where $P = 1/2\pi$
L = inductance of crystal, H
C = capacitance of crystal, F
C' = capacitance between electrodes, F

Write a computer program to calculate and print the values of the resonant frequencies.

Sample data

$L = 130$ H
$C = 0.025$ pF
$C' = 3.2$ pF

3-8. The Grashof number is a dimensionless parameter used in making heat transfer calculations involving free convection (see Prob. 1-60).

Write a FORTRAN program to evaluate and print the Grashof number for free convection in a pipe (where the inside diameter is the characterizing dimension) given the properties of the fluid and the temperature of the pipe wall.

Sample data

Fluid = air
Fluid temperature = 250°F
Density = 0.0489 lb-m/ft^3
Viscosity = 1.669×10^{-5} lb-m/ft-s
Coefficient of expansion = 0.00367, 1/°F
Pipe temperature = 800°F
Diameter = 1 ft

3-9. Write a FORTRAN program which will calculate $N!$ by Stirling's formula:

$$N! \approx \sqrt{2\pi N}\left(\frac{N}{e}\right)^N$$

3-10. In a surveyor's transit there are two horizontal "stadia" lines that intercept a segment of rod above and below the line of sight indicated by the cross hairs:

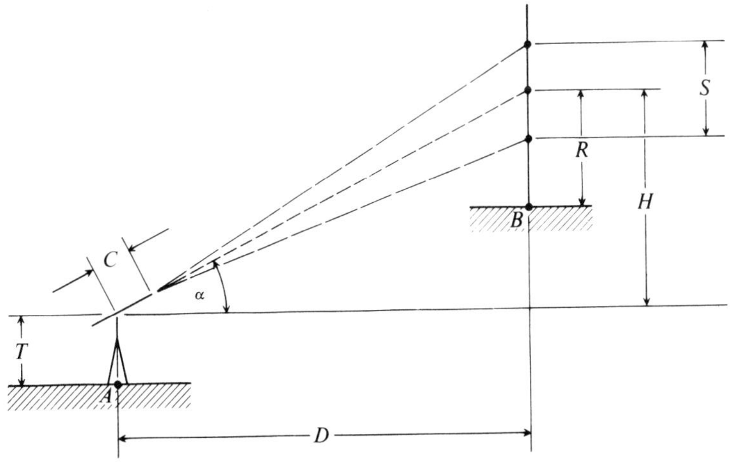

* J. Millman and C. C. Halkias, "Electronic Devices and Circuits," McGraw-Hill, 1967, p. 535.

The distances D and H are given by:

$$D = 100S \cos^2 \alpha + C \cos \alpha$$

$$H = 50S \sin 2\alpha + C \sin \alpha$$

Write a FORTRAN program that will accept the measured quantities R, S, and T in feet, α in degrees and minutes, and the instrument constant C; and will print the horizontal distance between the points A and B as well as the difference in their elevations.

Sample data

Set	R	S	T	α	C
1	6.00	2.35	5.00	10°23′	1.15
2	4.75	1.97	5.29	14°39′	0.98

3-11. A tapped capacitor can be used to provide impedance transformation, for example, to match a tuned circuit to a power amplifier. The necessary formulas are:

(1) $$C = \frac{1}{(2\pi f)^2 L}$$

(2) $$R_{par} = Q^2 R_{eff}$$

(3) $$C_2 = C \sqrt{\frac{R_{par}}{R_{amp}}}$$

(4) $$C_1 = \frac{CC_2}{C_2 - C}$$

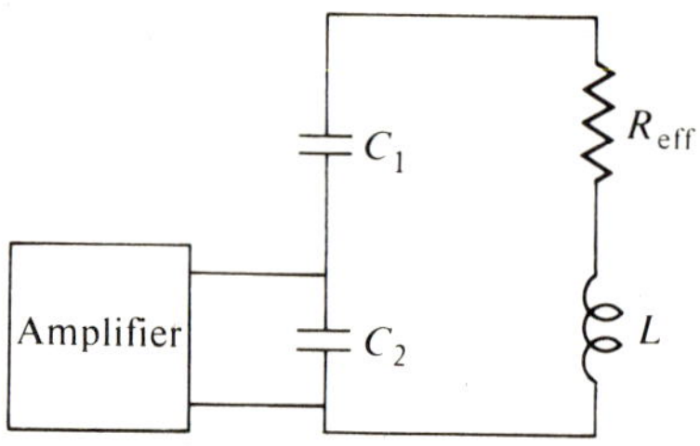

where C = capacitance required to achieve parallel resonance with inductance L at frequency f, F
f = resonant frequency, Hz
L = inductance, H
Q = Q of inductance L at frequency f
R_{eff} = effective series resistance of inductance L, ohms
R_{par} = effective parallel resistance of inductance L, ohms
C_1 = that part of C above the tap, F
C_2 = that part of C which is below the tap and which is connected to the amplifier, F
R_{amp} = optimum value of load resistance for the amplifier, ohms

Write a program that will calculate and print the values of C_1 and C_2, given values for f, L, Q, R_{amp}, and R_{eff}.

Sample data

f = 3300 Hz
L = 0.0028 H
Q = 60
R_{amp} = 150 ohms
R_{eff} = 0.91 ohm

3-12. The torque transmitted through a drive shaft that is delivering a specified horsepower at constant rotational speed is given by

$$T = \frac{(\text{hp})(63{,}000)}{N}$$

where T = torsional moment, lb-in
hp = horsepower
N = rotational speed, rev/min

The formula for selecting a shaft of appropriate size is given in Prob. 1-64.

Write a program that will accept values for the horsepower to be delivered, the shaft speed, and the allowable shear stress of the shaft material, and will print the required shaft diameter.

Sample data

Set	hp	*N*	*S*
1	600	85	6000
2	5	1800	6000
3	5	3600	6000

3-13. Modify the program written for Prob. 3-12 so the input data as well as the required shaft diameter are part of the computer output. Use FORMAT statements that will provide appropriate labels for all the output values. (This technique of "echoing" the input data along with the computed results is good programming practice.)

3-14.* The magnetic field intensity H_3 at the center of a spherical shell made of ferromagnetic metal of permeability μ is given by

$$H_3 = \frac{9H_0 \dfrac{\mu}{\mu_0}}{K}$$

$$K = \left(1 + \frac{2\mu}{\mu_0}\right)\left(2 + \frac{\mu}{\mu_0}\right) - 2\left(\frac{a}{b}\right)^3 \left(\frac{\mu}{\mu_0} - 1\right)^2$$

where H_3 = magnetic intensity at the center of the shell, A/m
H_0 = magnetic intensity external to the shell, A/m
μ = permeability of the shell, H/m
μ_0 = permeability of free space, H/m
a = inner radius of the shell, m
b = outer radius of the shell, m

Write a FORTRAN program that will read values of the two permeabilities and the ratio of the shell radii, and will print the value of the ratio H_3/H_0.

Sample data

$\mu_0 = 4\pi \times 10^{-7}$
$\mu = 500\mu_0$
$a/b = 0.98$

3-15. In determining the overall heat transfer coefficient of a parallel flow heat exchanger it is necessary to know the log mean temperature difference (see Prob. 1-70).

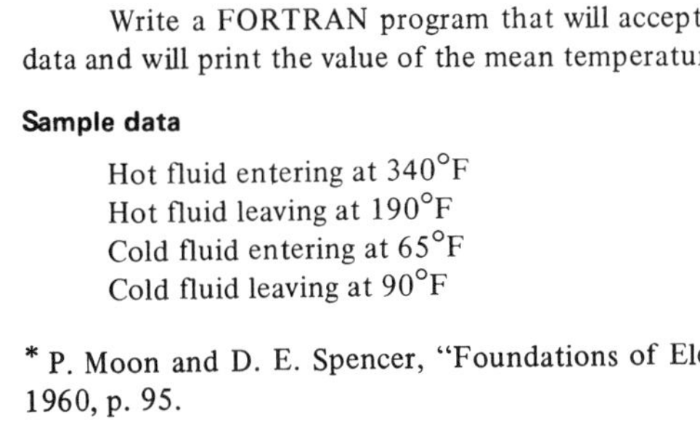

Write a FORTRAN program that will accept the four significant fluid temperatures as input data and will print the value of the mean temperature difference as output.

Sample data

Hot fluid entering at 340°F
Hot fluid leaving at 190°F
Cold fluid entering at 65°F
Cold fluid leaving at 90°F

* P. Moon and D. E. Spencer, "Foundations of Electrodynamics," Van Nostrand, Princeton, N.J., 1960, p. 95.

3-16.* The Laplace formula is an approximation to the binomial probability formula and is given by

$$P(k) = \frac{1}{\sqrt{2\pi npq}} e^{-(k-np)^2/2npq}$$

where k = number of successes in n trials
n = number of trials
p = probability of success on each trial
q = probability of failure on each trial
$P(k)$ = probability of k successes in n trials

Write a computer program that will accept values for k, n, p, and q, and will calculate and print the value of $P(k)$.

Sample data

$k = 15$ $n = 20$ $p = 0.5$ $q = 0.5$

3-17. According to the Richardson-Dushman equation, the current density due to electrons escaping from a metallic surface at temperature T is

$$J = AT^2 e^{-E_w/KT}$$

where J = current density, A/m^2
A = a constant for a particular metal, A/m^2 °K^2
E_w = work function for a particular metal, eV
K = Boltzmann constant = 8.62×10^{-5} eV/°K
T = absolute temperature, °K

Write a FORTRAN program which will calculate J in accordance with the Richardson-Dushman equation and will printout the value in E format.

Sample data

Set	*Material*	A	E_w	T
1	Tungsten	60.2×10^{-4}	4.52	300°K
2	Tungsten	60.2×10^{-4}	4.52	2500°K

3-18. The venturi meter formula given in Prob. 1-62 expresses the flow rate of water in terms of the differential pressure head between the main pipe and the venturi throat, $h_1 - h_2$. This pressure difference (in feet of water) is usually observed directly with a differential manometer. Revise the formula accordingly and write a FORTRAN program to read-in the pressure difference in feet of water, the meter coefficient, and the meter dimensions expressed as diameter of the inlet pipe and diameter at the throat, and calculate and print the value of the flow rate. (Have the program label the output and show the units of the flow rate.)

Sample data

Pressure drop = 6 ft of water
Meter coefficient = 0.98
Inlet pipe diameter = 8 in
Venturi throat diameter = 3.5 in

3-19. The polytropic expansion of an ideal gas is governed by the following laws:

$$T_2 = T_1 \left(\frac{P_2}{P_1}\right)^{(n-1)/n}$$

$$W = \frac{wRT_1}{M(n-1)} \left[1 - \left(\frac{P_2}{P_1}\right)^{(n-1)/n}\right]$$

* P. Beckmann, "Probability in Communication Engineering," Harcourt, Brace, New York, 1967, p. 35.

where the subscripts 1 and 2 refer to the initial and final conditions, respectively, and

T = absolute temperature, °R
P = absolute pressure, lb/in^2 absolute
n = polytropic exponent
W = work, ft-lb
R = universal gas constant = 1545.3 ft-lb/lb-mol/°R
w = weight of gas, lb
M = molecular weight of gas

Write a FORTRAN program to calculate and printout the final temperature and the work done during a slow expansion of w lb of gas at T_1°R from P_1 to P_2 lb/in^2 absolute. The physical properties of the gas that are required as input for the program are the polytropic exponent n and the molecular weight M.

Sample data

Gas = air
$n = 1.3$
$M = 28.97$
$w = 0.25$ lb
$T_1 = 1200$°R
$P_1 = 100$ lb/in^2
$P_2 = 20$ lb/in^2

3-20. In Sec. 3.5 formulas were given for calculating the performance of a single-stage transistor amplifier in terms of the H parameters. In this problem we will consider the performance of a two-stage transistor amplifier, with each stage using the same type of transistor. The analysis is limited to the case of midband frequencies, that is, frequencies at which reactances may be neglected. A block diagram of the system is shown below:

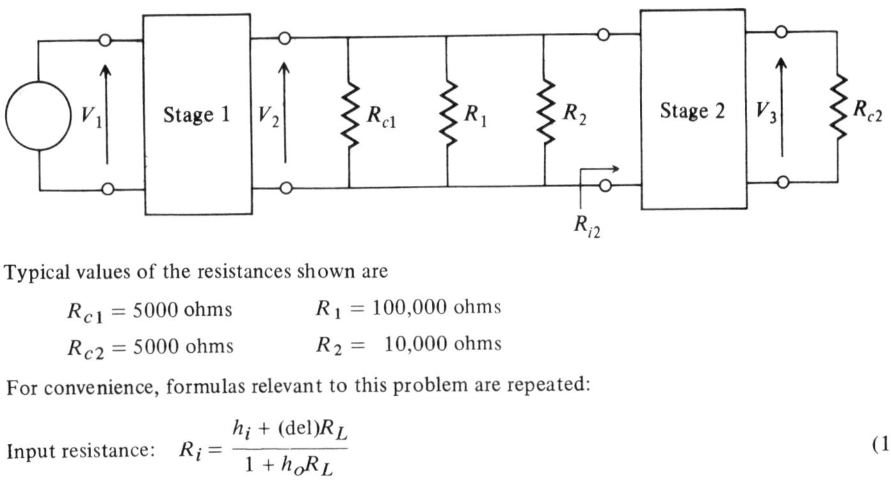

Typical values of the resistances shown are

$R_{c1} = 5000$ ohms $\qquad R_1 = 100{,}000$ ohms

$R_{c2} = 5000$ ohms $\qquad R_2 = 10{,}000$ ohms

For convenience, formulas relevant to this problem are repeated:

Input resistance: $$R_i = \frac{h_i + (\text{del})R_L}{1 + h_o R_L} \tag{1}$$

Voltage gain: $$A_v = \frac{-h_f R_L}{-h_i + (\text{del})R_L} \tag{2}$$

where del $= h_i h_o - h_f h_r$. It should be stressed that the value of R_L that must be used in these equations may be the resistance of one resistor or the equivalent resistance of several resistors in parallel.

The voltage gain of *each stage* can be calculated from Eq. (2), using appropriate values for R_L. For the *second stage*, the effective load resistance is R_{c2} (5000 ohms). Therefore, to calculate the voltage gain of stage 2, A_{v2}, a value of 5000 ohms should be used. The effective load resistance of stage 1 is R_{c1} in parallel with three other resistances: R_1, R_2, and R_{i2}:

$$R_{L1} = R_1 \| R_2 \| R_{i2} \| R_{c1} \tag{3}$$

where the symbol "||" denotes "in parallel with." Therefore, to calculate the voltage gain of stage 1, A_{v1}, the value of resistance given by Eq. (3) should be used. Finally, the overall gain A_v of the system is

$$A_v = A_{v1}A_{v2} \tag{4}$$

Write a program to calculate and print the values of A_{v1}, A_{v2}, and A_v. To test the program, use the values of the typical resistances given above and the following values for the H parameters:

$h_i = 1500$ ohms $\quad h_r = 6 \times 10^{-4}$

$h_f = 50$ $\quad h_o = 4 \times 10^{-5}$ mhos

3-21. The block diagram of the two-stage amplifier of Prob. 3-20 showed two resistors denoted R_1 and R_2. These two resistors, and a third resistor R_e (not shown in the diagram), establish the quiescent operating point of the transistor and are called the self-bias circuit. S, the stabilization factor, is a figure of merit for the self-bias circuit. The equation for S for a common emitter amplifier is

$$S = \frac{(1+\beta)(1+R_b/R_e)}{(1+\beta+R_b/R_e)} \tag{1}$$

where β = large-signal current gain of a common emitter transistor
$R_b = R_1 \,||\, R_2$, ohms
R_e = emitter resistor, ohms

In order to design the self-bias circuit, the following parameters must be specified: S, β, V_{CC}, V_{CE}, V_{BE}, I_C, and R_C. The first two parameters have already been defined. The remaining parameters are

V_{CC} = collector supply voltage, V
V_{CE} = collector to emitter voltage, V
V_{BE} = base to emitter voltage, V
I_C = collector current, A
R_C = collector resistor, A
I_B = base current, A

The equations for the design of the self-bias circuit are

$$R_e = \frac{V_{CC} - V_{CE}}{I_C} - R_C \tag{2}$$

$$I_B = \frac{I_C}{\beta} \tag{3}$$

$$R_b = \frac{(1+\beta)(S-1)}{(1+\beta-S)} R_e \tag{4}$$

$$R_2 = \frac{R_b V_{CC}}{I_B R_b + V_{BE} + (I_B + I_C)R_e} \tag{5}$$

$$R_1 = \frac{R_2[I_B R_b + V_{BE}(I_B + I_C)R_e]}{V_{CC} - [I_B R_b + V_{BE} + (I_B + I_C)R_e]} \tag{6}$$

Write a program which calculates and prints values of R_1, R_2, and R_e and test it using the following data:

$\beta = 100$ $\quad V_{BE} = 0.6$ $\quad V_{CC} = 22.5$ $\quad R_C = 5000$ $\quad V_{CE} = 12$ $\quad I_C = 0.002$ $\quad S = 4$

3-22. For a rectangular strain rosette in which the three gage readings are Q_1, Q_2, and Q_3:

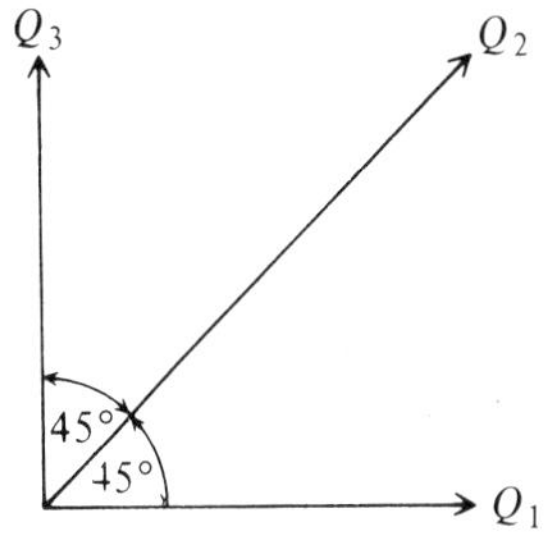

The principal stresses are given by

$$S_{max} = A + B$$

$$S_{min} = A - B$$

where $A = \dfrac{E(1-K)}{2(1-\mu)}(Q_1 + Q_3)$

$$B = \frac{E(1+K)}{2(1+\mu)}\sqrt{(Q_1 - Q_2)^2 + (Q_2 - Q_3)^2}$$

and E = modulus of elasticity, lb/in^2
μ = Poisson's ratio
K = transverse sensitivity of gage
Q_i = corrected gage reading, μin/in

Write a computer program that will read-in values for E, μ, K, and the corrected gage readings and will printout the values of the principal stresses.

Sample data

Set	E	μ	K	Q_1	Q_2	Q_3
1	29.5×10^6	0.31	0.03	+610	+308	−195
2	10.1×10^6	0.30	0.03	+140	−55	+765

4
Using the Computer to Make Decisions

In the previous chapter the computer was used to carry out various calculations. If this were its only function, it would be little more than a super slide rule. The power and utility of the computer is greatly extended by its ability to make decisions through the use of *conditional transfer statements.* In this chapter we will discuss this technique.

4.1 FLOW OF CONTROL

Under normal conditions the computer executes the FORTRAN statement instructions in the same sequence as they appear in the source program. The normal sequence can be altered by the use of *transfer statements* which direct the computer to some other part of the program for the next instruction by referencing another statement.

Transfer statements are classified as either *unconditional* or *conditional* transfers. The fundamental difference between the two classes of transfer statements can be summarized as follows:

> Each time an unconditional transfer statement is encountered in the normal flow of control, it produces a transfer or jump that was *specified when the source program was written.*
>
> Each time a conditional transfer statement is executed, it produces a transfer that is *chosen,* from among several specified alternatives, *during the execution of the object* program.

The *unconditional* transfer statements discussed are:

1. GO TO (Section 4.2)
2. CALL (Section 10.7)
3. RETURN (Section 10.4)

The *conditional* transfer statements discussed are:

1. *Arithmetic* IF (Section 4.3)
2. *Computed* GO TO (Section 4.4)
3. *Assigned* GO TO (Section 4.6)
4. *Logical* IF (Section 9.10)

4.2 GO TO STATEMENT

The GO TO statement is an executable statement of the form

GO TO n

where n is a statement number (integer constant).

This statement causes the computer to transfer control to the referenced statement. After transfer has occurred, the computer proceeds with the normal sequential execution of instructions from the point to which transfer has been made. Note that the referenced statement can be *anywhere* in the program. It does *not* necessarily have to be among those statements that *follow* the GO TO statement.

Some examples of the GO TO statement are

GO TO 26
GO TO 100

4.3 ARITHMETIC IF STATEMENT

The *arithmetic* IF statement is an executable statement of the form

IF(*exp*) n_1, n_2, n_3

where *exp* is any arithmetic or double precision expression (see Secs. 1.16 and 8.5) and n_1, n_2, n_3 are statement numbers.

This statement permits transfer of control to any one of *three* referenced statements depending on whether the value of the expression is *negative, positive,* or *zero.*

1. If the value is *negative,* control passes to the statement whose number is n_1.
2. If the value is *zero,* control passes to the statement whose number is n_2.
3. If the value is *positive,* control passes to the statement whose number is n_3.

This statement also makes it possible to compare two numbers to determine which is the larger one. For example, if the expression in the parentheses is (A – B):

If B is greater than A, the value is negative.

If B is equal to A, the value is zero.

If B is less than A, the value is positive.

For each of the conditions listed, transfer to an appropriate part of the program could be made.

Some examples of the arithmetic IF statement are

IF(1000 – K) 10,50,10

IF(Z) 6,2,100

Note that a test for Z = 0.0 may fail, even though a "hand" calculation of Z shows that it is zero. This is due to round-off errors that may occur in the computer when decimal numbers are coded.

4.4 COMPUTED GO TO STATEMENT

The *computed* GO TO statement is an executable statement of the form

GO TO ($n_1, n_2, n_3, \ldots, n_m$),$v$

where n_1, n_2, ..., n_m is a list of statement numbers and v is an unsigned and unsubscripted *integer* variable indicating which statement number in the list (reading from left to right) is being referenced.

This statement transfers control to any one of *several* referenced statements depending on the current value of the indicator v. For example, the statement

GO TO (6,14,25,3,10),K

will transfer control to:

Statement number 6 if K = 1

Statement number 14 if K = 2

Statement number 25 if K = 3

Statement number 3 if K = 4

Statement number 10 if K = 5

The value of the indicator may not exceed the number of statement numbers in the reference list and may not be set equal to zero.

Examples of computed GO TO statements are

GO TO (48,27),INDEX
GO TO (10,20,30,40),M

4.5 ASSIGN STATEMENT

The ASSIGN statement has the following general form:

ASSIGN n TO v

where n is a statement number (integer constant) and v represents an unsubscripted integer variable.

The ASSIGN statement is used in connection with the *assigned* GO TO statement (see Sec. 4.6) to set the value of the indicator at one of the optional statement numbers.

This statement *cannot* be used to:

1. Initialize a variable at some value
2. "Input" some value and store it in the location designated by the variable

Examples of ASSIGN statements are

ASSIGN 22 to J

ASSIGN 3 to LABEL

4.6 ASSIGNED GO TO STATEMENT

The *assigned* GO TO statement is an executable statement of the form:

GO TO v, $(n_1, n_2, n_3, \ldots, n_m)$

where v is an unsigned and unsubscripted *integer* variable and $n_1, n_2, \ldots, n_m$ is a list of statement numbers.

When this statement is encountered in a program, transfer is made to the statement whose value was *last assigned* to the variable v by an ASSIGN statement (see Sec. 4.5). The list of statement numbers indicates all the *permissible values* that can be assigned to v by ASSIGN statements elsewhere in the program.

This statement accomplishes essentially the same action as the computed GO TO. The major difference is that this statement requires that the indicator be changed to the *actual value* of the statement number to be referenced, while the computed GO TO requires that the indicator be adjusted to show the *position* of the referenced statement number in the list of numbers.

Examples of assigned GO TO statements are

GO TO MARK,(6,10,12,37)

GO TO NUMBER,(54,8,32,33,1)

The permissible values of MARK are 6, 10, 12, 37. The permissible values of NUMBER are 1, 8, 32, 33, 54.

4.7 FLOW-CHARTING DECISIONS

The flow-chart decision block shown below is used to indicate that a choice of *alternative paths* is to be made by the program.

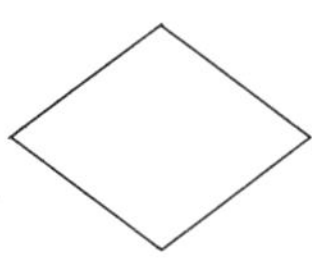

The decision-making operation is usually executed by one of the following FORTRAN statements:

1. *Arithmetic* IF (Section 4.3)
2. *Computed* GO TO (Section 4.4)
3. *Assigned* GO TO (Section 4.6)
4. *Logical* IF (Section 9.10)

In this text, the expression *interrogated* by the computer is shown *inside* the decision block, and the *conditions* controlling the alternative exit paths are shown above or to the right of the path lines exiting from the decision block. For example:

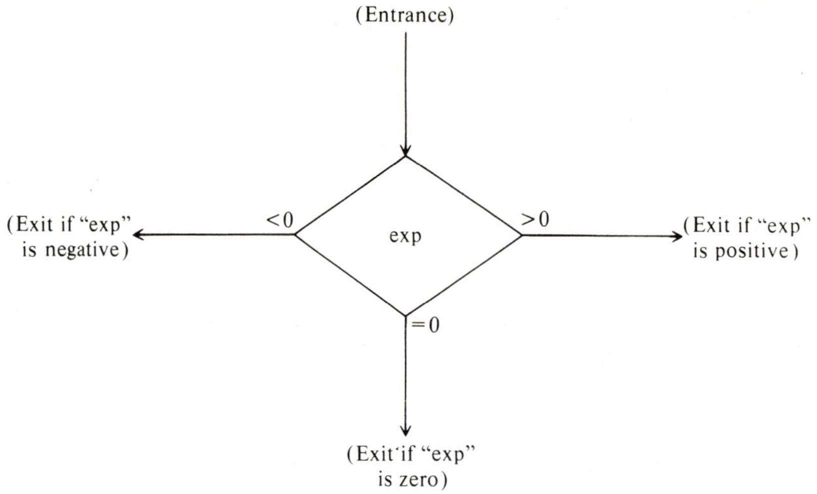

or:

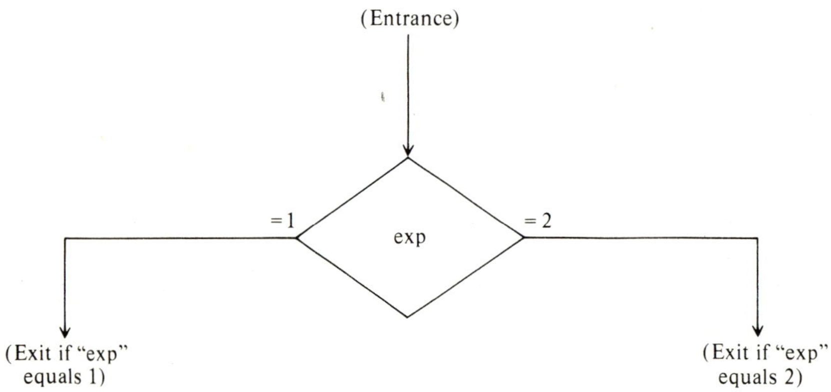

4.8 WATER TEMPERATURE FOR CONCRETE MIX[1]

During cold weather (below 40°F) concrete should be poured at temperatures between 70 and 100°F to prevent freezing of the mix before the concrete has set. The easiest way for the field engineer to control the mixture temperature is by adjusting the temperature of the water added. A simple heat balance on the materials added to the concrete makes it possible to obtain the following expression for the required water temperature:

[1] F. H. Kellogg, "Construction Methods and Machinery," Prentice-Hall, New York, 1954, pp. 340–342.

$$T_w = \frac{T_{\text{mix}}[(W_a + W_c)/5] + W_m + W_w - (T_a W_a + T_c W_c)/5 - T_m W_m}{W_w}$$

The symbols used in the equation are defined as follows.

Symbol	*Name*	*Meaning*	*Units*
T_w	TW	Required temperature of mixing water	°F
	TMIX	Desired temperature of the mixture	°F
T_a	TA	Temperature of the dry aggregate	°F
T_c	TC	Temperature of the dry cement	°F
T_m	TM	Temperature of moisture in the aggregate	°F
W_a	WA	Weight of dry aggregate	lb
W_c	WC	Weight of dry cement	lb
W_w	WW	Weight of mixing water	lb
W_m	WM	Weight of moisture in aggregate	lb

A flow chart of the procedure for calculating the water temperature is shown in Fig. 4.8.1 and includes a decision block that compares the computed value of water temperature with 170., the *maximum permissible temperature* of water to be added to the mixture. Above this temperature, the chemical reaction which produces setting in the concrete proceeds too rapidly and flash setting occurs. The flow chart shows that if the water temperature *exceeds the limit*, a "message" to that effect will be printed. If the temperature is *within the allowable range*, the calculated value is printed.

Figure 4.8.2 shows the FORTRAN program for accomplishing the plan outlined in the flow chart. The decision block has been programmed as an IF statement on line

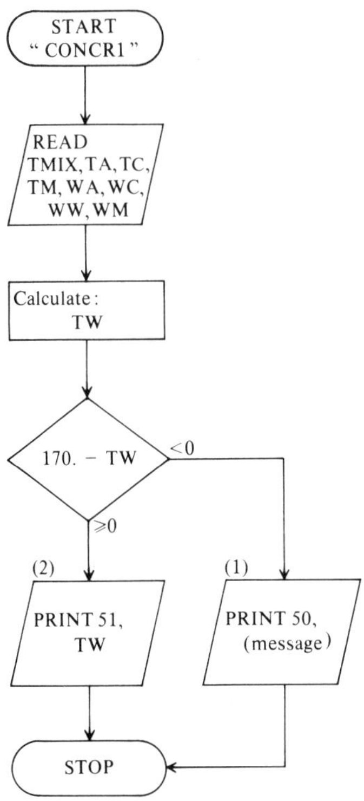

Fig. 4.8.1 Flow chart of program CONCR1.

```
00001C      CØNCR1
00002C      PRØGRAM TØ CALCULATE THE REQUIRED WATER TEMPERATURE
00003C      FØR MIXING CØNCRETE DURING CØLD WEATHER
00010       READ,TMIX,TA,TC,TM,WA,WC,WW,WM
00020       TW=(TMIX*((WA+WC)/5.+WM+WW)-(TA*WA+TC*WC)/5.-TM*WM)/WW
00030       IF(170.-TW)1,2,2
00040     1 PRINT 50,TW
00050    50 FØRMAT(27HREQUIRED WATER TEMPERATURE(F6.1,
00051&      15H ) IS EXCESSIVE/30HAGGREGATE MUST BE PREHEATED ØR/
00052&      38HDESIRED TEMPERATURE ØF MIXTURE REDUCED)
00060       STØP
00070     2 PRINT 51,TW
00080    51 FØRMAT(29HREQUIRED WATER TEMPERATURE ISF6.1,
00081&      19H DEGREES FAHRENHEIT)
00090       STØP
00100       END
```

Fig. 4.8.2 Program CONCR1.

```
INPUT:00010
?85.,40.,55.,40.,1410.,282.,125.,25.

REQUIRED WATER TEMPERATURE( 209.1 ) IS EXCESSIVE
AGGREGATE MUST BE PREHEATED ØR
DESIRED TEMPERATURE ØF MIXTURE REDUCED
STØP
```

(*a*)

```
INPUT:00010
?72.,40.,55.,40.,1410.,282.,125.,25.

REQUIRED WATER TEMPERATURE IS 158.3 DEGREES FAHRENHEIT
STØP
```

(*b*)

Fig. 4.8.3 Trial runs for CONCR1.

00030. Statement 1 on line 00040 causes the computer to print the *warning message* indicating an excessive temperature. Statement 2 on line 00070 prints the calculated value of the water temperature. Both FORMAT statements in this program used *continuation lines* which were indicated by the ampersand immediately following the line numbers (see Sec. 3.8). The slashes used in FORMAT statement 50 (lines 00050, 00051, and 00052) will produce the message on successive lines (see Sec. 2.11).

Two runs of this program are shown in Fig. 4.8.3: part (*a*) shows the output when the required water temperature is above the limiting value; part (*b*) shows the normal output.

4.9 A SIMPLE SORTING PROCESS

One of the tedious jobs that can be performed rapidly by a computer is the task of arranging a list of numbers into ascending or descending order. The problem is fundamentally one of comparing two values, deciding which has the greater magnitude, and taking appropriate action to arrange the values in the desired order.

To illustrate the process of sorting a list in descending order, let us assume that two storage locations in the computer are named X1 and X2 and each location contains a value of *unknown* magnitude. In order to test our program, we will make provision for entering values to be stored in X1 and X2, but if this sorting routine were only a part of some larger program, we could arrive at the "sorting" point without knowing the values. Our problem is to program the computer to select the value of greatest

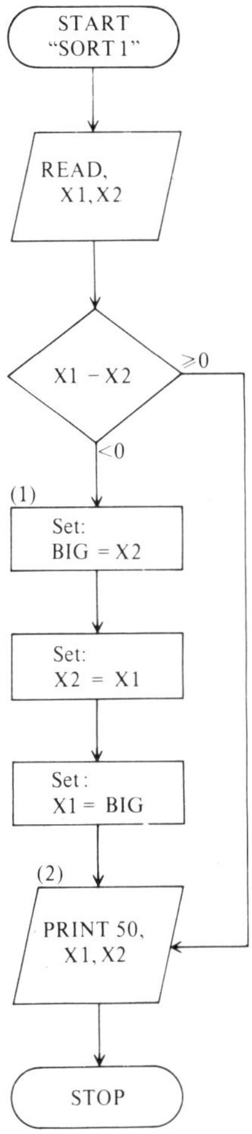

Fig. 4.9.1 Flow chart of program SORT1.

algebraic magnitude and store it in location X1. As part of the same operation the computer must also store the smaller value in location X2. As a check on the performance of our program we will have the computer printout the contents of X1 followed by the contents of X2.

Figure 4.9.1 is a flow chart outlining the necessary programming steps to accomplish our objective. If the value of (X1 − X2) is positive, the values are already stored in the desired order and the next step is to printout the values.

If the value of (X1 − X2) is exactly zero, the contents of the two locations must be equal in magnitude; therefore there is no "larger" or "smaller" value and again the next step is to printout the values presently in storage.

If the value of (X1 − X2) is negative, the larger value is stored in X2 and we must swap the contents of locations X1 and X2. This exchange can be accomplished by using a temporary storage location (called BIG in this program). Note that a "copy" of the contents of X2 is placed in BIG, at the same time erasing whatever was formerly stored in BIG. The original contents of X2 are then erased, and a copy of the value originally stored in X1 is placed in location X2. Finally, the original contents of X1 are erased and are replaced with a copy of the value presently stored in BIG.

A FORTRAN program for this sorting procedure is given in Fig. 4.9.2. Several

```
C     SØRT1
C     THIS PRØGRAM ARRANGES TWØ STØRED VALUES IN DESCENDING ORDER.
      READ,X1,X2
      IF(X1-X2) 1,2,2
    1 BIG=X2
      X2=X1
      X1=BIG
    2 PRINT 50,X1,X2
   50 FØRMAT(5X4HX1 =F10.5/5X4HX2 =F10.5)
      STØP
      END
```

```
INPUT:00010
?15.,63.

     X1 =  63.00000
     X2 =  15.00000

INPUT:00010
?15.,15.

     X1 =  15.00000
     X2 =  15.00000

INPUT:00010
?26.,-8.

     X1 =  26.00000
     X2 =  -8.00000

INPUT:00010
?-26.,8.

     X1 =   8.00000
     X2 = -26.00000

INPUT:00010
?-26.,-8.

     X1 =  -8.00000
     X2 = -26.00000

INPUT:00010
?4.,0.

     X1 =   4.00000
     X2 =   0.00000

INPUT:00010
?-4.,0.

     X1 =   0.00000
     X2 =  -4.00000
```

Fig. 4.9.2 Program SORT1 and trial runs.

combinations of values have been used as input data to test the operation of the program.

The reader should note that one additional storage location (BIG) is required to perform the sorting operation. The values being sorted are returned to the same storage locations in which they were originally held, but the arrangement, or order, has been altered. The same procedure applies if the list to be sorted contains a great many values rather than the two used here.

4.10 BEAM DEFLECTION (SINGLE POINT)

A simple beam with a uniform cross section, loaded with a single concentrated force, is shown in Fig. 4.10.1. The deflection of this beam at an arbitrary position x is given by[1]

$$y = \frac{-Pbx}{6EIL}\,[2L(L-x) - b^2 - (L-x)^2] \qquad (4.10.1)$$

for the region $0 \leqslant x \leqslant a$ and by

$$y = \frac{-Pa(L-x)}{6EIL}\,[2Lb - b^2 - (L-x)^2] \qquad (4.10.2)$$

for the region $a \leqslant x \leqslant L$.

The symbols are defined as follows.

Formula symbol	*Computer name*	*Meaning*	*Units*
y	Y	Deflection at location x	in
x	X	Position on the beam (measured from the origin)	in
P	P	Applied load	lb
L	XL	Length of beam	in
a	A	Distance to load application point (from left end)	in
b	B	Distance to load application point (from right end)	in
E	E	Modulus of elasticity	lb/in^2
I	XI	Second moment of cross section	in^4

A flow chart for one method of solution for this problem is shown in Fig. 4.10.2. The number in the parentheses at the upper left of some of the flow-chart symbols is the statement number of the program instruction represented by the symbol.

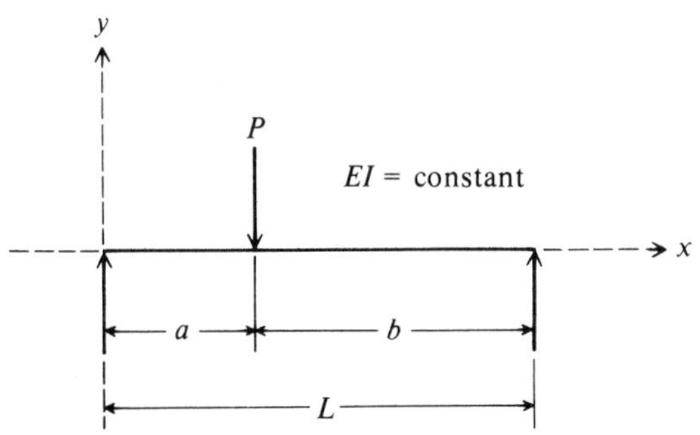

Fig. 4.10.1 Simple beam loading.

[1] H. A. Rothbart, "Mechanical Design and Systems Handbook," McGraw-Hill, New York, 1964, **15**, p. 37

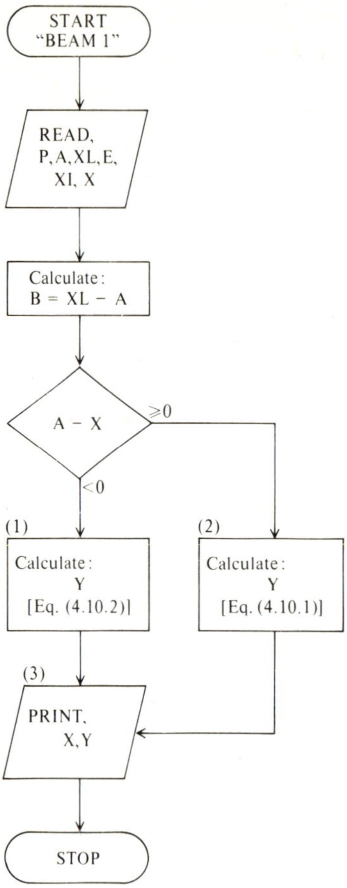

Fig. 4.10.2 Flow chart of program BEAM1.

A program for carrying out the diagrammed steps is shown in Fig. 4.10.3. In this program the decision as to which equation should be used is made by the IF statement on line 00030. The program actually compares the value of X that was read-in with the value of A which locates the applied load. If the point whose deflection is to be determined lies to the right of the applied load, X must be greater than A and the argument of the IF statement will be negative. The program will then branch to statement 1 and Eq. (4.10.2) will be used for evaluating Y. If the point X is to the left of the applied load, the IF statement argument will be positive and the program will jump to statement 2 and Eq. (4.10.1) will be used for evaluating Y. If the point being

```
00001C    BEAM1
00002C    THIS PRØGRAM CALCULATES SIMPLE BEAM DEFLECTIØN PRØDUCED
00003C    BY A SINGLE CØNCENTRATED LØAD.
00010     READ,P,A,XL,E,XI,X
00020     B=XL-A
00030     IF(A-X) 1,2,2
00040   1 Y=-P*A*(XL-X)*((2.*XL-B)*B-(XL-X)**2)/(6.*E*XI*XL)
00050     GØ TØ 3
00060   2 Y=-P*B*X*(2.*XL*(XL-X)-B*B-(XL-X)**2)/(6.*E*XI*XL)
00070   3 PRINT,X,Y
00080     STØP
00090     END
```

Fig. 4.10.3 Program BEAM1.

```
INPUT:00010
?500 65 154 3E7 2.5 46.2

     46.2000          -.4053
STØP


INPUT:00010
?500 65 154 3E7 2.5 107.8

     107.8000         -.3761
STØP
```

Fig. 4.10.4 Trial runs for BEAM1.

examined happens to coincide with the point of load application (i.e., the deflection *under* the load is being sought), X will be equal to A and the argument of the IF statement will be zero. This condition will also produce a jump to statement 2.

In the expression on line 00040, that part of Eq. (4.10.2) requiring the evaluation of $2Lb - b^2$ has been written in the factored form (2.*XL–B)*B. While this is not essential for proper operation of the program, it produces a slightly more efficient program and is preferred by many programmers.

If the program branches to statement 1, then, after Y has been evaluated, it is necessary to skip over the intervening steps and jump to the PRINT statement. This jump is accomplished by the GO TO statement on line 00050. If statement 2 on line 00060 is used for evaluating Y, no jump is required because the next statement in the normal flow of control is the PRINT statement.

The results of two runs (one for a location to the left of the load and one to the right) are shown in Fig. 4.10.4. In Sec. 5.11 we shall look at this beam problem again and discuss a method of programming the computer to calculate the deflection at several points.

4.11 PIPE SIZE SELECTION

In Sec. 3.8 we discussed a program for calculating the optimum pipe size based on economy of operation. The result obtained was the required inside diameter of the pipe and was simply printed as a numerical value. The engineer then selected the correct nominal pipe size. In this section we will extend that program so the result printed by the computer *will be the nominal pipe size.* The procedure for accomplishing this requires that a series of decisions be made by the computer after the required inside diameter has been calculated. In each of these decisions the *required diameter* is compared with the *actual inside diameter* of one of the standard pipe sizes. If the required diameter is *smaller*, then, as explained in Sec. 3.8, the *next smaller* standard pipe size is printed as the result. If the required diameter is *larger* than the actual inside diameter of the given pipe, the program advances and makes a similar comparison with the *next larger* standard size. The procedure is continued until the correct nominal pipe size has been determined.

Figure 4.11.1 shows a flow chart outlining the procedure. The reader will note that only pipe sizes from 1 to 4 in were included. This abridgement of the table of standard pipe sizes was deliberate and was done to shorten the program. The effect, as can be seen from the flow chart, is to specify a 1-in standard steel pipe if the required diameter is less than 1.380 in, and to merely indicate the required diameter if it should be greater than 5.047 in. If the required diameter lies between 1.380 and 5.047 in, the

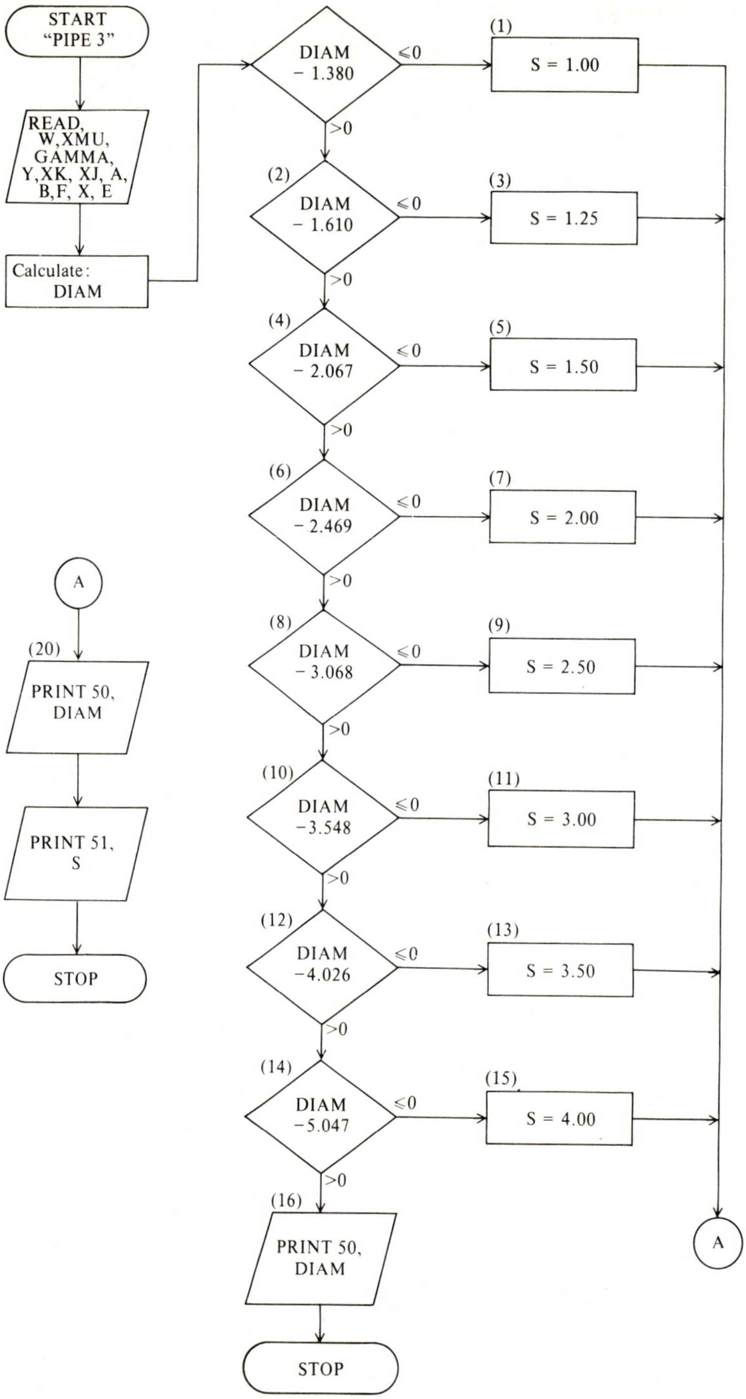

Fig. 4.11.1 Flow chart of program PIPE3.

next smaller nominal pipe size will be specified in the output of the computer.

The FORTRAN program is shown in Fig. 4.11.2. Lines 00010 through 00025 are identical with those of PIPE1 in Fig. 3.8.2 and merely evaluate the required

```
00001C      PIPE3
00002C      PRØGRAM TØ CALCULATE THE ØPTIMUM PIPE SIZE FØR FLUID
00003C      FLØW BASED ØN MINIMUM ØPERATING CØST. (PRØGRAM GIVES
00004C      NØMINAL PIPE SIZE REQUIRED.)
00010       READ,W,XMU,GAMMA,Y,XK,XJ,A,B,F,X,E
00020       DIAM=(W/1000.)**0.448*XMU**0.025/GAMMA**0.315
00025&       *(0.0232*Y*XK*(1.+XJ)/((A+B)*(F+1.)*X*E))**0.158
00030       IF(DIAM-1.380)1,1,2
00040     1 S=1.00
00050       GØ TØ 20
00060     2 IF(DIAM-1.610)3,3,4
00070     3 S=1.25
00080       GØ TØ 20
00090     4 IF(DIAM-2.067)5,5,6
00100     5 S=1.50
00110       GØ TØ 20
00120     6 IF(DIAM-2.469)7,7,8
00130     7 S=2.00
00140       GØ TØ 20
00150     8 IF(DIAM-3.068)9,9,10
00160     9 S=2.50
00170       GØ TØ 20
00180    10 IF(DIAM-3.548)11,11,12
00190    11 S=3.00
00200       GØ TØ 20
00210    12 IF(DIAM-4.026)13,13,14
00220    13 S=3.50
00230       GØ TØ 20
00240    14 IF(DIAM-5.047)15,15,16
00250    15 S=4.00
00260       GØ TØ 20
00270    16 PRINT 50,DIAM
00280    50 FØRMAT(14HREQUIRED I.D.=F8.4/)
00290       STØP
00300    20 PRINT 50,DIAM
00310       PRINT 51,S
00320    51 FØRMAT(3HUSEF5.2,19H INCH STANDARD PIPE/)
00330       STØP
00340       END
```

Fig. 4.11.2 Program PIPE3.

diameter. The series of decisions shown in the flow chart (Fig. 4.11.1) has been programmed as a set of arithmetic IF statements starting on line 00030 and extending to line 00240. Note that in this program we have an example of one FORMAT statement serving several output operations (the PRINT statements on lines 00270 and 00300 refer to FORMAT statement 50 on line 00280).

In Fig. 4.11.3 we have the results of a run made using the same data as were used in Fig. 3.8.3. Because of the *table search* made in our *programmed pipe table,* the printed results of this program are more readily usable by the engineer. The principle demonstrated here can be used with any table of standard sizes (e.g., wire gages, beam sizes, insulation thicknesses, etc.) and, where employed, makes a program more effective as a tool for *computer-aided design.*

```
INPUT:00010
?19000.,98.,78.8,8500.,0.011,0.3,0.1,0.02,2.,0.15,0.4

REQUIRED I.D.=  2.2885

USE 2.00 INCH STANDARD PIPE

STØP
```

Fig. 4.11.3 Trial run for PIPE3.

4.12 CHANGING COMPLEX NUMBERS FROM RECTANGULAR TO POLAR FORM[1]

In the steady-state analysis of electric circuits with sinusoidal excitation, the use of complex numbers greatly simplifies calculations. In carrying out such calculations, it is necessary to transform complex numbers from rectangular to polar form, and from polar to rectangular form. In this section, we are concerned only with the transformation of a complex number $a + jb$ to polar form.[2] The equivalence of these representations of a complex number is indicated by Eq. (4.12.1).

$$a + jb = r\underline{/\beta} \tag{4.12.1}$$

where $r = a^2 + b^2$

$$\beta = \tan^{-1} \frac{b}{a}$$

The arctangent function is multivalued and it is necessary to place β in the correct quadrant.

The program of Fig. 4.12.1 carries out this transformation. In the program, the quantities a, b, r and β are represented by the names A, B, R and BETA, respectively. The operation of this program is probably best appreciated by study of the flow chart shown in Fig. 4.12.2. However, analysis of any of the trial runs in Fig. 4.12.3 is also

```
C     PØLAR
C     THIS PRØGRAM CØNVERTS CØMPLEX NUMBERS FRØM RECTANGULAR
C     TØ PØLAR FØRM.
   30 READ,A,B
      R=SQRT(A**2+B**2)
      IF(A) 2,3,4
    3 IF(B) 5,6,7
    5 BETA=270.
      GØ TØ 1
    6 PRINT 20
   20 FØRMAT(22HDEGENERATE CASE:A=B=0.)
      GØ TØ 30
    7 BETA=90.
      GØ TØ 1
    2 IF(B) 8,9,10
    8 BETA=180.+57.296*ATAN(ABS(B/A))
      GØ TØ 1
    9 BETA=180.
      GØ TØ 1
   10 BETA=180.-57.296*ATAN(ABS(B/A))
      GØ TØ 1
    4 IF(B) 11,12,13
   11 BETA=360.-57.296*ATAN(ABS(B/A))
      GØ TØ 1
   12 BETA=0.
      GØ TØ 1
   13 BETA=57.296*ATAN(ABS(B/A))
    1 PRINT 23,R,BETA
   23 FØRMAT(14HRADIUS VECTØR=F8.3,7H ANGLE=F8.3,8H DEGREES/)
      GØ TØ 30
      END
```

Fig. 4.12.1 Program POLAR.

[1] W. H. Hayt, Jr., and J. E. Kemmerly, "Engineering Circuit Analysis," McGraw-Hill, New York, 1962, p. 287.

[2] In accordance with customary engineering practice, we denote $\sqrt{-1}$ by the letter j. Mathematicians usually use the letter i.

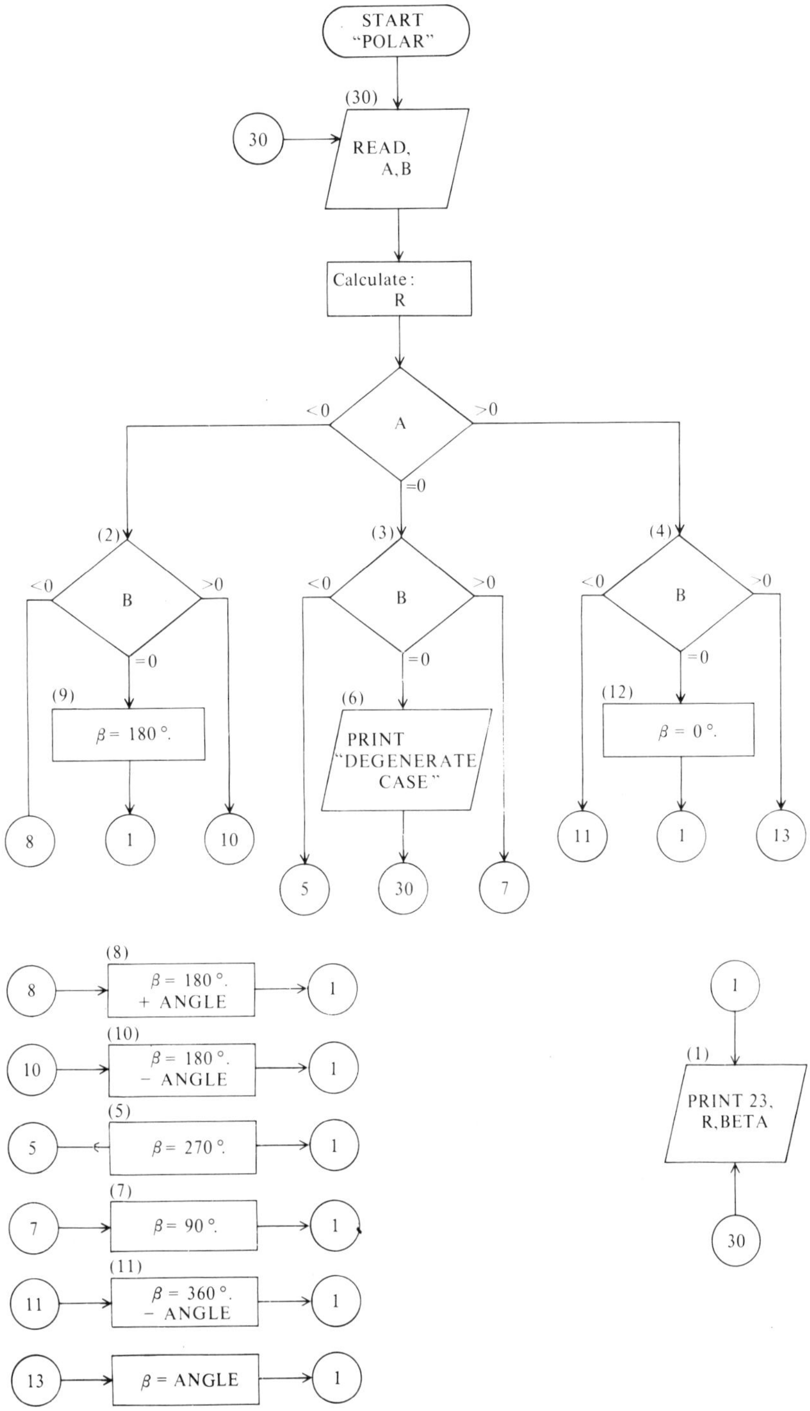

Fig. 4.12.2 Flow chart of program POLAR.

helpful. Let us consider the complex number $-5 + j2$, which was entered in the sixth trial run. In this case, $A = -5.$ and $B = +2.$ On line 00020 the radius vector R is

```
INPUT:00010
 ?0.,0.

DEGENERATE CASE:A=B=0.
INPUT:00010
 ?2.1,0.

RADIUS VECTØR=   2.100 ANGLE=    0.000 DEGREES

INPUT:00010
 ?0.,2.1

RADIUS VECTØR=   2.100 ANGLE=   90.000 DEGREES

INPUT:00010
 ?-2.1,0.

RADIUS VECTØR=   2.100 ANGLE= 180.000 DEGREES

INPUT:00010
 ?0.,-2.1

RADIUS VECTØR=   2.100 ANGLE= 270.000 DEGREES

INPUT:00010
 ?-5.,2.

RADIUS VECTØR=   5.385 ANGLE= 158.199 DEGREES

INPUT:00010
 ?4.,-3.

RADIUS VECTØR=   5.000 ANGLE= 323.130 DEGREES
```

Fig. 4.12.3 Trial runs for POLAR.

calculated. Next, as A is negative, the IF statement on line 00030 causes the program to branch to statement 2 on line 00120 (jumping over the intermediate statements). B is positive, and the IF statement on line 00120 causes the program to branch to statement 10 on line 00170. There, the value of the angle corresponding to $\tan^{-1} |B/A|$ is found, converted to degrees, and subtracted from 180°. Thus, the angle BETA lies in the second quadrant. The transfer statement GO TO 1 causes the computer to print the answer: $5.385\underline{/158.199^\circ}$ The other trial runs may be analyzed in a similar fashion.

The reader may have observed that the expression 57.296*ATAN (ABS(B/A)) appears four times. It is possible to represent this expression by use of an arithmetic statement function which will be discussed in Chap. 10. For example, one could write the statement ANG(B,A) = 57.296*ATAN(ABS(B/A)) at the beginning of a program. In the remainder of the program, the shorter expression ANG(B,A) would be used. It may also be useful to mention that the program discussed in this section is used, with minor modification, in programs in Secs. 4.13, 6.6, and 8.11.

Finally, it should be noted that it is certainly possible to write a simpler version of this program by the use of complex arithmetic, which will be discussed in Chap. 8.

4.13 THE CHARACTERISTIC IMPEDANCE OF A TRANSMISSION LINE[1]

In Sec. 4.12 we discussed a program for converting complex numbers from rectangular to polar form. As an illustration of this conversion, we will now discuss the calculation of the characteristic impedance Z_0 of a transmission line. Z_0 is calculated from the parameters of the line

$$Z_0 = \sqrt{\frac{R + j\omega L}{G + j\omega C}} \tag{4.13.1}$$

where R = ohms/loop-mi
L = H/loop-mi
G = mhos/mi
C = F/mi

When R and G are small relative to $j\omega L$ and $j\omega C$, respectively, Z_0 approaches the value $+\sqrt{L/C}$. (The positive value of the square root is used because the expression $\sqrt{L/C}$ has the dimensions of resistance and, therefore, is positive.) This case is important at high frequencies, but we will consider the more general form of Z_0.

By multiplying the numerator and denominator of the fraction under the radical in Eq. (4.13.1) by $G - j\omega C$, we obtain

$$Z_0 = \sqrt{\frac{RG + \omega^2 LC + j\omega(LG - RC)}{G^2 + (\omega C)^2}}$$

The complex fraction can be separated into real and imaginary parts:

$$Z_0 = \sqrt{\frac{RG + \omega^2 LC}{G^2 + (\omega C)^2} + j\,\frac{\omega(LG - RC)}{G^2 + (\omega C)^2}} \tag{4.13.2}$$

The complex expression under the radical is in rectangular form, $A + jB$, and the latter can be converted to polar form $R\underline{/\beta}$. Therefore, we can express Z_0 as

$$Z_0 = \sqrt{A + jB} = \sqrt{R\underline{/\beta}} \tag{4.13.3}$$

where

$$A = \frac{RG + \omega^2 LC}{G^2 + (\omega C)^2} \tag{4.13.4}$$

$$B = \frac{\omega(LG - RC)}{G^2 + (\omega C)^2} \tag{4.13.5}$$

$$R = A^2 + B^2 \tag{4.13.6}$$

$$\beta = \tan^{-1}\frac{B}{A} \tag{4.13.7}$$

[1] Charles A. Holt, "Introduction to Electromagnetic Fields and Waves," Wiley, New York, 1963, p. 487.

The roots of a complex number are given by

$$(R/\underline{\beta})^{1/n} = R^{1/n}/\underline{\theta}$$

where $\theta = \dfrac{\beta + 2\pi K}{n}$

$$K = 0, 1, 2, \ldots, n-1$$

In Eq. (4.13.3), $n = 2$, and, therefore, K is limited to the two values 0 and 1. In the computer program which follows, we use $K = 1$, and, therefore, $\theta = \beta/2 + \pi$. This choice is dictated by the fact that Z_0 should approach $+\sqrt{L/C}$ as R and G approach zero.

In the program shown in Fig. 4.13.1, the characteristic impedance of a transmission line, Z_0, is calculated in polar form. The line parameters R, L, G, and C are represented in the program by the names OM, XL, G, and C. The name OM was used instead of the conventional R, so that R could be retained as the name for the magnitude of the complex quantity $A + jB$, and XL was used instead of L so that inductance would be a floating point quantity. The radian frequency ω is denoted by W.

In the program of Fig. 4.13.1, the values of A and B are calculated from the parameters of the transmission line in accordance with Eqs. (4.13.4) and (4.13.5) (see

```
C      TRANS1
C      THIS PRØGRAM CALCULATES THE CHARACTERISTIC IMPEDANCE ØF
C      A TRANSMISSIØN LINE. THE PRØGRAM FØR RECTANGULAR TØ PØLAR
C      CØNVERSIØN DISCUSSED IN SECTIØN 4.12 IS USED AS PART ØF
C      THIS PRØGRAM.
       READ,ØM,XL,G,C,W
       DENØM=G**2+(W*C)**2
       A=(ØM*G+(W**2)*XL*C)/DENØM
       B=(W*(XL*G-ØM*C))/DENØM
       R=SQRT(A**2+B**2)
       ZØMAG=SQRT(R)
       IF(A) 2,3,4
     3 IF(B) 5,6,7
     5 BETA=270.
       GØ TØ 1
     6 PRINT 20
    20 FØRMAT(22HDEGENERATE CASE:A=B=0.)
       GØ TØ 25
     7 BETA=90.
       GØ TØ 1
     2 IF(B) 8,9,10
     8 BETA=180.+57.296*ATAN(ABS(B/A))
       GØ TØ 1
     9 BETA=180.
       GØ TØ 1
    10 BETA=180.-57.296*ATAN(ABS(B/A))
       GØ TØ 1
     4 IF(B) 11,12,13
    11 BETA=360.-57.296*ATAN(ABS(B/A))
       GØ TØ 1
    12 BETA=0.
       GØ TØ 1
    13 BETA=57.296*ATAN(ABS(B/A))
     1 THETA=BETA/2.+180.
       PRINT 23,ZØMAG,THETA
    23 FØRMAT(6HZØMAG=F6.1,7H ANGLE=F6.2//)
    25 STØP
       END
```

Fig. 4.13.1 Program TRANS1.

```
INPUT:00010
 ?4.,.003,1.E-06,1.5E-08,5000.

ZØMAG= 454.9 ANGLE=352.92

STØP
```

Fig. 4.13.2 Trial run for TRANS1.

lines 00020, 00030, 00040). The magnitude R of $A + jB$ is calculated on line 00050, and the magnitude of the characteristic impedance ZOMAG is calculated on line 00060. From line 00070 to line 00280, the programs of Fig. 4.12.1 and 4.13.1 are identical. In this part of the programs, the angle of the complex quantity is calculated. On line 00290 the phase angle of Z_0 (denoted by THETA) is calculated in accordance with the discussion above concerning roots of complex numbers. A sample run is shown in Fig. 4.13.2.

4.14 CHOOSING ONE OF SEVERAL PATHS

In the solution of engineering problems it is sometimes necessary to employ a particular sequence of calculations for one set of conditions, a second sequence for a second set of conditions, still a third sequence for a third set of conditions, and so on. If the conditions can be characterized by assigning a number to each set of conditions, all the special sequences of calculations can be placed in one program and, when the *characterizing number* is referred to, the program will select and follow the appropriate sequence.

There are two fundamental methods of using this programming device in the FORTRAN language. The first method employs the *computed* GO TO statement (see Sec. 4.4), and the second the *assigned* GO TO statement (see Sec. 4.6). In both methods the programmer must provide a means of *setting* the characterizing number for each execution of the program. The simplest means of accomplishing this is to enter the appropriate value at the time the other problem data are being read-in.

In the solution of problems requiring iteration, this same technique can be used to follow one path the first time through a *loop* (i.e., an iteration sequence); a *second* path the second, third, fourth, etc., times through the loop (until some condition is met); and a *third* path until some additional condition is satisfied; and so on.

To illustrate the method of programming the computer to follow a prescribed calculation sequence, we will use the *computed* GO TO statement. Let us consider the problem of calculating the effectiveness of a heat exchanger in which the fluid passages may be arranged in three ways. These cases are (see Fig. 4.14.1):

Case A Counter flow
Case B Parallel flow
Case C Cross flow

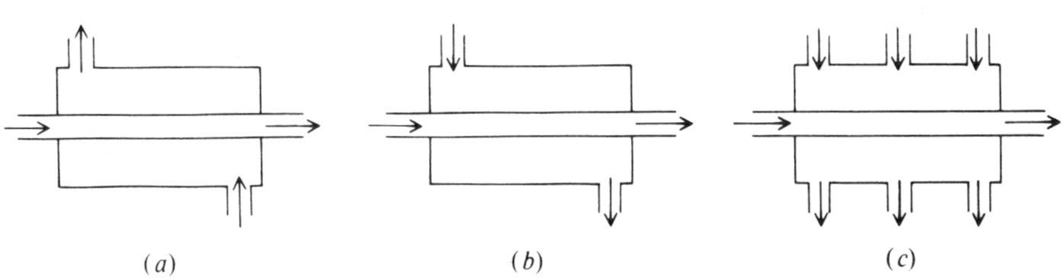

Fig. 4.14.1 Heat exchanger flow paths. (*a*) Counter flow; (*b*) parallel flow; (*c*) cross flow.

The basic equations needed to evaluate the effectiveness for each case are as follows[1] (see Table 4.14.1 for definitions of the symbols used):

1. The thermal *capacity rate* of the hot and cold fluids:

$$C_H = W_H C_{pH} \tag{4.14.1}$$

$$C_C = W_C C_{pC} \tag{4.14.2}$$

2. The *capacity rate ratio*:

$$R = \frac{C_{min}}{C_{max}} \tag{4.14.3}$$

where C_{min} and C_{max} are, respectively, the smaller and larger values of the capacity rates for the hot and cold fluids. Note that before this ratio can be evaluated, the program must designate either C_H or C_C as C_{max} and the remaining capacity rate as C_{min}. A variation of the simple sorting procedure described in Sec. 4.9 can be used to accomplish this.

3. *The number of heat transfer units* of the exchanger:

$$\text{NTU} = \frac{AU_{avg}}{C_{min}} \tag{4.14.4}$$

Table 4.14.1

Symbol	*Computer name*	*Meaning*	*Units*
. . .	INDEX	"Characterizing" number for calculation conditions	dimensionless
W_H	WH	Weight flow rate of hot fluid	lb/h
W_C	WC	Weight flow rate of cold fluid	lb/h
C_{PH}	SPHTH	Specific heat at constant pressure for hot fluid	Btu/lb °F
C_{PC}	SPHTC	Specific heat at constant pressure for cold fluid	Btu/lb °F
C_H	CH	Capacity rate of hot fluid	Btu/h °F
C_C	CC	Capacity rate of cold fluid	Btu/h °F
C_{min}	CMIN	*Smaller* value of C_H and C_C	Btu/h °F
C_{max}	CMAX	*Larger* value of C_H and C_C	Btu/h °F
R	R	Capacity rate ratio	dimensionless
NTU	TU	Number of heat transfer units	dimensionless
A	A	Heat transfer area	ft^2
U_{avg}	U	Average overall thermal conductance (based on A)	Btu/h ft^2 °F
ϵ	EFF	Effectiveness of heat exchanger	dimensionless
γ	GAM	An interim calculated value	dimensionless

The particular equations describing the *effectiveness* for each of the cases under consideration are as follows.

CASE A—COUNTER FLOW In this type of exchanger the two fluids move along parallel paths, separated by a conducting wall, and flow in *opposite* directions so that one fluid is entering at the point where the second fluid is leaving. The effectiveness is

$$\epsilon = \frac{1 - e^{-(\text{NTU})(1-R)}}{1 - Re^{-(\text{NTU})(1-R)}} \tag{4.14.5}$$

[1] W. M. Kays and A. L. London, "Compact Heat Exchangers," National Press, Palo Alto, Calif., 1955, pp. 7–12.

CASE B—PARALLEL FLOW For this case the fluid passages are the same as for the counterflow exchanger, but the two fluids move in the *same* direction, both entering at the same point and exiting at another point. The effectiveness can be calculated by

$$\epsilon = \frac{1 - e^{-(\mathrm{NTU})(1 + R)}}{1 + R} \tag{4.14.6}$$

CASE C—CROSS FLOW It will be assumed here that the cross flow arrangement is such that one fluid passes through the exchanger in an unmixed condition (e.g., by flowing through a bundle of parallel tubes) while the second fluid passes through in a mixed condition (e.g., by flowing *around* the tube bundle and at right angles to it). This assumption about the configuration of the exchanger gives rise to two algebraic expressions for the effectiveness; the particular expression to be used depends on whether the unmixed fluid has the larger or smaller capacity rate.

Case C1 For cross flow where the unmixed fluid has the *larger* capacity rate:

$$\epsilon = 1 - e^{-\gamma/R} \tag{4.14.7}$$

where $\gamma = 1 - e^{-(\mathrm{NTU})R}$ (4.14.8)

Case C2 For cross flow where the unmixed fluid has the *smaller* capacity rate:

$$\epsilon = \frac{1 - e^{-\gamma R}}{R} \tag{4.14.9}$$

where $\gamma = 1 - e^{-\mathrm{NTU}}$ (4.14.10)

Although the subcases, C1 and C2, are described in terms of the magnitude of the capacity rate of the unmixed fluid, the program user will generally not know in advance which capacity rate, that of the mixed or unmixed fluid, is the larger one. Therefore the program must be designed so this comparison is made by the computer and the correct subcase selected for evaluation of the effectiveness. The comparison is easily made if the unmixed fluid is specified. For example, if it is known that the hot fluid is passing through the tubes and is therefore designated the unmixed fluid, we only need to compare C_{H} [evaluated according to Eq. (4.14.1)] and C_{max} (previously determined in the program as a prerequisite for the evaluation of the rate ratio R) to determine if we are to follow the method described in subcase C1 or in subcase C2. This means that the program user must not only characterize the calculation as one for cross flow, but must *also* indicate whether the hot fluid or the cold fluid is passing through the tube bundle.

In summary, the cases that are to be characterized by an assigned number are as follows.

Number	*Case*	*Description*
1	A	Counter flow
2	B	Parallel flow
3	C1	Cross flow (*hot* fluid unmixed with $C_{\mathrm{H}} > C_{\mathrm{C}}$)
3	C2	Cross flow (*hot* fluid unmixed with $C_{\mathrm{H}} < C_{\mathrm{C}}$)
4	C1	Cross flow (*cold* fluid unmixed with $C_{\mathrm{C}} > C_{\mathrm{H}}$)
4	C2	Cross flow (*cold* fluid unmixed with $C_{\mathrm{C}} < C_{\mathrm{H}}$)

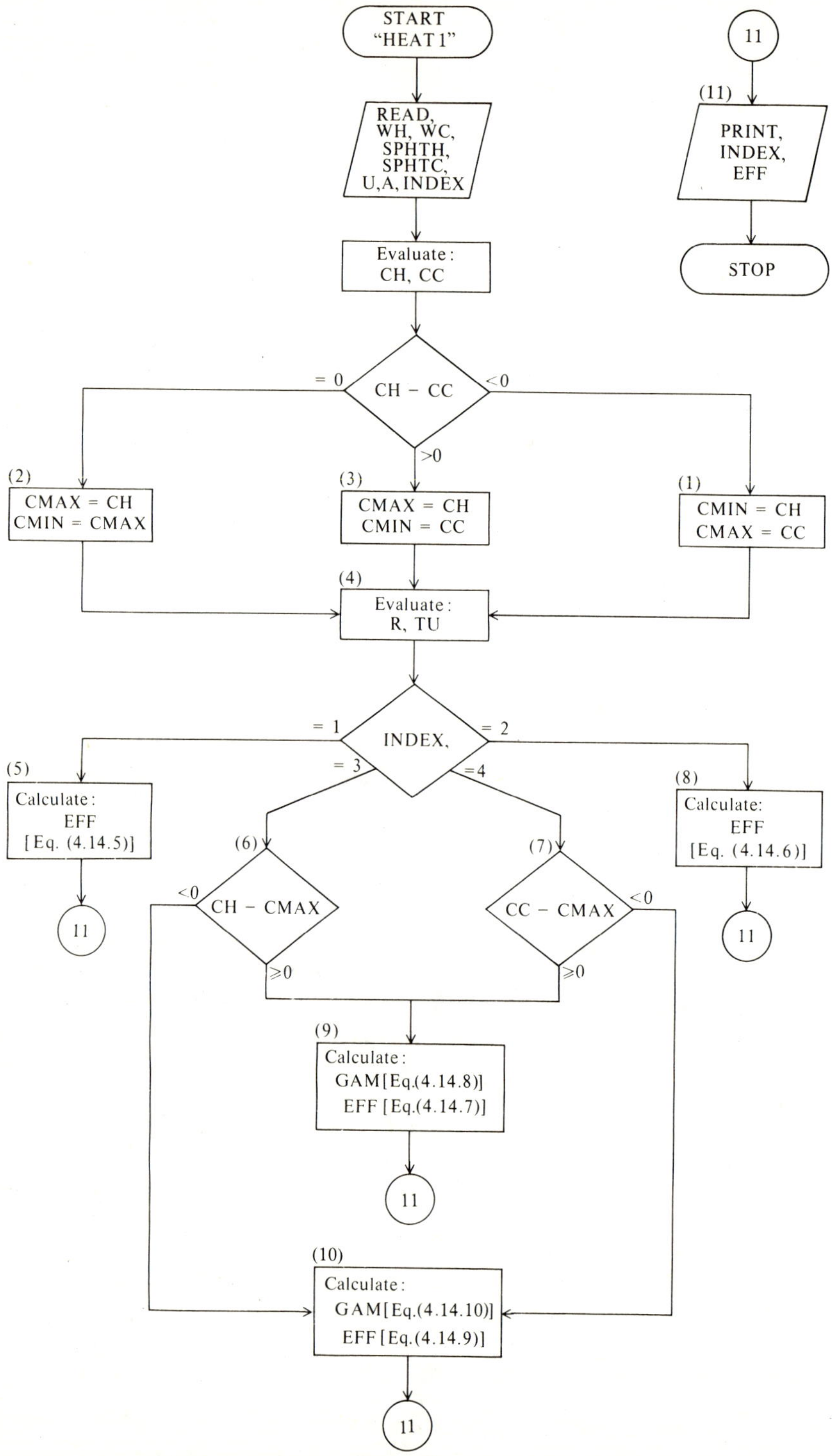

Fig. 4.14.2 Flow chart of program HEAT1.

See Table 4.14.1 for definitions of the symbols and names used in programming this problem. A flow chart outlining the logic for the program is shown in Fig. 4.14.2. As usual, the *arithmetic* IF statements in the program are represented by decision boxes showing alternative exits for conditions of negative, zero, and positive. The decision box containing the name INDEX represents, in this program, a *computed* GO TO statement with four possible exits. The particular exit path that will be followed is determined by the value stored in the location named INDEX at that time.

Figure 4.14.3 shows the FORTRAN program for carrying out the procedure diagrammed in the flow chart.

The IF statement on line 00040 makes a comparison of the relative magnitudes of the values stored in locations CH and CC. Since we are not actually sorting these values and returning them to the same storage locations (as was done in Sec. 4.9), the branching that is indicated merely assigns the value of the larger capacity rate to CMAX and the value of the smaller to CMIN. The possibility that CH and CC will be exactly equal (in which event CMAX and CMIN will also be identical) is accounted for by statement 2 on line 00080 and the following unnumbered statement on line 00090. Line 00080, which "sets" the value of CMAX, could also be written 2 CMAX = CC because this statement will be accessed only for the special case in which CH = CC.

The computed GO TO statement on line 00150 is used to choose the intended path. If the value stored in INDEX is 3 (i.e., the characterization number for cross flow with the hot fluid unmixed) the program will branch to statement 6, the *third* statement number in the list.

Statement 6 on line 00210 makes a comparison between CH and CMAX for the

```
00001C       HEAT1
00002C       THIS PRØGRAM CALCULATES THE THERMØDYNAMIC EFFECTIVENESS ØF
00003C       A HEAT EXCHANGER FØR SEVERAL ØPTIØNAL FLØW ARRANGEMENTS.
00010        READ,WH,WC,SPHTH,SPHTC,U,A,INDEX
00020        CH=WH*SPHTH
00030        CC=WC*SPHTC
00040        IF(CH-CC) 1,2,3
00050      1 CMIN=CH
00060        CMAX=CC
00070        GØ TØ 4
00080      2 CMAX=CH
00090        CMIN=CMAX
00100        GØ TØ 4
00110      3 CMAX=CH
00120        CMIN=CC
00130      4 R=CMIN/CMAX
00140        TU=A*U/CMIN
00150        GØ TØ(5,8,6,7),INDEX
00160      5 GAM=EXP(-TU*(1.-R))
00170        EFF=(1.-GAM)/(1.-R*GAM)
00180        GØ TØ 11
00190      8 EFF=(1.-EXP(-TU*(1.+R)))/(1.+R)
00200        GØ TØ 11
00210      6 IF(CH-CMAX) 10,9,9
00220     10 GAM=1.-EXP(-TU)
00230        EFF=(1.-EXP(-GAM*R))/R
00240        GØ TØ 11
00250      9 GAM=1.-EXP(-TU*R)
00260        EFF=1.-EXP(-GAM/R)
00270        GØ TØ 11
00280      7 IF(CC-CMAX) 10,9,9
00290     11 PRINT,INDEX,EFF
00300        STØP
00310        END
```

Fig. 4.14.3 Program HEAT1.

```
INPUT:00010
?1250.,875.,1.,.7,180.,6.,1

         1          0.7408

INPUT:00010
?1250.,875.,1.,.7,180.,6.,2

         2          0.6226

INPUT:00010
?1250.,875.,1.,.7,180.,6.,3

         3          0.6929

INPUT:00010
?1250.,875.,1.,.7,180.,6.,4

         4          0.6810
```

Fig. 4.14.4 Trial runs for HEAT1.

purpose of determining whether Eq. (4.14.7) or Eq. (4.14.9) should be used for evaluating the effectiveness of the cross-flow exchanger. It should be noted that since there are only *two* capacity rates, CH and CC, it follows that CH is *either* the larger one, CMAX, *or* the smaller one, CMIN, and therefore the argument (CH − CMAX) can only be negative or equal to zero. However, all arithmetic IF statements require three reference numbers, and therefore statement 9 was referenced for the condition that the argument be positive even though this was recognized as an impossible event.

Figure 4.14.4 shows several test runs of the program. In each run the same problem parameters were used, but the characterizing number was changed to show the effect of varying the type of heat exchanger. This is not representative of the way in which this program would normally be used, but serves to demonstrate its operation.

PROBLEMS

4-1. Modify the program outlined in the flow chart of Fig. 4.10.2 to provide for the possibility of a user inadvertently reading-in a value of X that does not lie on the beam (that is, $L < x < 0$).

4-2. Write a program that will accept any three lengths and store them in locations A, B, and C. Have the program determine whether or not a triangle can be formed from the given lengths, and, if a triangle *can* be formed, whether or not it is a right triangle.

The output of the program should be one of the following three messages:

1. TRIANGLE CANNOT BE FORMED
2. TRIANGLE CAN BE FORMED
3. RIGHT TRIANGLE IS FORMED

Sample data

Set	*A*	*B*	*C*
1	7	2	4
2	12	13	5
3	3	9	6
4	8	4	10

4-3. Given a general quadrilateral in which the lengths of the four sides and the angle at one corner are specified:

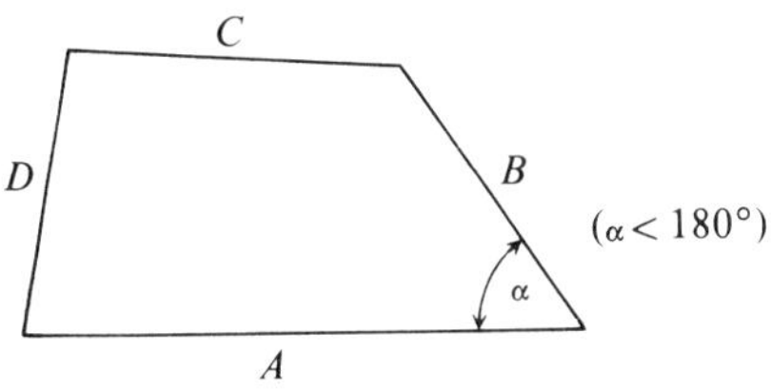

write a FORTRAN program that will read-in the specified values and printout the given data *and* the maximum area that can be enclosed by the figure. The program should check the given data for consistency; that is, the quadrilateral must close.

Sample data

Set	*A*	*B*	*C*	*D*	α
1	8	6	7	4	30°
2	12	5	6	7	90°
3	10	8	3	5	60°

4-4. Given the system of simultaneous equations

$$a_1x + b_1y = c_1$$
$$a_2x + b_2y = c_2$$

use Cramer's rule to solve for the two variables in terms of the coefficients and constants, and write a FORTRAN program that will read-in values for the coefficients and constants and will print the values of the variables.

Note that if the determinant of the system is zero, the equations are either dependent or inconsistent. Have the computer printout an appropriate message if the system determinant is zero.

Sample data

Set	a_1	b_1	c_1	a_2	b_2	c_2
1	+2	−3	+ 5	+2	+5	+13
2	+2	−7	+15	−2	+7	+ 4
3	+3	−1	+ 9	+5	+6	− 8
4	+3	−1	+ 9	−9	+3	−27

4-5. Revise the procedure outlined in the flow chart of Fig. 4.9.1 to sort in *ascending* order, that is, to place the larger value in X2 and the smaller value in X1.

4-6. Write a FORTRAN program to arrange the contents of two storage locations A and B so that A will contain the greater absolute value and B will contain the smaller absolute value. Have the program read-in a pair of test values and printout the results after the sorting process has been completed.

Sample data

Set	*A*	*B*
1	−26.6	14.2
2	143.8	76.9
3	65.2	−99.2
4	47.5	83.6

4-7. In stress analysis the algebraically largest of the three principal stresses is designated S1 and the algebraically smallest is labeled S3. The third stress is called S2.

Write a FORTRAN program to arrange the contents of three storage locations S1, S2, and S3 so that S1 will contain the largest algebraic value and S3 the smallest algebraic value.

Test the program by reading-in one of the following sets of values and having the computer printout the values in their proper order.

Sample data

Set	*S1*	*S2*	*S3*
1	−0.7	26.9	−12.3
2	4.3	8.9	16.4
3	6.3	−14.1	11.8

4-8. The velocity of sound transmission in a gas is given by

$$C = \sqrt{kg_cRT}$$

where C = sound velocity, ft/s
k = ratio of specific heats
g_c = gravitational constant, ft/s^2
R = gas constant, ft/°R
T = absolute temperature, °R

Write a FORTRAN program that will contain the physical properties of five gases so that the sonic velocity at a specified temperature can be calculated for any one of the gases. The input to the program is to be a number indicating which gas is to be selected and the gas temperature in degrees Fahrenheit. Design the program to print the results and automatically return to the READ statement for a new set of data.

Sample data

Number	*Gas*	*R*	*k*
1	Air	53.34	1.400
2	CO_2	35.10	1.285
3	He	386.0	1.66
4	N_2	55.15	1.400
5	O_2	48.28	1.395

Use the program to compare the velocities of sound in air and in helium at 75°F.

4-9. At high frequencies *short* sections of transmission lines are often used as circuit elements. Such sections may be used in either short-circuited or open-circuited form.

Write a program which will calculate the length l of a lossless, short-circuited transmission line with a specified characteristic impedance that is to be used as an inductor of value L at a particular frequency. The following formulas define the circuit conditions:

$$2\pi fL = Z_0 \tan\theta \qquad (1)$$

$$\theta = \frac{2\pi l}{\lambda} \qquad (2)$$

$$\lambda = \frac{c}{f} \qquad (3)$$

where f = frequency, Hz
L = inductance, H
Z_0 = characteristic impedance, ohms
l = length of line, m
λ = wavelength, m
c = 3×10^8, m/s

Sample data

$f = 75 \times 10^6$ Hz
$L = 7 \times 10^{-6}$ H
$Z_0 = 300$ ohms

4-10. If the transmission line described in Prob. 4-9 is open circuited, Eqs. (2) and (3) given there are still valid. However, Eq. (1) must be replaced by

$$2\pi fL = -Z_0 \cot\theta \qquad (4)$$

Write a program that will calculate the required length of line l for this case and test it with the sample data given in Prob. 4-9.

4-11. Problem 4-9 dealt with a short-circuited transmission line and Prob. 4-10 dealt with an open-circuited line. Write a FORTRAN program which gives the user the option of making calculations for either a short-circuited or an open-circuited line.

4-12. In an RLC circuit the impedance is given by

$$Z = R + j\left(\omega L - \frac{1}{\omega C}\right)$$

Referring to the discussion in Sec. 4.12, write a computer program to read-in values of R, L, C, and ω, and print the value of the impedance Z in polar form.

Sample data

Set	R, ohms	L, H	C, F	ω, rad/s
1	100	0.01	1×10^{-6}	5×10^{3}
2	100	0.01	1×10^{-6}	5×10^{4}
3	10	0.001	1×10^{-10}	5×10^{6}

4-13. In calculating the impedance of an RLC circuit, if C happens to be equal to zero, the term $1/\omega C$ is not defined in algebra and in a computer calculation would produce an error message.

Modify the program written as a solution to Prob. 4-12 to allow for the possibility that the input value of C is equal to zero.

Sample data

$R = 100$ ohms
$L = 0.01$ H
$\omega = 5 \times 10^3$ rad/s

4-14. The combustion of 1 mol of propane with the theoretical amount of air is represented by the following chemical equation:

$$C_3H_8 + 5(O_2 + 3.76N_2) \rightarrow 4H_2O + 3CO_2 + 18.8N_2$$

The number of moles of product is 4 + 3 + 18.8 = 25.8, and the composition of the products of combustion (by volume) is

$$\%\ H_2O = \frac{4}{25.8}\,100 = 15.5$$

$$\%\ CO_2 = \frac{3}{25.8}\,100 = 11.6$$

$$\%\ N_2 = \frac{18.8}{25.8}\,100 = 72.9$$

If propane is burned with X percent of the theoretical air, the combustion equations (with X expressed as a decimal) are

$$C_3H_8 + 5(X)(O_2 + 3.76N_2) \rightarrow 4H_2O + 3(X)CO_2 + 3(1 - X)CO + 18.8(X)N_2 \qquad (\text{for } X \leqslant 1)$$

and

$$C_3H_8 + 5(X)(O_2 + 3.76N_2) \rightarrow 4H_2O + 3CO_2 + 5(X - 1)O_2 + 18.8(X)N_2 \qquad (\text{for } X \geqslant 1)$$

Write a FORTRAN program that will accept the percentage of theoretical air used in the combustion process and will print the percentage composition of the products.

Sample data

Set	X, %
1	92
2	100
3	130

5
The DO Loop

In this chapter we will discuss *iterative processes* and illustrate them with several examples. The general technique of programming the computer to perform an iterative process will be demonstrated here, but the major emphasis will be placed on the use of the DO loop, a FORTRAN-supplied procedure, for achieving this purpose. All the problems solved in this chapter employ *normal exits* (see Sec. 5.4) from the DO loop. In the next chapter we will extend our discussion to include *exits other than normal.*

5.1 ITERATIVE PROCESSES

An algorithm is a procedure for solving a problem. To be effective, an algorithm must be unambiguous, must take into consideration all alternatives that may be presented, and must produce an answer in a finite number of steps.

In many algorithms it is necessary to perform a given calculation many times, using a different value for a variable each time. Executing such an iterative or recursive procedure with a computer program is called *looping.*

For example, the following program segment could be used to evaluate 20!:

```
     FACT = 1.                 (a)
     X = 1.                    (a)
   1 FACT = FACT*X             (b)
     IF(20.-X) 2,2,3           (c)
   3 X = X + 1                 (d)
     GO TO 1                   (e)
   2 PRINT 50, FACT            (f)
```

Step *a* *Initializes* the values of FACT and X.
Step *b* *Calculates* (start of the loop).
Step *c* *Tests* the current value of X (i.e., the value during *this* loop). Note that if X is equal to 20. the result is printed.
Step *d* *Increments* the value of X by 1.
Step *e* *Returns* control to the start of the loop to allow calculation to continue.
Step *f* Is an output statement and is called into operation when the calculation has been completed.

As a further example, consider the following algorithm for calculating the square root of a number A.

$$X_{i+1} = \frac{X_i + A/X_i}{2}$$

Assuming the value of A is already stored in the computer, this procedure can be programmed as follows:

```
   X = A/2.                          (a)
 1 XNEW = (X + A/X)/2.               (b)
   IF(ABS (X - XNEW) - .01) 2,2,3    (c)
 3 X = XNEW                          (d)
   GO TO 1                           (e)
 2 PRINT 50, XNEW                    (f)
```

Step *a* *Initializes* the first estimate of the square root.
Step *b* *Calculates* the next approximation of the root using the procedure of the algorithm.
Step *c* *Tests*[1] the *latest* estimate of the root by comparing it to the *previous* estimate. If successive estimates are within ±0.01 of each other, the *current* (latest) estimate is printed.
Step *d* *Replaces* the current value of X with the value of the *latest* estimate. (This step is necessary because the algorithm calculates succeeding estimates on the basis of the latest estimate.)
Step *e* *Returns* control to the start of the loop.
Step *f* Is an output statement that will print the result when the test in step *c* has been *satisfied.*

It should be noted that both of the above examples utilized the looping process. In the factorial calculation, the loop would be circuited exactly 20 times. In the square root calculation, the loop would be circuited an unspecified number of times until the calculation *test* is satisfied.

The DO loop is a provision for the FORTRAN language that makes it possible to *automatically* produce the steps described in the above examples. The programs for solving the problems in this chapter illustrate the use of the DO loop and the *nested* DO loop. In all of these programs, exit from the loop is *normal*; that is, the loop executes the prescribed number of cycles before the main program continues. (This point is explained in detail in Sec. 5.4.)

[1] This is not the *best* test that could be devised for this program but it is a simple one and serves to illustrate the looping procedure.

5.2 DO STATEMENT

The DO statement is an executable statement having one of the following forms:

DO n $i = n_1, n_2, n_3$

or

DO n $i = n_1, n_2$

where

- n is the statement number of the *last* statement in the loop (the *range* of the DO extends from the first statement following the DO statement down to the statement numbered n).
- i is a nonsubscripted integer variable whose value is incremented in successive loops. It is called the *index.*
- n_1, n_2, n_3 are unsigned integer constants or nonsubscripted integer variables and are called the *index parameters.* All the parameters must have values greater than zero.
- n_1 is the *initial value* of *i.*
- n_2 is the maximum value that i may assume and is called the *test value* (n_2 must be greater than or equal to n_1).
- n_3 is the amount i is increased for each successive loop. It is called the *increment.* (*Note*: If n_3 does not appear in the list, it is understood to be equal to 1.)

The DO statement automatically accomplishes all the steps of looping with the exception of the calculations which must be specified within the *range* of the DO.

Examples of DO statements are

```
DO 16 I = 1,12
DO 30 MARK = 6,18,3
```

It should be noted that the same index name, for example, I, may be used for as many DO loops as desired, provided that none of the loops are *nested* (see Sec. 5.14).

5.3 RESTRICTIONS ON THE DO STATEMENT

The following rules apply when a DO statement is used in a FORTRAN program:

1. The range of the DO must *not* contain any statement that redefines the index.
2. The range of the DO must *not* contain any statements that redefine the index parameters when they are written as integer variables in the source program. That is, n_1, n_2, and n_3, must be "set" before the DO loop is entered.
3. The *first statement* in the range of a DO *must be an executable statement.*
4. The *last statement* in the range of a DO (i.e., statement n) *must not be a transfer statement.*[1] Where such a statement would normally be the last statement, the range is extended with a CONTINUE statement (see Sec. 5.5).

[1] An exception to this rule is the *logical* IF statement (see Sec. 9.10). While the logical IF is truly a transfer statement, it *may* be used as the last statement in the range of a DO. If the argument of the logical IF is .TRUE., control passes to *stat 1*; when the value of the argument is .FALSE., control passes to the first statement in the range of the DO (providing incrementation of the index does not produce a value greater than the test value).

5. It is possible to *nest* DO loops providing the *inner* DO is entirely contained within the *outer* DO. There is no limitation on the number that may be nested. The names used for the indexes of an inner DO loop must differ from that of the outer loop and from each other.
6. Transfer of control *out of* the range of a DO *is* permitted.
7. Transfer of control *into* the range of a DO *is not* permitted. The only exception to this rule occurs when a subprogram (see Sec. 10.2) is called as part of the calculations in the range and the transfer into the loop is actually the *return* from the subprogram.
8. Transfer of control *within* the range of a DO *is* permitted. In a nest of DO's it is possible to transfer from an inner loop to an outer loop because the transfer is *out* of the *inner* loop and entirely *within* the *outer* loop. The reverse process is not permitted.

5.4 OPERATION OF THE DO LOOP

When a DO statement (see Sec. 5.2) is encountered in a program, the computer automatically executes the following steps:

1. *Initializes*: sets i equal to n_1.
2. *Computes*: follows the instructions in the range of the DO.
3. *Increments*: increases the present value of i by n_3.
4. *Tests*: compares the value of i with n_2.
   a. If i is *less than or equal to* n_2, control returns to the first statement in the range of the DO.
   b. If i is *greater than* n_2, the looping process is stopped and control passes to the statement immediately following the last statement in the range of the DO. This procedure is called a normal exit from the DO. The requirements of the DO have been satisfied.

If exit from a DO is *other* than normal, the current value of the index at the time of exit is retained and may be used later in the program. If exit from a DO *is* normal, the value of the index will be *greater than* the test value because the sequence of events following the last statement in a DO is

1. Increment
2. Test
3. Either exit or loop

To illustrate the application of the DO statement, let us consider the problem that was posed in Sec. 5.1, that is, the evaluation of 20!. Using a DO loop to evaluate this factorial would require the following statements:

```
      FACT = 1.                    (a)
      DO 1 N=1,20                  (b)
      X = N                        (c)
    1 FACT = FACT*X                (d)
      PRINT 50, FACT               (e)
```

Step *a* Initializes FACT.
Step *b* Initiates the DO loop (defines the index parameters and the range).

Step *c* Transforms N into floating point mode.
Step *d* Calculates.
Step *e* Prints the results. When the DO is *satisfied* (i.e., when N = 21 in this illustration), control passes to the PRINT statement (i.e., the *next* statement). This is a *normal exit* from the DO loop.

In this example, the value of the *index* (after transformation to real mode) was *actually used* in the calculation. This use of the index is *not* a requirement, as is shown in the following illustration.

Now let us examine the use of a DO loop to calculate the square root of A, as discussed in Sec. 5.1. Before we can write the program, we must decide what the *maximum* number of iterations will be before a suitable result is obtained. Let us choose 1000 as a very conservative estimate. We will assume that location A already contains the number whose square root is required. The program would then be:

```
     X = A/2.                              (a)
     DO 3 J=1,1000                         (b)
     XNEW = (X + A/X)/2.                   (c)
     IF(ABS (X–XNEW)–.01) 2,2,3            (d)
   3 X = XNEW                              (e)
   2 PRINT 50, XNEW                        (f)
```

Step *a* Initializes the value of X at the first estimate of the square root.
Step *b* Initiates the DO loop.
Step *c* Calculates the next estimate according to the algorithm.
Step *d* Tests the latest estimate. If the test is met, control is transferred *out* of the DO loop (an *abnormal exit*) and the result is printed, step *f*. If the test fails, control is transferred to statement 3, which is *within* the DO.
Step *e* Replaces the current value of X with the latest estimate. This is the last statement in the range of the DO, and at this point J will be incremented. If J is less than or equal to 1000, the control will transfer to the start of the range. If J is greater than 1000, a *normal exit* will be accomplished and the current value of XNEW will be printed whether or not it is the correct result.

In the above example the index of the DO was used only as a *counter.* It is assumed that the algorithm will give a correct result in less than 1000 iterations. As a matter of interest, we might have printed the value stored in J along with the value in XNEW. This would tell us the number of iterations that were performed before the process converged.

5.5 CONTINUE STATEMENT

The CONTINUE statement is classified as an executable statement, but it does not produce any commands in the object program. It is sometimes referred to as a "dummy" statement because it produces no action other than to pass control to the next statement in the normal flow of control (see Sec. 4.1). It can appear anywhere in a program without altering the normal sequence of actions.

The CONTINUE statement may be used as a reference point for one of the *transfer* statements (see Sec. 4.1), but it is most frequently used in connection with DO loops to create a final statement in the *range* of the DO to which a statement number may be assigned (see rule 4 in Sec. 5.3).

5.6 FLOW CHARTING DO LOOPS

To clearly mark the beginning and end of each DO loop in a flow chart, the authors have employed the following symbols:

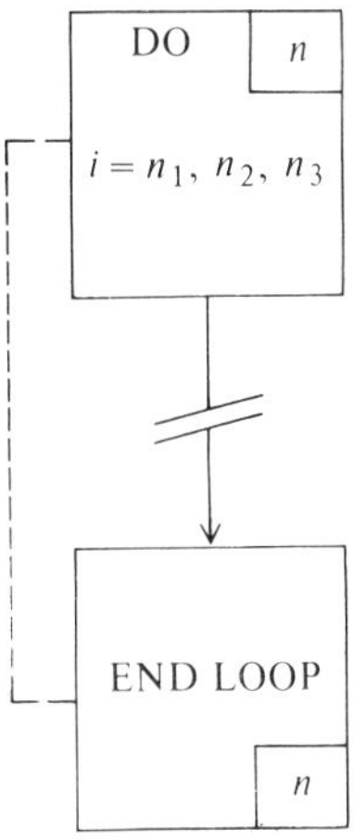

The *beginning block* symbolizes a FORTRAN DO statement. The *end block* symbolizes the point where the DO loop automatically increments, tests, and either loops or exits. It may either follow the *processing block* (see Sec. 1.27) containing the last executable statement in the range of the DO, or represent the location of a CONTINUE statement that was used as the last statement in the range of the DO.

In this text, a dotted line connector has been used to bracket all the program steps contained in a DO loop. This procedure accentuates the iterative nature of the DO loop and also serves to clarify the repetitive operation of nested DO loops.

5.7 COMPOUND INTEREST TABULATION

The single-investment, compound interest formula is

$$V = P\left(1 + \frac{i}{m}\right)^{mn} \tag{5.7.1}$$

The symbols have the following definitions.

Formula symbol	*Computer name*	*Meaning*	*Units*
V	V	Value of principal at the end of the investment period	\$
P	P	Principal invested	\$
i	. . .	Annual interest rate	decimal
. . .	XI	Annual interest rate	%
m	M	Number of times interest is compounded each year	
	XM	M expressed as a real	
n	N	Number of years principal is invested (assumed to be an integer)	years
	K	DO loop index	years

Equation (5.7.1) can be used as it is to calculate the value of the principal at the end of any investment period. However, the purpose of this exercise is to *tabulate* the value at the end of each year, for a specified number of years. The tabulation actually

```
00001C      INTER1
00002C      THIS PRØGRAM TABULATES THE ANNUAL VALUE ØF "P" DØLLARS
00003C      CØMPØUNDED "M" TIMES A YEAR FØR "N" YEARS AT A NØMINAL
00004C      INTEREST RATE ØF "XI" PERCENT.
00010       READ,P,XI,M,N
00020       XM=M
00030       DØ 10 K=1,N
00040       V=P*(1.+XI/100./XM)**(K*M)
00050    10 PRINT,K,V
00060       STØP
00070       END
```

Fig. 5.7.1 Program INTER1.

requires repeated evaluation of the formula with different values of N, and a simple method of accomplishing the repetition is by use of the DO loop. In the program INTER1, shown in Fig. 5.7.1, the formula is contained in a DO loop (lines 00030, 00040, and 00050) that is to be executed N times.

The interest rate is not necessarily an integer value and therefore, I, which normally implies an integer mode, has been avoided, and the interest was expressed in real mode by assigning its contents to a real location designated XI. It is desirable to use M as an integer in the exponentiation process indicated on line 00040 (see Sec. 1.15). However, a real value of M, for use in computing the interest rate in decimal form, was obtained by effecting a change of mode on line 00020. It should be noted that in Eq. (5.7.1), the nominal annual interest rate i is expressed as a decimal, but, in the computer program, XI is expressed in percent and the decimal equivalent for each interest period is calculated within the program. A sample run for this program is shown in Fig. 5.7.2.

During program execution, the DO loop evaluates and prints the value of V a total of N times. Note that each time line 00040 is executed, the value of (1.+XI/100./XM) is determined anew, despite the fact that none of the terms in this expression is changed during successive iterations. Whenever a condition such as this occurs, the program can be made to run more efficiently by placing, *outside* the DO loop, a statement that will evaluate the unchanging expression. This procedure is illustrated in the revised program INTER2, shown in Fig. 5.7.3. The reader's attention is called to the insertion of the statement on line 00030 and the revision shown on line 00050. As the interest rate XI is not used again in this program, it was not necessary to use an additional location for storing the value of the expression. If the original contents of location XI had been required subsequently in the program, the revised lines could have been written as

```
00030     FAC = 1.+XI/100./XM
00050     V   = P*FAC**(K*M)
```

```
INPUT:00010
?100,5.5,4,10

          1        105.6145
          2        111.5442
          3        117.8068
          4        124.4211
          5        131.4067
          6        138.7845
          7        146.5765
          8        154.8060
          9        163.4975
         10        172.6771
STØP
```

Fig. 5.7.2 Trial run for INTER1.

```
      INTER2
      THIS PRØGRAM TABULATES THE ANNUAL VALUE ØF "P" DØLLARS
      CØMPØUNDED "M" TIMES A YEAR FØR "N" YEARS AT A NØMINAL
      INTEREST RATE ØF "XI" PERCENT.
      READ,P,XI,M,N
      XM=M
      XI=1.+XI/100./XM
      DØ 10 K=1,N
      V=P*XI**(K*M)
   10 PRINT,K,V
      STØP
      END
```

Fig. 5.7.3 Program INTER2.

```
INPUT:00010
?100,5.5,4,10

         1       105.6145
         2       111.5442
         3       117.8068
         4       124.4211
         5       131.4067
         6       138.7845
         7       146.5765
         8       154.8060
         9       163.4975
        10       172.6771
STØP
```

Fig. 5.7.4 Trial run for INTER2.

A run for this program, using the same problem data as before, is shown in Fig. 5.7.4. The reader can see that the results are the same as those shown in Fig. 5.7.2.

5.8 CALCULATION OF THE FACTORIAL

Calculation of a factorial is frequently required in problems concerning probability. For example, the calculation of the number of combinations of k objects selected from n objects, all different, requires three factorials.[1] A segment of a program for calculating 20! was discussed in Sec. 5.1.

A similar, but slightly more general, program for calculating M! is shown in Fig. 5.8.2. In this program, which is diagrammed in Fig. 5.8.1, the numerical value of M! is stored in location PROD. In order to demonstrate the sequence of iterations performed by the DO loop in making the calculations, a PRINT statement for printing the current value of the variable PROD is included in the DO loop. Normally, the PRINT statement would be outside the DO loop, and only the answer would be printed. This would result in faster execution of the program.

In the first trial run shown in Fig. 5.8.3, 5! was calculated. After this calculation, the transfer statement on line 00100 caused the program to branch back to statement 2 on line 00010, thus giving the user the option of obtaining a new factorial calculation. After two additional factorial calculations, execution was halted by typing the command STOP when the next input request was made. Note that in the calculation of 1! the loop is executed only once.

[1] B. W. Lindgren and G. W. McElrath, "Introduction to Probability and Statistics," 2d ed., Macmillan, New York, 1969, p. 7.

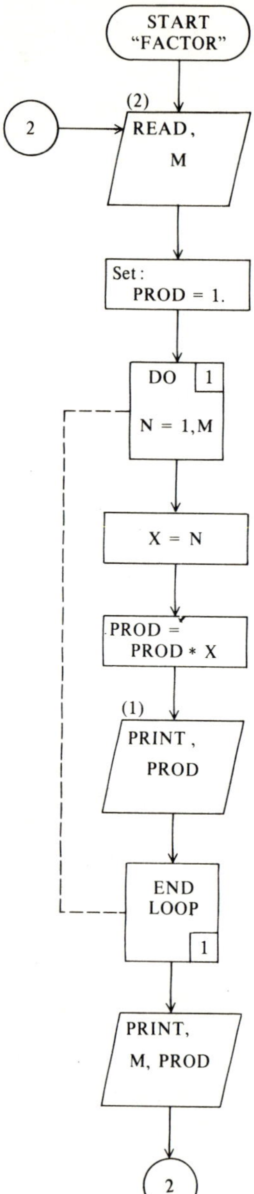

Fig. 5.8.1 Flow chart of program FACTOR.

```
C     FACTØR
C     THIS PRØGRAM ILLUSTRATES ØNE WAY ØF CALCULATING
C     M FACTØRIAL.
    2 READ,M
      PRØD=1.
      DØ 1 N=1,M
      X=N
      PRØD=PRØD*X
    1 PRINT 8,PRØD
    8 FØRMAT(E15.8)
      PRINT 9,M,PRØD
    9 FØRMAT(I3,11H FACTØRIAL=E15.8//)
      GØ TØ 2
      END
```

Fig. 5.8.2 Program FACTOR.

```
INPUT:00010
?5

0.10000000E+01
0.20000000E+01
0.60000000E+01
0.24000000E+02
0.12000000E+03
 5 FACTØRIAL= 0.12000000E+03


INPUT:00010
?1

0.10000000E+01
 1 FACTØRIAL= 0.10000000E+01


INPUT:00010
?7

0.10000000E+01
0.20000000E+01
0.60000000E+01
0.24000000E+02
0.12000000E+03
0.72000000E+03
0.50400000E+04
 7 FACTØRIAL= 0.50400000E+04


INPUT:00010
?STØP
```

Fig. 5.8.3 Trial runs for FACTOR.

5.9 THE HYPERGEOMETRIC PROBABILITY DISTRIBUTION[1]

The hypergeometric probability distribution involves nine factorial calculations. This probability distribution concerns selection without replacement from a population of N objects, M of which are of one type, and $N-M$ of which are of another type. (For example, N may consist of M men and $N-M$ women.) A random sample of size NS is drawn from the population and not replaced. The problem is to calculate the probability $P(K)$ that there are K of type M in the sample. A flow chart of the program is shown in Fig. 5.9.1.

The solution to this problem is:

$$P(K) = \frac{\binom{M}{K}\binom{N-M}{NS-K}}{\binom{N}{NS}} = \frac{\dfrac{M!}{K!\,(M-K)!}\,\dfrac{(N-M)!}{(NS-K)!\,(N-M-NS+K)!}}{\dfrac{N!}{NS!\,(N-NS)!}} \tag{5.9.1}$$

The program for this calculation is shown in Fig. 5.9.2 and a sample run follows in Fig. 5.9.3. The major part of the program is given to the calculation of the nine factorials required in Eq. (5.9.1). Each of these calculations parallels the factorial calculations illustrated in Fig. 5.8.2. For convenience, the names F1, . . . , F9 were chosen to represent the various factorials. For example, F1 represents $M!$, and F2 represents $K!$ Also, note that it is permissible to use the same label, I, as the index of each of the DO loops for calculating the factorials *because the DO loops are*

[1] *Ibid.*, p. 60.

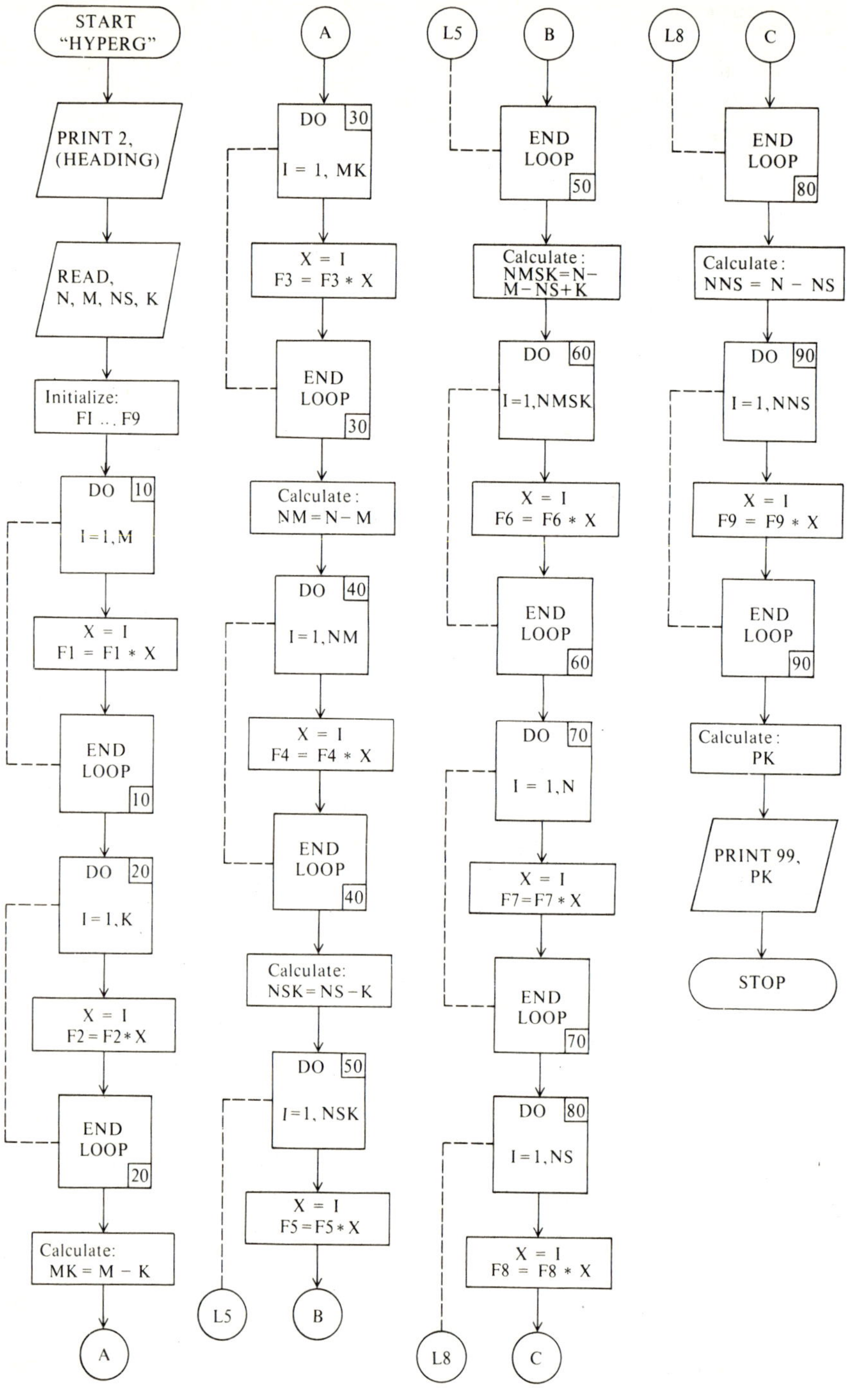

Fig. 5.9.1 Flow chart of program HYPERG.

independent of each other. However, when DO loops are nested (see Sec. 5.14), the label of the index for each loop must be different. The variables F1, . . . , F9 are initialized at a value of 1. with a multiple assignment statement which appears on line

```
00001C    HYPERG
00002C    THIS IS A PRØGRAM WHICH CALCULATES HYPERGEØMETRIC
00003C    PRØBABILITIES. N IS THE TØTAL PØPULATIØN. M IS THE NUMBER
00004C    ØF ØNE KIND. NM IS THE NUMBER ØF THE ØTHER KIND. NS IS
00005C    THE SAMPLE SIZE. K IS THE NUMBER ØF M IN NS.
00010     PRINT 2
00020   2 FØRMAT(29HENTER VALUES FØR N,M,NS,AND K)
00030     READ,N,M,NS,K
00040     F1=F2=F3=F4=F5=F6=F7=F8=F9=1.
00050     DØ 10 I=1,M
00060     X=I
00070  10 F1=F1*X
00080     DØ 20 I=1,K
00090     X=I
00100  20 F2=F2*X
00110     MK=M-K
00120     DØ 30 I=1,MK
00130     X=I
00140  30 F3=F3*X
00150     NM=N-M
00160     DØ 40 I=1,NM
00170     X=I
00180  40 F4=F4*X
00190     NSK=NS-K
00200     DØ 50 I=1,NSK
00210     X=I
00220  50 F5=F5*X
00230     NMSK=N-M-NS+K
00240     DØ 60 I=1,NMSK
00250     X=I
00260  60 F6=F6*X
00270     DØ 70 I=1,N
00280     X=I
00290  70 F7=F7*X
00300     DØ 80 I=1,NS
00310     X=I
00320  80 F8=F8*X
00330     NNS=N-NS
00340     DØ 90 I=1,NNS
00350     X=I
00360  90 F9=F9*X
00370     PK=(F1/(F2*F3))*(F4/(F5*F6))/(F7/(F8*F9))
00380     PRINT 99,PK
00390  99 FØRMAT(5HP(K)=F6.4//)
00400     STØP
00410     END
```

Fig. 5.9.2 Program HYPERG.

```
ENTER VALUES FØR N,M,NS,AND K
INPUT:00030
 ?30,20,5,3

P(K)=0.3600


STØP
```

Fig. 5.9.3 Trial run for HYPERG.

00040. This is a convenience which is *not* permitted on all computers. It is possible to make this program more compact by use of a function subprogram and this will be discussed in Sec. 10.14.

5.10 THE BIRTHDAY PROBLEM[1]

Another illustration of the use of the DO loop is provided by a program for the solution of the classic "birthday problem." If there are n people in a room, what is the probability that two or more have the same birthday, assuming 365 days in a year? Let the probability of the event "no two people have the same birthday" be denoted by $P(E)$. It follows from probability theory that

$$P(E) = \frac{365 \cdot 364 \cdot 363 \cdots (365 - n + 1)}{365^n} \tag{5.10.1}$$

The solution to the problem is:

$$P(\text{two or more people have the same birthday}) = 1. - P(E) \tag{5.10.2}$$

Although Eq. (5.10.1) is correct, it is not convenient for computer calculation because the numerator and denominator each will be large numbers for even relatively small values of n. For example, for $n = 12$, the denominator of the fraction in Eq. (5.10.1) would be $365^{12} \approx 5.59 \times 10^{30}$. Thus, for large values of n, there is the possibility of overflow.[2] A better way of calculating $P(E)$ is shown in Eq. (5.10.3). In this equation, $P(E)$ is expressed as the product of n fractions. The first fraction has a value of unity, and the succeeding fractions are progressively smaller.

$$P(E) = \frac{365}{365} \cdot \frac{364}{365} \cdot \frac{363}{365} \cdots \frac{(365 - n+1)}{365} \tag{5.10.3}$$

The program for solving this problem is shown in Fig. 5.10.1. N is the specified number of people in a room. $P(E)$ is represented by the name POFE. POFE is calculated within the DO loop which begins on line 00030. In the loop, the index I is used to generate values of the real variable T (line 00040). T causes RATIO to vary, and POFE is generated by RATIO. For example, when I = 2, T = 1., RATIO = 364./365., and POFE = 364./365. When I = 3, T = 2., RATIO = 363./365., and POFE = (364./365.)(363./365.). Thus, $P(E)$ is calculated in accordance with Eq. (5.10.3). The initial value of the index I has been set at 2 because a value of I = 1

```
00001C      BIRTH
00002C      THIS PRØGRAM CALCULATES THE PRØBABILITY THAT IN A RØØM
00003C      ØF N PEØPLE AT LEAST TWØ HAVE THE SAME BIRTHDAY.
00010       READ,N
00020       PØFE=1.
00030       DØ 1 I=2,N
00040       T=I-1
00050       RATIØ=(365.-T)/365.
00060     1 PØFE=PØFE*RATIØ
00070       PRØB=1.-PØFE
00080       PRINT 5,PRØB
00090     5 FØRMAT(12HPRØBABILITY=F6.4//)
00100       STØP
00110       END
```

Fig. 5.10.1 Program BIRTH.

[1] W. Feller, "An Introduction to Probability Theory and Its Applications," 2d ed., Wiley, New York, 1957, vol. 1, p. 31.

[2] In the computer used for this program, the largest permissible value of a real number is $0.578960444 \times 10^{77}$.

```
INPUT:00010
 ?12

PRØBABILITY=0.1670

STØP
```

Fig. 5.10.2 Trial run for BIRTH.

would result in RATIO = (365./365.) = 1. In the trial run shown in Fig. 5.10.2, the number 12 was entered as the value of N. For this value of N, the probability that 2 or more people in a room of 12 people have the same birthday was found to be 0.167. A flow chart of the program is shown in Fig. 5.10.3.

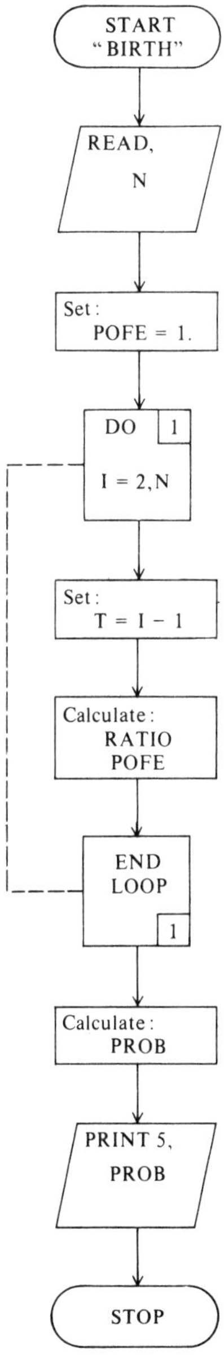

Fig. 5.10.3 Flow chart of program BIRTH.

5.11 BEAM DEFLECTION (SEVERAL POINTS)

The simple beam deflection problem that we examined in Sec. 4.10 can be solved, using a DO loop, to give the value of the deflection at *several* equally spaced points along the beam rather than for the *single* arbitrary point that was used earlier. The information obtained in this manner could be used to plot the deflection curve of the beam.

Equations (4.10.1) and (4.10.2), as applied to the sketch in Fig. 4.10.1, are again employed in writing the program. Names used in the program are the same as those described in Sec. 4.10 with the following additional symbols:

Computer name	*Meaning*
N	Number of points at which deflection is to be determined
SECTS	Number of segments into which beam is divided by the N points
I	DO loop index (used to select points consecutively)
POINT	Real equivalent of I

A flow chart for one method of solution is shown in Fig. 5.11.1. Within the DO loop, the points at which the deflection is to be determined are chosen consecutively, and the values of X (i.e., the actual positions on the beam) are calculated as fractional parts of the beam length. Once the value of X has been calculated, the program proceeds just as it did in Sec. 4.10 when the value of X was part of the input data.

When the value of X and its corresponding deflection Y have been printed, the index of the loop is incremented and a new position on the beam is chosen. After the DO loop has been satisfied, the program stops.

The FORTRAN program for carrying out this plan is entitled BEAM2 and is shown in Fig. 5.11.2. This program is a revision of the earlier one shown as Fig. 4.10.3. Note the change in line 00010 that makes it possible to specify *how many* deflection points will be evaluated rather than *which* point, as was done formerly. The DO loop starts on line 00040. The statement on line 00050 changes the mode of the index I to a real value, POINT, so the latter can be used in the arithmetic statement on line 00060 without mixing modes. The remainder of the program is written exactly as it was in Chap. 4, but it should be noted that in this version statement 3, line 00110, is the last statement in the DO loop and the computer will not proceed to the STOP command until the DO has been satisfied. Figure 5.11.3 shows two trial runs made for the same load but with different numbers of evaluation points.

Figure 5.11.4 is a program called BEAM3, which is a modification of BEAM2. The changes made include the calculation of a factor FAC on line 00030 and its use in the two deflection equations on lines 00090 and 00110. This alteration is not necessary but will produce a more efficient program by removing the evaluation of a program constant from the DO loop. In the program BEAM1 in Fig. 4.10.3 there would be no advantage in evaluating FAC separately because in *that* program *either* line 00040 *or* line 00060 is executed only *once*.

BEAM3 also contains a change on line 00120 in that the PRINT statement now references an output FORMAT (see Sec. 2.3). Line 00130 gives the FORMAT statement which includes *H, F,* and *X* specifications (see Secs. 2.6, 2.9 and 2.10). The sample run in Fig. 5.11.5 shows that although the output is arranged differently, the numerical values produced by both of these programs are the same.

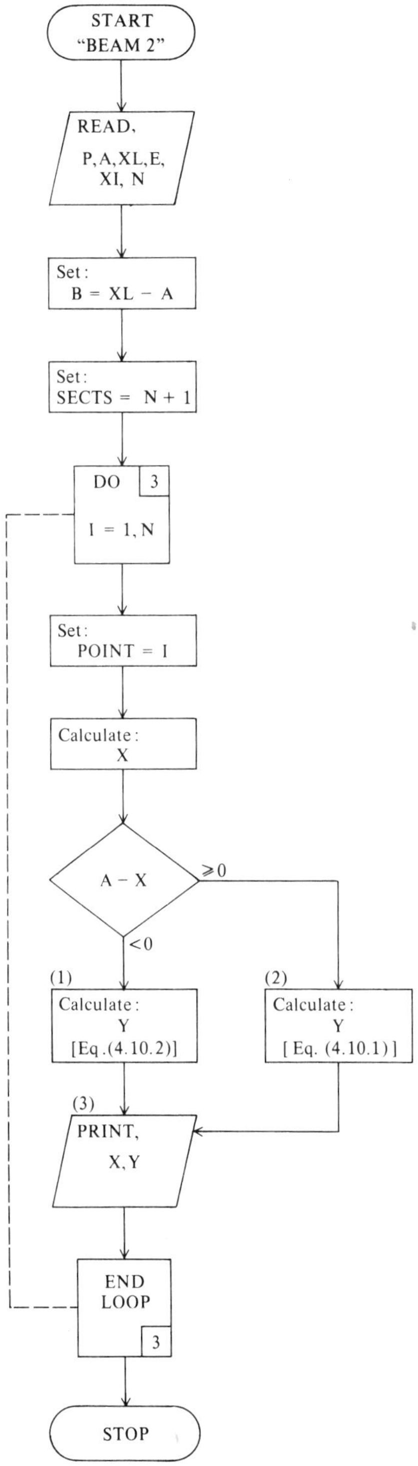

Fig. 5.11.1 Flow chart of program BEAM2.

```
C      BEAM2
C      THIS PRØGRAM CALCULATES SIMPLE BEAM DEFLECTIØN,AT SEVERAL
C      PØINTS,PRØDUCED BY A SINGLE CØNCENTRATED LØAD.
       READ,P,A,XL,E,XI,N
       B=XL-A
       SECTS=N+1
       DØ 3 I=1,N
       PØINT=I
       X=PØINT*XL/SECTS
       IF(A-X) 1,2,2
     1 Y=-P*A*(XL-X)*((2.*XL-B)*B-(XL-X)**2)/(6.*E*XI*XL)
       GØ TØ 3
     2 Y=-P*B*X*(2.*XL*(XL-X)-B*B-(XL-X)**2)/(6.*E*XI*XL)
     3 PRINT,X,Y
       STØP
       END
```

Fig. 5.11.2 Program BEAM2.

```
INPUT:00010
?500,65,154,3E7,2.5,10

      14.0000        -.1402
      28.0000        -.2699
      42.0000        -.3784
      56.0000        -.4552
      70.0000        -.4899
      84.0000        -.4790
      98.0000        -.4295
     112.0000        -.3492
     126.0000        -.2456
     140.0000        -.1267
STØP


INPUT:00010
?500,65,154,3E7,2.5,15

       9.6250        -.0970
      19.2500        -.1907
      28.8750        -.2774
      38.5000        -.3538
      48.1250        -.4165
      57.7500        -.4621
      67.3750        -.4870
      77.0000        -.4897
      86.6250        -.4724
      96.2500        -.4376
     105.8750        -.3876
     115.5000        -.3252
     125.1250        -.2527
     134.7500        -.1726
     144.3750        -.0876
STØP
```

Fig. 5.11.3 Trial runs for BEAM2.

So far we have discussed programs that evaluate the deflection of a simple beam at a *single specified point* and at *any number of equally spaced points.* In Sec. 6.5 we will examine a computer method for locating the point of *maximum deflection* and the value of that deflection.

```
C    BEAM3
C    THIS PRØGRAM CALCULATES SIMPLE BEAM DEFLECTIØN,AT SEVERAL
C    PØINTS, PRØDUCED BY A SINGLE CØNCENTRATED LØAD.
     READ,P,A,XL,E,XI,N
     B=XL-A
     FAC=-P/(6.*E*XI*XL)
     SECTS=N+1
     DØ 3 I=1,N
     PØINT=I
     X=PØINT*XL/SECTS
     IF(A-X) 1,2,2
   1 Y=FAC*A*(XL-X)*((2.*XL-B)*B-(XL-X)**2)
     GØ TØ 3
   2 Y=FAC*B*X*(2.*XL*(XL-X)-B*B-(XL-X)**2)
   3 PRINT 50,X,Y
  50 FØRMAT(6HAT X =F8.3,3X3HY =F8.4)
     STØP
     END
```

Fig. 5.11.4 Program BEAM3.

```
INPUT:00010
?500,65,154,3E7,2.5,15

AT X =   9.625   Y =  -.0970
AT X =  19.250   Y =  -.1907
AT X =  28.875   Y =  -.2774
AT X =  38.500   Y =  -.3538
AT X =  48.125   Y =  -.4165
AT X =  57.750   Y =  -.4621
AT X =  67.375   Y =  -.4870
AT X =  77.000   Y =  -.4897
AT X =  86.625   Y =  -.4724
AT X =  96.250   Y =  -.4376
AT X = 105.875   Y =  -.3876
AT X = 115.500   Y =  -.3252
AT X = 125.125   Y =  -.2527
AT X = 134.750   Y =  -.1726
AT X = 144.375   Y =  -.0876
STØP
```

Fig. 5.11.5 Trial run for BEAM3.

5.12 SPECIFIC PERMEABILITY OF A POROUS MEDIUM[1]

The specific permeability k of a porous medium consisting of particles of various sizes and shapes is an important parameter in the study of the movement of groundwater and in the design of wells. The following formula is used to calculate specific permeability:

$$k = \frac{1}{m\ \{[(1-p)^2/p^3]\ [\Sigma(\theta_j/100)\ (P_j/d_j)]^2\}} \tag{5.12.1}$$

where k = specific permeability of mixture, mm^2
p = porosity (fraction of the mixture occupied by interstices)
m = packing factor
θ_j = particle shape factor
P_j = percent by weight of material j in the medium
d_j = diameter of the particles of material j in the medium, mm

[1] R. K. Linsley and J. B. Franzini, "Water Resources Engineering," McGraw-Hill, New York, 1964, p. 82.

If all the particles of the mixture have the same shape factor, then Eq. (5.12.1) simplifies to

$$k = \frac{1}{m\,\{[(1-p)^2/p^3](\theta/100)^2\,[\Sigma(P_j/d_j)]^2\}} \tag{5.12.2}$$

Let us consider the following problem. We have a mixture of two types of sand, type A and type B. How does the specific permeability of the mixture vary as the proportions of A and B in the mixture are varied? The following data concerning the sand are given:

$d_a = 0.08$ mm $\quad \theta = 6.5 \quad m = 5$

$d_b = 0.40$ mm $\quad p = 0.34$

Because the particles of each type of sand are assumed to have the same shape factor, Eq. (5.12.2) is used. A program for carrying out the necessary calculations is shown in Fig. 5.12.1. The following table defines the symbols used in the program:

Formula symbol	*Computer name*	*Meaning*	*Units*
d_a	DA	Particle diameter (type A)	mm
d_b	DB	Particle diameter (type B)	mm
θ	THETA	Shape factor	
p	P	Porosity	
m	XM	Packing factor	
P_a	PWA	Percentage by weight of type A	
P_b	PWB	Percentage by weight of type B	
k	XK	Specific permeability	mm^2

```
C       HYDRA1
C       THIS PRØGRAM CALCULATES THE SPECIFIC PERMEABILITY IN
C       SQUARE MM. ØF A MIXTURE ØF TWØ MATERIALS. THE MATERIALS
C       ARE DENØTED "A" AND "B" AND THEIR PRØPØRTIØNS BY WEIGHT IN
C       THE MIXTURE ARE DENØTED BY PWA AND PWB. THE MEAN DIAMETER
C       ØF THE PARTICLES ØF A IS .08 MM. AND THE MEAN DIAMETER ØF
C       THE PARTICLES ØF B IS .40 MM.THE SHAPE FACTØR ØF THE
C       PARTICLES ØF EACH MATERIAL IS ASSUMED TØ BE 6.5,AND THE
C       PØRØSITY ØF THE MIXTURE IS ASSUMED TØ BE .34.
        THETA=6.5
        P=.34
        XM=5.
        PWA=-20.
        PWB=120.
        F=XM*(1.-P)**2*(THETA/100.)**2/P**3
        DA=.08
        DB=.40
        PRINT 5
      5 FØRMAT(4X,"PWA",4X,"PWB",4X,"SPEC.PERMEABILITY")
        DØ 1 I=1,6
        PWA=PWA+20.
        PWB=PWB-20.
        XK=1./(F*(PWA/DA+PWB/DB)**2)
      1 PRINT 9,PWA,PWB,XK
      9 FØRMAT(I6,2X,I6,2X,E15.4/)
        STØP
        END
```

Fig. 5.12.1 Program HYDRA1.

```
    PWA      PWB      SPEC.PERMEABILITY
     0       100         0.6834E-04

    20        80         0.2109E-04

    40        60         0.1010E-04

    60        40         0.5912E-05

    80        20         0.3874E-05

   100         0         0.2734E-05

STØP
```

Fig. 5.12.2 Trial run for HYDRA1.

On lines 00010, 00020, and 00030, values are assigned to the shape factor, the porosity, and the packing factor, respectively; and on lines 00070 and 00080 values are assigned to the diameter of the particles of sand A and of sand B, respectively. On lines 00040 and 00050 the percentages by weight of sands A and B, PWA and PWB, respectively, are initialized. The calculation of the specific permeability XK is carried out with the DO loop which begins on line 00110. At each iteration, PWA is increased by 20 per cent and PWB is decreased by 20 per cent, so that PWA is increased in steps from 0 to 100 per cent and PWB is decreased in steps from 100 to 0 per cent. For example, when I = 1, PWA is 0 per cent and PWB is 100 per cent. The value of XK corresponding to these proportions of sand is calculated and printed, and then the program branches back to the beginning of the DO loop. The results of the run are shown in Fig. 5.12.2.

5.13 OXYGEN DEFICIT IN A POLLUTED STREAM[1]

The calculation of the variation with time of the dissolved oxygen in a polluted stream is important in water-resources engineering. In this section, we will briefly discuss this topic and summarize a method for calculating D_t, the oxygen deficit in a polluted stream.

Organic matter in sewage decomposes through chemical and bacterial action. In this process, free oxygen is consumed, and the sewage is deoxygenated. A standard procedure for determining the rate of deoxygenation of sewage involves diluting a sewage sample with water containing a known amount of dissolved oxygen and determining the loss in oxygen after the mixture has been maintained at a temperature of 20°C for a period of 5 days. This loss is called the biochemical oxygen demand (hereafter denoted BOD) for 5 days. The BOD for a period of 20 days at a temperature of 20°C is called the first-stage demand and denoted Λ_{20} (the subscripts indicate the temperature: 20°C). For any temperature, T, the first-stage demand Λ can be calculated by the formula

$$\Lambda = \Lambda_{20}(0.02T + 0.6) \tag{5.13.1}$$

As previously mentioned, when sewage is discharged into a stream, oxygen is consumed in the decomposition of organic matter. At the same time, oxygen is

[1] R. K. Linsley and J. B. Franzini, "Water Resources Engineering," McGraw-Hill, New York, 1964, p. 548.

absorbed from the air. However, deoxygenation and reoxygenation take place, in general, at different rates. Usually, reoxygenation lags behind deoxygenation; and the dissolved oxygen decreases with time, reaches a minimum, and then increases. As the dissolved oxygen decreases, an oxygen deficit is said to occur. When the dissolved oxygen is at a minimum, the oxygen deficit is at a maximum.

The oxygen deficit of the polluted stream can be calculated from

$$D_t = \frac{K_d \Lambda_m}{K_r - K_d} (10^{-K_d t} - 10^{-K_r t}) + D_0 \times 10^{-K_r t} \tag{5.13.2}$$

where D_t = oxygen deficit of the stream at time t, mg/l
K_d = coefficient of deoxygenation
K_r = coefficient of reoxygenation
D_0 = initial oxygen deficit, mg/l
Λ = first stage BOD of polluted stream, mg/l
t = elapsed time (when $t = 0$, $D_t = D_0$), days

We will calculate D_t by use of Eq. (5.13.2) and will assume the following data:

1. The temperature of the mixture of sewage and stream = 17.6°C. This temperature will be denoted T_m.
2. $K_r = 0.2$ (at temperature T_m).
3. $K_{d_{20}} = 0.1$ (K_d at 20°C).
4. The biochemical oxygen demand of the stream $(\text{BOD})_R$, above the point at which the sewage discharges into it, is zero.
5. The biochemical oxygen demand of the sewage $(\text{BOD})_S$ is 145 mg/l.
6. $D_0 = 1.3$ mg/l.
7. The rate of flow of the stream Q_R = 23.9 million gal/day.
8. The rate of flow of the sewage Q_S = 3.5 million gal/day.

Before we can calculate D_t, we have to perform the following preliminary calculations:

1. Calculate K_d at temperature T_m by use of the following equations:
$$K_d = (K_{d_{20}})(1.047)(T_m - 20) \tag{5.13.3}$$
2. Calculate the BOD of the mixture of sewage and stream, $(\text{BOD})_m$, by use of the following equation:

$$(\text{BOD})_m = \frac{(\text{BOD})_S Q_S + (\text{BOD})_R Q_R}{Q_S + Q_R} \tag{5.13.4}$$

3. Calculate Λ_{20}, the first-stage BOD of the mixture at 20°C:

$$\Lambda_{20} = \frac{(\text{BOD})_m}{0.68} \tag{5.13.5}$$

If Λ_{20}, as given by Eq. (5.13.5), is substituted into Eq. (5.13.1), we obtain the first-stage BOD at temperature T_m:

$$\Lambda_m = \Lambda_{20}(0.02 T_m + 0.6)$$

$$\Lambda_m = \frac{(\text{BOD})_m}{0.68}(0.02 T_m + 0.6) \tag{5.13.6}$$

```
C     SANIT1
C     THIS PRØGRAM CALCULATES THE ØXYGEN DEFICIT IN A PØLLUTED
C     STREAM AS A FUNCTIØN ØF TIME.
      READ,CR,QSTR,QSEW,CD20,TMIX,BØDSTR,BØDSEW,DO
      CD=CD20*1.047**(TMIX-20.)
      BØDMIX=(BØDSTR*QSTR+BØDSEW*QSEW)/(QSTR+QSEW)
      FBØD=BØDMIX*(.02*TMIX+.6)/.68
      PRINT,"          T             DT"
      DØ 1 I=1,35
      Z=I
      T=Z/10.
      A=10.**(-CR*T)
      DT=((CD*FBØD)/(CR-CD))*(10.**(-CD*T)-A)+DO*A
    1 PRINT,T,DT
      STØP
      END
```

Fig. 5.13.1 Program SANIT1.

With the data assumed above, and the use of Eqs. (5.13.2) to (5.13.4), and (5.13.6), we can predict the effect of the sewage on the oxygen content of the stream by evaluating D_t for several increments of time after the addition of the sewage.

```
INPUT:00010
?.2,23.9,3.5,.1,17.6,0.,145.,1.3

          T             DT
       0.1000         1.7587
       0.2000         2.1863
       0.3000         2.5842
       0.4000         2.9541
       0.5000         3.2974
       0.6000         3.6156
       0.7000         3.9099
       0.8000         4.1816
       0.9000         4.4320
       1.0000         4.6621
       1.1000         4.8732
       1.2000         5.0661
       1.3000         5.2419
       1.4000         5.4016
       1.5000         5.5460
       1.6000         5.6760
       1.7000         5.7924
       1.8000         5.8960
       1.9000         5.9875
       2.0000         6.0676
       2.1000         6.1369
       2.2000         6.1961
       2.3000         6.2458
       2.4000         6.2866
       2.5000         6.3190
       2.6000         6.3435
       2.7000         6.3605
       2.8000         6.3707
       2.9000         6.3743
       3.0000         6.3718
       3.1000         6.3636
       3.2000         6.3502
       3.3000         6.3317
       3.4000         6.3086
       3.5000         6.2812
STØP
```

Fig. 5.13.2 Trial run for SANIT1.

A program for carrying out these calculations is shown in Fig. 5.13.1. The quantities used in the program are defined as follows.

Formula symbol	*Computer name*
T_m	TMIX
K_d	CD
$K_{d_{20}}$	CD20
D_0	DO
D_t	DT
Q_S	QSEW
Q_R	QSTR
$(\mathrm{BOD})_S$	BODSEW
$(\mathrm{BOD})_R$	BODSTR
$(\mathrm{BOD})_m$	BODMIX
Λ_{20}	FBOD

First, the data listed above (items 1 through 7) are entered. Then, the coefficient of deoxygenation at temperature T_m is calculated in accordance with Eq. (5.13.3) (line 00020). Next, $(\mathrm{BOD})_m$ and Λ_{20} are calculated in accordance with Eqs. (5.13.4) and (5.13.6), respectively (lines 00030 and 00040). The oxygen deficit D_t is calculated as a function of time for a period of 3.5 days by means of the DO loop which begins on line 00060. The trial run shown in Fig. 5.13.2 indicates that the maximum oxygen deficit is *approximately* 6.37 mg/l, and occurs at approximately 2.90 days.

5.14 NESTED DO LOOPS

Figure 5.14.1 shows a tutorial program which has one nested DO loop. The index of the main loop is I. The initial value of I is 1, and I attains a maximum value of 2 with increments of 1. The index of the nested loop is J. The initial value of J is 1, and J attains a maximum value of 4 with increments of 1. As noted in Sec. 5.2, it is necessary to use a different name for the index in the nested loop.

```
C     NEST
C     THIS PRØGRAM ILLUSTRATES A NESTED DØ LØØP.
      IØTA=0.
      DØ 4 I=1,2
      DØ 4 J=1,4
      K=I*J
      IØTA=IØTA+K
    4 PRINT 9,I,J,K,IØTA
    9 FØRMAT(2HI=I3,2X,2HJ=I3,2X,2HK=I3,2X,5HIØTA=I4)
      STØP
      END
```

Fig. 5.14.1 Program NEST.

```
I=  1  J=  1  K=  1  IØTA=   1
I=  1  J=  2  K=  2  IØTA=   3
I=  1  J=  3  K=  3  IØTA=   6
I=  1  J=  4  K=  4  IØTA=  10
I=  2  J=  1  K=  2  IØTA=  12
I=  2  J=  2  K=  4  IØTA=  16
I=  2  J=  3  K=  6  IØTA=  22
I=  2  J=  4  K=  8  IØTA=  30
STØP
```

Fig. 5.14.2 Trial run for NEST.

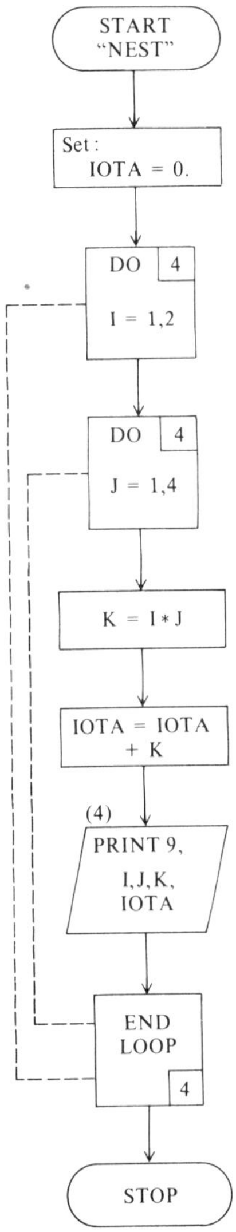

Fig. 5.14.3 Flow chart of program NEST.

```
C     STABLE
C     THIS PRØGRAM CALCULATES THE STABILIZATIØN FACTØR S ØF A
C     CØMMØN EMITTER TRANSISTØR STAGE WITH SELF BIAS. S IS
C     CALCULATED AS A FUNCTIØN ØF BETA AND THE RATIØ ØF
C     RB TØ RE.
      DØ 2 I=1,5
      BETA=10*I
      PRINT 20,BETA
   20 FØRMAT(/5HBETA=F5.1//6H RB/RE8X1HS)
      RATIØ=0.01
      DØ 2 J=1,5
      RATIØ=10.*RATIØ
      S=(1.+BETA)*(1.+RATIØ)/(1.+BETA+RATIØ)
    2 PRINT 90,RATIØ,S
   90 FØRMAT(F7.1,4XF5.1)
      STØP
      END
```

Fig. 5.14.4 Program STABLE.

```
BETA= 10.0

 RB/RE          S
    0.1       1.1
    1.0       1.8
   10.0       5.8
  100.0      10.0
 1000.0      10.9

BETA= 20.0

 RB/RE          S
    0.1       1.1
    1.0       1.9
   10.0       7.5
  100.0      17.5
 1000.0      20.6

BETA= 30.0

 RB/RE          S
    0.1       1.1
    1.0       1.9
   10.0       8.3
  100.0      23.9
 1000.0      30.1

BETA= 40.0

 RB/RE          S
    0.1       1.1
    1.0       2.0
   10.0       8.8
  100.0      29.4
 1000.0      39.4

BETA= 50.0

 RB/RE          S
    0.1       1.1
    1.0       2.0
   10.0       9.2
  100.0      34.1
 1000.0      48.6
STØP
```

Fig. 5.14.5 Trial run for STABLE.

At the outset, I assumes a value of 1, and remains at that value while J successively assumes values of 1, 2, 3, and 4. During each passage through the nested loop, the variables K and IOTA are recalculated. In this program, the indices I and J perform two functions: They control the number of iterations *and* their numeric values are used in the calculations. The output of this program is in Fig. 5.14.2. The nested DO loop was executed eight times. In general, *when exit is normal*, the nested loop is executed $(I_{max})(J_{max})$ times, where (I_{max}) is the maximum value of I, and (JMAX) is the maximum value of J. Notice that the outer loop and the nested loop terminate on the same statement. A flow chart of this program is shown in Fig. 5.14.3. An example of a DO loop with two loops nested within it is to be found in Sec. 7.7.

Figure 5.14.4 shows a program for calculating the stabilization factor S of a transistor amplifier.[1] S is given by

$$S = \frac{(1+\beta)(1+R_B/R_E)}{1+\beta+R_B/R_E} \tag{5.14.1}$$

[1] J. Millman and C. C. Halkias, "Electronic Devices and Circuits," McGraw-Hill, New York, 1968, p. 271.

where β = BETA = large-signal current gain of a common emitter transistor stage
R_B, R_E = RB,RE = resistors which form the self-bias network of the stage, ohms

BETA is set at a particular value, for example, 10, and then S is calculated as a function of the ratio of the resistance RB to the resistance RE. The quantity RB/RE is denoted by the name RATIO. For each value of BETA, S is calculated with RATIO assuming successive values of 0.1, 1.0, 10.0, 100.0, and 1000.0.

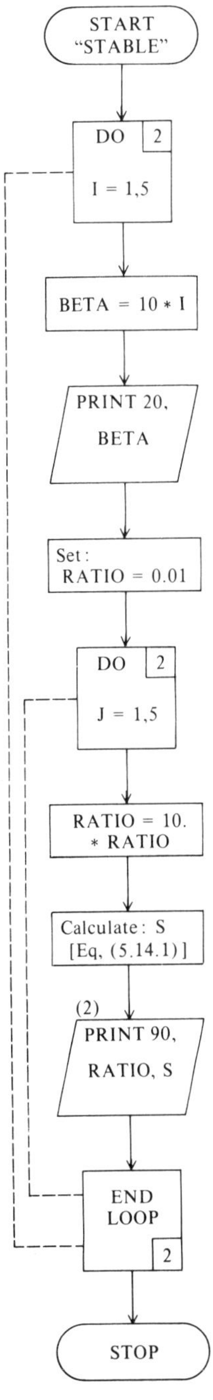

Fig. 5.14.6 Flow chart of program STABLE.

The sequence described above is accomplished by two DO loops, one nested within the other. The *main DO loop* (index I) extends from line 00010 through line 00090, inclusive, and the *nested DO loop* (index J) extends from line 00060 through 00090, inclusive. Let us consider a cycle of operation. Index I is set at its initial value of 1. The assignment statement on line 00020 converts the integer expression 10*I to real mode and stores it in BETA as 10. RATIO is initialized at a value of 0.01. The nested DO loop is then executed five times. During these five iterations, BETA remains at a value of 10., and RATIO successively assumes values of 0.1, 1.0, 10.0, 100.0, and 1000.0. Note that in the *main* loop the index is used as part of the calculation; in the *nested* loop the index is used as a counter.

In the second cycle, I is set at 2, and BETA is assigned a value of 20. RATIO is again initialized at a value of 0.01. The nested DO loop is again executed five times, with BETA remaining at 20., while RATIO successively assumes values of 0.1, 1.0, 10.0, 100.0, and 1000.0. There are five cycles in all, and in the completion of these cycles the nested loop is executed 25 times.

The program results are shown in Fig. 5.14.5, and a flow chart of the procedure is given in Fig. 5.14.6.

PROBLEMS

5-1. Write a program to produce a temperature conversion table for converting integral values of degrees Fahrenheit to the equivalent value in degrees Celsius, degrees Kelvin, and degrees Rankine. The conversion equations are

$$C = \frac{5(F - 32)}{9}$$
$$R = F + 459.67$$
$$K = C + 273.15$$

where F = temperature in degrees Fahrenheit
C = temperature in degrees Celsius
R = temperature in degrees Rankine
K = temperature in degrees Kelvin

Have the program print headings for each column in the table.

Sample data

Set

1 from 28 to 54°F in steps of 1°F
2 from 450 to 950°F in steps of 50°F
3 from −50 to 250°F in steps of 10°F

5-2. In FORTRAN, *integer division* returns the greatest integer value as the answer, that is, the decimal portion of the quotient is truncated. This fact makes it possible to program expressions whose predetermined value will recur every *i*th cycle. For example, if K is a cycle counter and I represents the period of the desired action, the following FORTRAN expression will only be equal to zero when K is an integer multiple of I:

K/I*I − K

Using the principle described above, modify the program of Prob. 5-1 so that a blank line will be printed after every three lines of calculated results.

5-3. The sine of an angle expressed in radians can be evaluated using N terms of the following series:

$$\sin x = \sum_{n=1}^{n=N} (-1)^{n+1} \frac{x^{(2n-1)}}{(2n-1)!}$$

Write a FORTRAN program employing a DO loop to execute the summation of the N terms.

The program should accept as input a value of X in degrees and a value of N indicating the number of terms to be used.

Sample data

Set	*X, degrees*	*N*
1	30	3
2	30	20
3	30	5

5-4. Write a FORTRAN program that will compute and print the sum of m consecutive integers, starting with any positive integer n.

Sample data

Set	m	n
1	50	42
2	23	107
3	110	36

5-5. The work done on a piston during a reversible isothermal expansion process with an ideal gas is given by

$$W = P_1 V_1 \ln \frac{P_1}{P_2}$$

where W = work, ft-lb
P = absolute pressure, lb/ft^2 absolute
V = volume, ft^3

and the subscripts 1 and 2 refer to the initial and final conditions, respectively.

Write a program that will accept the initial gas conditions and will print a table of values of the final pressure and the expansion work for a specified range of final pressures.

Sample data

$P_1 = 100$ lb/in^2 absolute $\quad V_1 = 30$ ft^3
P_2 = is to range from 70 to 30 lb/in^2 absolute in steps of 5 lb/in^2 absolute.

Note: The program must be designed to make the required adjustment in the dimensions of the factors.

5-6. The Poisson probability function is

$$P(K) = \frac{\lambda^K e^{-\lambda}}{K!}$$

Write a FORTRAN program that will accept values for K and will print the probability of K events occurring for a given lambda.

Sample data

Set	K	λ
1	5	1
2	5	2
3	8	2

5-7. Write a FORTRAN program to calculate and print a tabulation of the values of tan α for positive integral values of α, expressed in degrees, starting at α_1 and ending at α_2 in steps of α_3.

Sample data

Set	α_1	α_2	α_3
1	40°	50°	1°
2	65°	85°	5°
3	0°	90°	10°

5-8. The equation for the velocity of an electron which starts with zero velocity and is accelerated by a potential difference V is

$$\mu' = c\sqrt{1 - \frac{1}{(1 + \mu^2/2c^2)^2}}$$

where μ' = velocity *with* relativistic correction, m/s
c = speed of light = 3×10^8, m/s
μ = velocity *without* relativistic correction, m/s

The equation for μ is

$$\mu = \sqrt{2eV/m_0}$$

where e = electron charge $\approx 1.6 \times 10^{-19}$, C
m_0 = rest mass of electron $\approx 9.11 \times 10^{-31}$, kg
V = potential difference, V

Write a program which calculates μ and μ' as V varies from 2000 to 10,000 V in steps of 500 V.

Suggestion: Use E format for printing the results. Why?

5-9. The equation for the volt-ampere characteristic of an ideal *p-n* junction diode is:

$$I = I_s[e^{eV/kt} - 1]$$

where I = net junction current, A
I_s = saturation current, A
e = electron charge $\approx 1.6 \times 10^{-19}$, C
k = Boltzmann constant = 1.381×10^{-23}, J/°K
T = junction temperature, °K
V = voltage applied across junction, V

When V is positive, the diode is said to be *forward biased*, and I is positive. When V is negative the diode is said to be *reverse biased*, and I is negative. Write a program which calculates I as a function of V as V varies from -0.2 to 0.5 V, with $I_s = 1 \times 10^{-6}$ A and T = 300°K.

5-10. The binomial probability function is

$$P(K) = \frac{N!}{K!(N-K)!} p^K q^{N-K}$$

where

$$p + q = 1 \qquad \text{for } 0 < p \leqslant 1$$

Write a program that will accept values of p and N and will print a table of probabilities as K varies from 0 to N.

Sample data

Set	*p*	*N*
1	0.25	10
2	0.50	10
3	0.75	10

5-11. The Freundlich adsorption equation can be expressed as

$$X = \exp\,[\ln X_1 + C\,(\ln Y - \ln Y_1)]$$

where $X = \dfrac{R}{A}$

$$Y = \frac{100 - R}{100}$$

A = adsorbent concentration, ppm
R = contaminant removed by A, %

and the subscript 1 refers to conditions at a known point.

The equation constant C can be evaluated with the following formula if conditions at two points are known:

$$C = \frac{\ln\,(R_1A_2/R_2A_1)}{\ln\,[(100 - R_1)/(100 - R_2)]}$$

Write a computer program that will read-in conditions at two known points, and, by varying R in a loop, will calculate values for A and print the results in a table.

Sample data
Two tests were made on a sample of water containing suspended material. The water was treated with an adsorbing agent and the following observations made:

Adsorbent concentration A, ppm	*Solids removed from suspension R*, %
5	70
10	93

Use the computer to determine the adsorbent concentration that will be required to remove 95.0, 95.5, 96.0, . . . , 99.5 per cent of the solids in suspension.

5-12. In Prob. 4-8 the sonic velocity in one of several selected gases was calculated for a given temperature. Revise the program so that the input data, in addition to the code number of the gas, specifies a range of temperatures (and the step size between the extremes of the range) at which the sonic velocity is to be computed.

The output of the program should be a table showing the name of the gas, the temperatures in the range, and the velocities at each of the temperatures.

Sample data
Set 1: Gas = carbon dioxide.

Temperature range is from 40 to 135°F in steps of 5°F.

Set 2: Gas = helium.

Temperature range is from 72 to 102°F in steps of 3°F.

5-13. The current in an ideal (zero resistance) circuit is given by the following equation:

$$I = \frac{V}{|\omega L - 1/\omega C|}$$

where I = current, A
V = voltage, V
ω = radian frequency, rad/s
L = inductance, H
C = capacitance, F

For given values of V, L, and C, there is a value of ω at which the current becomes infinite.

Write a FORTRAN program that will increment ω over a range of values and print the values of ω and I at each step. When

$$\left|\omega L - \frac{1}{\omega C}\right| \leqslant \epsilon$$

(where ϵ = an arbitrary small number) have the computer print the value of ω and the message, CURRENT IS INFINITE. The program should then resume looping operations until the test value is reached.

Input to the program should be: V, L, C, ϵ.

Sample data

V = 10.0 V
L = 0.001 H
$C = 1. \times 10^{-9}$ F
Lower limit of $\omega = 10^4$ rad/s
Upper limit of $\omega = 10^8$ rad/s

5-14. In Prob. 3-14 a formula was given for calculating the ratio of magnetic intensity at the interior of a shell to the magnetic intensity outside the shell. This ratio was denoted by H_3/H_0. Write a computer program to calculate H_3/H_0 as the ratio of the shell radii, a/b, assumes values of 0.99, 0.98, 0.97. For each of the radii ratios, calculate the magnetic intensity ratios for shell permeabilities of $500\mu_0$, $1000\mu_0$, and $1500\mu_0$.

5-15. Revise the program written as a solution to Prob. 4-14 so it will print a table showing the composition of the products of combustion for various percentages of theoretical air.

The program should read-in values for the lower and upper limits of the range of theoretical air percentages, as well as the steps in the range at which calculations are to be made.

Use a FORMAT statement for printing the results and arrange for the computer to print an appropriate table heading.

Sample data
Lower limit = 90%
Upper limit = 120%
Steps = 5%

6
Iterative Solutions

In the previous chapter the DO loop was used to carry out a prescribed number of iterations. There are circumstances under which it is inconvenient, or not possible, to specify the desired number in advance. Instead, iteration is halted when a condition specified in a test statement has been satisfied. The test can be performed by either the *arithmetic* IF statement or the *logical* IF statement discussed in Secs. 4.3 and 9.10, respectively. As described earlier, the iterative process itself may be directed by either a DO statement or by a program loop designed by the programmer. Both approaches are demonstrated in this chapter.

6.1 VAN DER WAAL'S EQUATION

As a first illustration of an iterative solution we will discuss a program which solves Van der Waal's equation for the volume of a gas when the temperature and pressure are known.[1] For 1 mol of gas, Van der Waal's equation is

$$P + \frac{A}{V^2}(V - B) = RT \tag{6.1.1}$$

where P = pressure, atm
T = absolute temperature, °K
R = ideal gas constant = 0.08205 atm/°K mol
V = volume, l
A,B = constants for a particular gas

[1] L. G. Bassett, S. C. Bunce, et al., "Principles of Chemistry," Prentice-Hall, Englewood Cliffs, N.J., 1966, p. 188.

```
C      VAN1
C      THIS PRØGRAM CALCULATES GAS VØLUME BY USE ØF THE VAN DER
C      WAAL EQUATIØN. CØNSTANTS ARE FØR CØ2. "C" IS IN
C      DEGREES CELSIUS AND "P" IS IN ATMØSPHERES.
       READ,C,P
       T=273.+C
       A=3.59
       B=0.0427
       R=0.08205
       F3=R*T
       DØ 5 I=1,10000
       X=I
       V=X/1000.
       F1=P+A/V**2
       F2=V-B
       IF(F1*F2-F3) 5,6,6
     5 CØNTINUE
       PRINT,"V IS GREATER THAN 10 LITERS"
       STØP
     6 PRINT,
       PRINT,"V=",V," LITERS"
       PRINT,F1*F2
       PRINT,F3
       STØP
       END
```

Fig. 6.1.1 Program VAN1.

```
INPUT:00010
 ?32,0.75

V IS GREATER THAN 10 LITERS
STØP


INPUT:00010
 ?32,7.5

V=          3.2340   LITERS
       25.0302
       25.0252
STØP


INPUT:00010
 ?32,75

V=          0.1120   LITERS
       25.0306
       25.0252
STØP
```

Fig. 6.1.2 Trial runs for VAN1.

In the program under discussion, we wish to find the volume of a mole of CO_2 at a specified temperature in degrees Celsius and at a specified pressure in atmospheres. For CO_2, the values of the parameters A and B are

$$A = 3.59 \text{ atm l}^2 \text{ mol}^{-2}$$
$$B = 0.0427 \text{ l mol}^{-1}$$

The computer names used in the program are the same as the symbols used in Eq. (6.1.1). In addition, the following computer names are used:

Formula symbol	*Computer name*
$P+\dfrac{A}{V^2}$	F1
$V-B$	F2
RT	F3

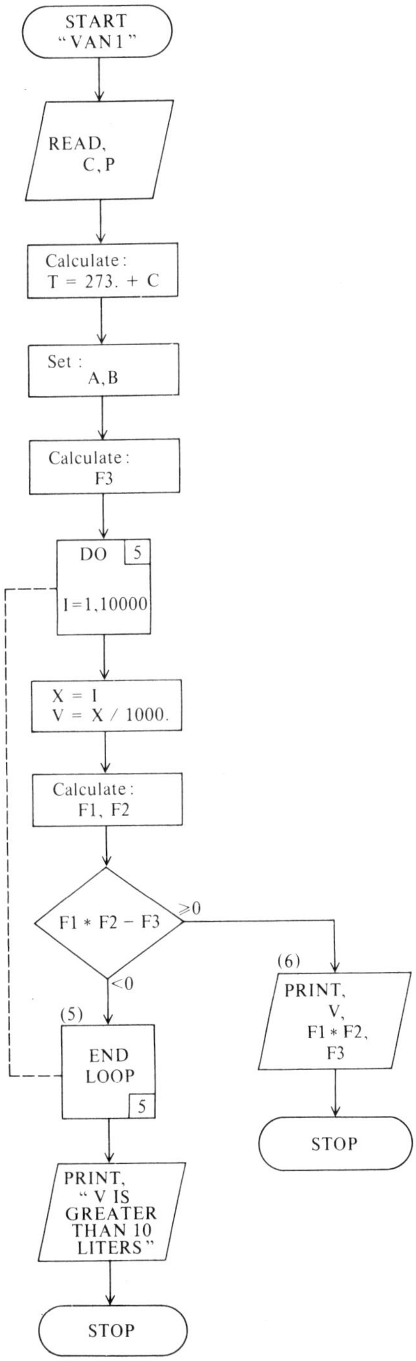

Fig. 6.1.3 Flow chart of program VAN1.

The program and several trial runs are shown in Figs. 6.1.1 and 6.1.2, respectively. A flow chart for the program is shown in Fig. 6.1.3. The value of the specified temperature in degrees Celsius and the pressure in atmospheres are entered into the program (line 00010). The temperature is converted to degrees Kelvin, and F3 is calculated (lines 00020 and 00060). Now let us consider the DO loop which extends from line 00070 to line 00130. In the first passage through the DO loop, I = 1, X = 1. and V = 0.001. F1 and F2 are evaluated with V = 0.001. Then the IF statement compares the value of the product of F1 and F2 with F3. As the product of F1 and F2 is negative, the product is less than F3 and the program branches back to the beginning of the DO loop. At each passage through the DO loop, V assumes a larger value (in steps of 0.001) until the product of F1 and F2 equals or exceeds the value of F3. At this point, the program branches to statement 6 (line 00160) and the unformatted PRINT statement causes a blank line to be "printed." Then the current value of V is printed. As a matter of interest, the value of the product of F1 and F2 and also the value of F3 were printed.

The reader should understand that this method of solution was employed only to illustrate the technique of computer iteration. The program could be designed to produce convergence in a shorter time by starting the loop with I = 1000.*B, but this would not affect the principle being demonstrated here.

6.2 RE-FORMING OF CYCLOHEXANE[1]

In calculating mass balances for chemical reactions, the engineer often encounters equations which are difficult or impossible to solve analytically. However, it is relatively simple to obtain *approximate* solutions of such equations by use of the digital computer. As an illustration, let us consider the re-forming of cyclohexane. (The conversion of hydrocarbons to other hydrocarbons of higher octane number is called *re-forming*.) An analysis on a pound-mole basis of the reaction for the re-forming of cyclohexane results in the following equation:

$$\frac{(4 + 3n_B)^3 n_B}{[(1 - n_B)/13](5 + 3n_B)^3} = 14.7 \qquad (6.2.1)$$

where n_B = benzene produced, lb-mol.

A flow chart for a procedure which finds an approximate solution to Eq. (6.2.1) is shown in Fig. 6.2.1. In the program listing, REACT1 in Fig. 6.2.2, the name B is used to denote the number of pound-moles of benzene [symbolized by n_B in Eq. (6.2.1)], and the name TEST is used to denote the left-hand side of Eq. (6.2.1) (see line 00040). The desired solution is obtained by means of a DO loop and an IF statement. For each iteration of the DO loop, the value of B is increased by 0.01 (see line 00030). Iteration continues until TEST is equal to or greater than 14.7 [the value on the right-hand side of Eq. (6.2.1)]. When this condition is attained, the program branches out of the DO loop to statement 1 and the current value of B is printed. This is the *approximate* solution. The trial run shown in Fig. 6.2.3 indicates that the approximate value of B is 0.65 lb-mol.

Let us now briefly examine the statements on lines 00070 and 00080. Occasionally, a procedure of the type used in this program fails either because the

[1] L. B. Andersen and L. A. Wenzel, "Introduction to Chemical Engineering," McGraw-Hill, New York, 1961, p. 203.

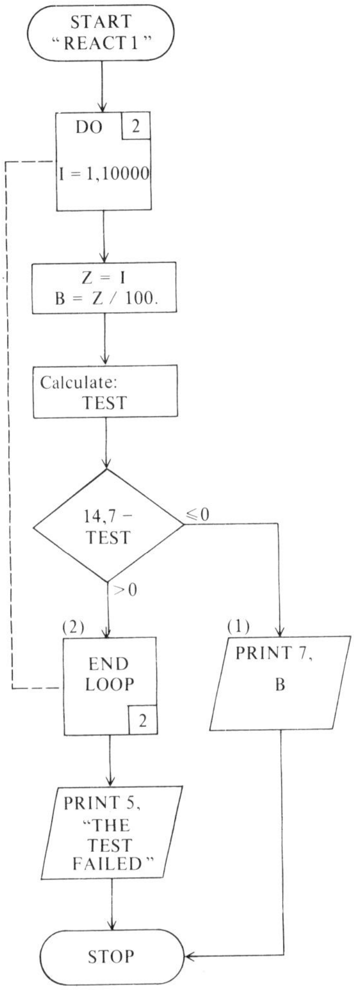

Fig. 6.2.1 Flow chart of program REACT1.

method does not converge to a solution or, more frequently, because the programmer has made some error in the logic of the program. It is advisable, therefore, for the programmer to build into the program a "safety valve" that will prevent endless

```
C      REACT1
C      THIS PRØGRAM CALCULATES THE LB MØLES ØF BENZENE PRØDUCED
C      IN THE REFØRMING ØF CYCLØHEXANE.
       DØ 2 I=1,10000
       Z=I
       B=Z/100.
       TEST=((4.+3.*B)**3*B)/((1.-B)/13.)/(5.+3.*B)**3
       IF(14.7-TEST) 1,1,2
     2 CØNTINUE
       PRINT 5
     5 FØRMAT("THE TEST FAILED")
       GØ TØ 6
     1 PRINT 7,B
     7 FØRMAT("THE NØ. ØF LB MØLE ØF BENZENE PRØDUCED="F8.4//)
     6 STØP
       END
```

Fig. 6.2.2 Program REACT1.

```
THE NØ. ØF LB MØLE ØF BENZENE PRØDUCED=  0.6500

STØP
```

Fig. 6.2.3 Trial run for REACT1.

cycling. In this program the DO loop itself provides just such a contingency escape route, and no more than 10,000 cycles are possible. As an example of a mistake in programming logic, consider what would happen if the positions of TEST and 14.7 were accidentally reversed in the IF statement on line 00050. In this event, 10,000 iterations of the DO loop would occur and the program would exit normally from the DO to the PRINT statement on line 00070. The warning message THE TEST FAILED would be printed. This would alert the programmer to the existence of an error in his program and would probably lead him to examine the IF statement on line 00050 and the arithmetic replacement statement on line 00040.

6.3 MAXIMUM OXYGEN DEFICIT IN A STREAM

In Sec. 5.13, we discussed a program which calculates the oxygen deficit in a polluted stream, D_t, as a function of time. In that program, time was increased in steps of 0.01 day. The trial run (Fig. 5.13.2) showed that the maximum oxygen deficit was approximately 6.3743 and occurred at about 2.9 days after the discharge of sewage began.

Let us assume that we now wish to determine *only* the maximum oxygen deficit and the time at which it occurs. Furthermore, we wish to determine these quantities to a greater accuracy than previously. We can achieve these goals by a few modifications of the program shown in Fig. 5.13.1. The revised program is shown in Fig. 6.3.1. In order to achieve greater accuracy, time is increased in steps of 0.001 day (line 00080). The approximate value of the maximum oxygen deficit is found by means of the IF statement on line 00110. Iteration continues until the current value of the oxygen deficit DT is equal to or less than the previously calculated value DTOLD. It also should be noted that the initial value of the index of the DO loop has been set at 2700, and the maximum value of the index has been set at 3000, because it is known from the earlier work that the answers being sought will lie in this range. The results of the run are shown in Fig. 6.3.2.

```
00001C |    SANIT2
00010  |    READ,CR,QSTR,QSEW,CD20,TMIX,BØDSTR,BØDSEW,DO
00020  |    CD=CD20*1.047**(TMIX-20.)
00030  |    BØDMIX=(BØDSTR*QSTR+BØDSEW*QSEW)/(QSTR+QSEW)
00040  |    FBØD=BØDMIX*(.02*TMIX+.6)/.68
00050  |    DTØLD=DØ
00060  |    DØ 1 I=2700,3000
00070  |    Z=I
00080  |    T=Z/1000.
00090  |    A=10.**(-CR*T)
00100  |    DT=((CD*FBØD)/(CR-CD))*(10.**(-CD*T)-A)+DO*A
00110  |    IF(DT-DTØLD) 2,2,1
00120  |  1 DTØLD=DT
00130  |    PRINT,"THE TEST FAILED."
00140  |    GØ TØ 7
00150  |  2 PRINT 5,T,DT
00160  |  5 FØRMAT("THE MAXIMUM DEFICIT AT T=",F9.4," DAYS IS",F9.4//)
00170  |  7 STØP
00180  |    END
```

Fig. 6.3.1 Program SANIT2.

```
INPUT:00010
?.2,23.9,3.5,.1,17.6,0.,145.,1.3

THE MAXIMUM DEFICIT AT T=    2.9090 DAYS IS    6.3743


STØP
```

Fig. 6.3.2 Trial run for SANIT2.

6.4 TIME OF FLIGHT OF AN ELECTRON[1]

In this section, a program is presented for calculating the time of flight of an electron from the cathode to the anode of a vacuum diode. The electrodes are parallel planes, and the electron is assumed to begin its flight with zero velocity. A generator, $V_m \sin \omega t$, is connected to the electrodes. The parameters are so chosen that the time of flight is less than a half period of the generator voltage.

The instantaneous distance y of the electron from the cathode is

$$y = \frac{e}{m} \frac{V_m}{\omega^2 D} (\theta - \sin\theta) \tag{6.4.1}$$

where e/m = electron charge to mass ratio
D = distance from cathode to anode, in m
$\theta = \omega t$

It is sufficiently accurate to replace $\sin\theta$ by the first three terms of its Taylor series representation about the origin. We then obtain

$$y = \frac{e}{m} \frac{V_m}{\omega^2 D}\left[\theta - \left(\theta - \frac{\theta^3}{3!} + \frac{\theta^5}{5!}\right)\right] = \frac{e}{m} \frac{V_m}{\omega^2 D}\left(\frac{\theta^3}{6} - \frac{\theta^5}{120}\right)$$

Let $P = (e/m)(V_m/\omega^2 D)$, and write the equation as

$$\frac{y}{P} - \left(\frac{\theta^3}{6} - \frac{\theta^5}{120}\right) = 0 \tag{6.4.2}$$

If y is set equal to the distance D from cathode to anode, the quantity D/P is a constant. By substituting successively increasing values of θ, we can find a value which satisfies the equation. This trial-and-error procedure is diagrammed in Fig. 6.4.1. It is accomplished by the joint action of the DO loop and the IF statement in the program of Fig. 6.4.2. In the program, the angle θ is represented by the name TH, the frequency ω by W, and the quantity within parentheses in Eq. (6.4.2) by the name THET. The constant 1.759E11 which appears on line 00020 is the ratio of the charge to the mass of an electron in the mks system.

The index I of the DO loop is incremented in steps of 1 from an initial value of 1. During each iteration, the current value of I is assigned to the variable B (line 00050), and values of TH are generated by dividing B by 10^{+8} (line 00060). The constant has been introduced (line 00030) to reduce computation time. If D/P were used in the IF statement (line 00080), it would be recalculated during each pass through the loop. For the data used in the trial run (Fig. 6.4.3) it took 7856 iterations to satisfy the equation (see lines 00150 and 00160 and the trial run). Thus if D/P had

[1] J. Millman and C. C. Halkias, "Electronic Devices and Circuits," McGraw-Hill, New York, 1968, p. 8.

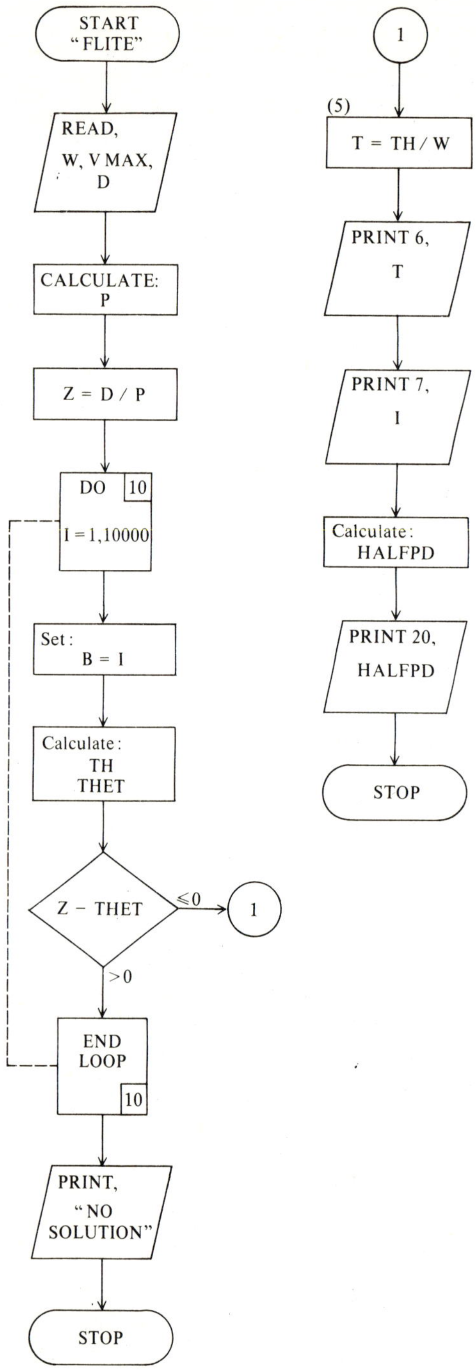

Fig. 6.4.1 Flow chart of FLITE.

been used in the IF statement, it would have been calculated 7856 times. The time of flight of the electron is equal to the value of θ at the last iteration divided by ω, the radian frequency. On line 00120 this calculation is expressed as T = TH/W.

```
C      FLITE
C      THIS PROGRAM CALCULATES THE TIME ØF FLIGHT ØF AN ELECTRØN
C      RELEASED WITH ZERØ INITIAL VELØCITY AT A PLATE ØF A
C      PARALLEL PLATE DIØDE.A SINUSØIDAL VØLTAGE GENERATØR IS
C      CØNNECTED TØ THE PLATES.
       READ,W,VMAX,D
       P=(1.759E11)*VMAX/((W**2)*D)
       Z=D/P
       DØ 10 I=1,10000
       B=I
       TH=B/1.E08
       THET=(TH**3)/6.-(TH**5)/120.
       IF(Z-THET) 5,5,10
    10 CØNTINUE
       PRINT,"NØ SØLUTIØN"
       STØP
     5 T=TH/W
       PRINT 6,T
     6 FØRMAT("THE TIME ØF FLIGHT IN SECØNDS="E15.8/)
       PRINT 7,I
     7 FØRMAT("THE NUMBER ØF ITERATIØNS="I6)
       HALFPD=3.1416/W
       PRINT 20,HALFPD
    20 FØRMAT("HALF PERIØD IN SECØNDS="E15.8//)
       STØP
       END
```

Fig. 6.4.2 Program FLITE.

```
INPUT:00010
?377.,1000.,.01

THE TIME ØF FLIGHT IN SECØNDS= 0.20838196E-06

THE NUMBER ØF ITERATIØNS=  7856
HALF PERIØD IN SECØNDS= 0.83331565E-02


STØP
```

Fig. 6.4.3 Trial run for FLITE.

6.5 BEAM DEFLECTION (MAXIMUM POINT)

The simple beam deflection problem discussed in Secs. 4.10 and 5.11 will be employed here as the basis for illustrating the use of several IF statements in an iterative solution. The answers being sought are the location and magnitude of the maximum deflection. The reader may recall that there are well-known formulas for determining these values. The method we are using here is actually an inefficient one for so elementary a problem. However, the technique that is used is a fundamental *maximum seeking procedure* that has application in optimization design.

The optimum seeking method under study is applicable to any unimodal function, i.e., any function having only one maximum (or minimum) in the region under investigation. The elastic curve of the simple beam under consideration falls into this category. The region being investigated is the entire length of the beam, and the function whose maximum is being sought is actually the *two* functions that describe the elastic curve, one in effect to the *left* of the applied load, and the second one in effect for the part of the beam to the *right* of the load.

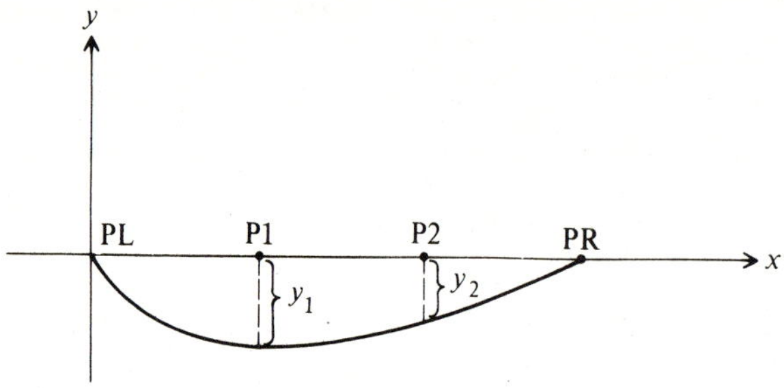

Fig. 6.5.1 Elastic curve of simple beam.

The search method itself is a simple process for reducing the size of the *region of uncertainty* in which the maximum is located. The procedure depends on successive *pairs* of simultaneous experiments being used to section the region and reduce its size.

Figure 6.5.1 shows the elastic curve of a simple beam on which two experiments (i.e., function evaluations) have been made at points P1 and P2. These points were chosen, according to the plan described below, within the region of uncertainty contained between the points PL and PR. Now, since the beam curve is the graph of a unimodal function, it is obvious that

1. If $|y_1| > |y_2|$
   then y_{max} must lie between PL and P2.
2. If $|y_1| < |y_2|$
   then y_{max} must lie between P1 and PR.
3. If $|y_1| = |y_2|$
   then y_{max} must lie between P1 and P2.

Therefore, for any of the three cases, a part of the beam can be eliminated from the search. With the knowledge obtained by the above analysis, the region of uncertainty can be reduced in size and two *new* experiments made which will lead to a further reduction. The process can be repeated until the region has been reduced to some acceptable size and no further significant improvement can be made by continuing.

In applying this procedure to our beam problem, the bounds of the region of uncertainty are defined as PL and PR. The points within the region at which the experiments are to be made are taken as the one-third points, so that for *any* region of uncertainty the X locations of these points are

$$P1 = PL + \tfrac{1}{3}(PR - PL)$$

and

$$P2 = PL + \tfrac{2}{3}(PR - PL)$$

The *original* region of uncertainty is defined as the entire beam, with PL = 0.0 and PR = XL (the beam length). After the first pair of experiments has been analyzed, the bounds of the region are redefined and two more experiments are made. Each time the bounds of the region are redefined, a test must be made to determine if the region has been reduced to the specified size. In preparing this program, the acceptable size

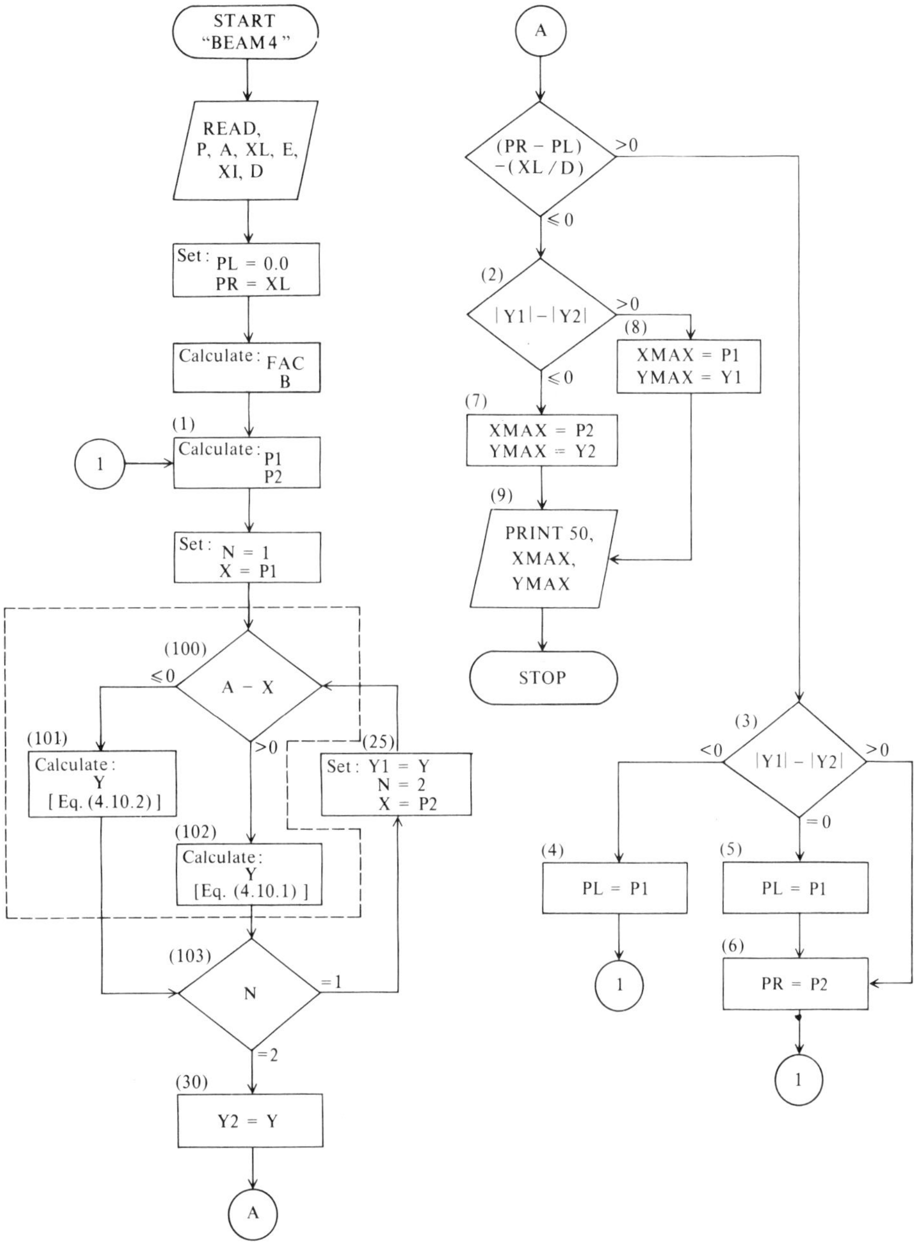

Fig. 6.5.2 Flow chart of program BEAM4.

for the region of uncertainty was set at some fractional part of the beam length rather than a predetermined distance. The fraction to be used is part of the input data and therefore the accuracy can be specified by the program user within the capabilities of the computer being employed.

Figure 6.5.2 shows a flow chart of the procedure to be programmed. The

```
C        BEAM4
C        PRØGRAM TØ CALCULATE THE MAXIMUM DEFLECTIØN PRØDUCED BY
C        A SINGLE CØNCENTRATED LØAD ACTING ØN A SIMPLE BEAM.
         READ,P,A,XL,E,XI,D
         PL=0.0
         PR=XL
         FAC=-P/(6.*E*XI*XL)
         B=XL-A
       1 P1=(2.*PL+PR)/3.
         P2=(PL+2.*PR)/3.
         N=1
         X=P1
     100 IF(A-X) 101,102,102
     101 Y=FAC*A*(XL-X)*((2.*XL-B)*B-(XL-X)**2)
         GØ TØ 103
     102 Y=FAC*B*X*(2.*XL*(XL-X)-B*B-(XL-X)**2)
     103 GØ TØ (25,30),N
      25 Y1=Y
         N=2
         X=P2
         GØ TØ 100
      30 Y2=Y
         IF((PR-PL)-(XL/D)) 2,2,3
       3 IF(ABS(Y1)-ABS(Y2)) 4,5,6
       4 PL=P1
         GØ TØ 1
       5 PL=P1
       6 PR=P2
         GØ TØ 1
       2 IF(ABS(Y1)-ABS(Y2)) 7,7,8
       7 YMAX=Y2
         XMAX=P2
         GØ TØ 9
       8 YMAX=Y1
         XMAX=P1
       9 PRINT 50,XMAX,YMAX
      50 FØRMAT(F10.4,E20.8//)
         STØP
         END
```

Fig. 6.5.3 Program BEAM4.

program symbols used have the same meanings as those given in Secs. 4.10 and 5.11, with the following additions:

Computer name	*Meaning*	*Units*
D	Divisor for beam length	
PL	Location of left end of region of uncertainty	in
PR	Location of right end of region of uncertainty	in
P1, P2	Location of points at which function is to be evaluated	in
N	Indicator for determining calculation path	
Y1, Y2	Deflection at P1, P2	in
XMAX	Location of maximum deflection	in
YMAX	Maximum deflection	in

```
INPUT:00010
?500,65,154,3E7,2.5,1E5

  73.3961        -.49118989E+00


STØP
```

Fig. 6.5.4 Trial run for BEAM4.

The section of the flow chart enclosed in the dotted line is the basic procedure for calculating the beam deflection and has been discussed in detail in Secs. 4.10 and 5.11. The reader should note that the technique of choosing alternative paths the first and second times through the deflection calculation, as discussed in Sec. 4.14, was used in evaluating Y1 and Y2.

The program corresponding to this flow chart is shown in Fig. 6.5.3 as BEAM4. After the values of Y1 and Y2 have been calculated, the program tests the size of the region of uncertainty by comparing it with the specified fraction of the beam length. If the region is too large, the bounds of the reduced region are redefined and the program returns to statement 1 and makes another pair of function evaluations. However, if the region is smaller than the preset fraction of the beam, the program chooses the greater value of Y1 and Y2, assigns this value and its corresponding location on the X axis to YMAX and XMAX, respectively, and prints the results. It should be observed that the assignment of these values to locations YMAX and XMAX is not really necessary in this program. That is, once the appropriate values have been selected, a properly constructed PRINT statement is all that is needed. For example, the following program statements starting at line 00280 would achieve the same printout without assigning the values:

```
 7 PRINT 50, P2, Y2
   STOP
 8 PRINT 50, P1, Y1
   STOP
50 FORMAT (F10.4, E20.8)
   END
```

However, if the program did not end at this point and the maximum values were to be used subsequently, it would be advisable to store the maximum values in designated storage locations. The data used for the sample run are the same as those used for the earlier beam deflection programs and the results are shown in Fig. 6.5.4.

6.6 INSERTION RATIO OF A CONSTANT-K FILTER[1]

The normalized insertion ratio of a two-section constant-K filter terminated at each end by fixed resistors of value $R = \sqrt{L/C}$ is

$$\frac{V_o}{V_s} = \frac{1}{(1 - 8x^2 + 8x^4) + j(4x - 10x^3 + 4x^5)} \tag{6.6.1}$$

In this equation x is normalized frequency, and

$$x = \omega/\omega_c$$

where ω_c is the cutoff frequency in radians per second. The signal attenuation produced by this filter is

$$\text{Signal attenuation} = -10 \log_{10} [1 + 4x^6(2x^2 - 1)^2] \quad \text{dB} \tag{6.6.2}$$

[1] O. S. Meixell, Constant-K Low and High Pass Filters, *RCA Rev.*, January 1941.

```
00001C      FILTR1
00002C      THIS PRØGRAM CALCULATES THE ATTENUATIØN AND PHASE SHIFT
00003C      ØF A TWØ SECTIØN CØNSTANT K FILTER. THE FILTER IS
00004C      TERMINATED BY CØNSTANT RESISTØRS ØF VALUE R.
00010       PRINT 99
00020    99 FØRMAT("NØRM.FREQ.          DECIBELS          PHASE,DEG.")
00030       DØ 14 I=1,200
00040       Z=I
00050       X=Z/10.-0.1
00060       IF(X-1.9) 97,97,98
00070    97 CØNTINUE
00080       DEC=-4.343*ALØG(1.+4.*X**6*(2.*(X**2)-1.)**2)
00090       B=4.*X-10.*X**3+4.*X**5
00100       A=1.-8.*X**2+8.*X**4
00110       IF(A) 2,3,4
00120     3 IF(B) 5,6,7
00130     5 BETA=270.
00140       GØ TØ 1
00150     6 PRINT 20
00160    20 FØRMAT("DEGENERATE CASE:A=B=0.")
00170       STØP
00180     7 BETA=90.
00190       GØ TØ 1
00200     2 IF(B) 8,9,10
00210     8 BETA=180.+57.296*ATAN(ABS(B/A))
00220       GØ TØ 1
00230     9 BETA=180.
00240       GØ TØ 1
00250    10 BETA=180.-57.296*ATAN(ABS(B/A))
00260       GØ TØ 1
00270     4 IF(B) 11,12,13
00280    11 BETA=360.-57.296*ATAN(ABS(B/A))
00290       GØ TØ 1
00300    12 BETA=0.
00310       GØ TØ 1
00320    13 BETA=57.296*ATAN(ABS(B/A))
00330     1 THETA=-BETA
00340    14 PRINT 33,X,DEC,THETA
00350    33 FØRMAT(3X,F6.2,9X,F8.3,8X,F10.3)
00360    98 STØP
00370       END
```

Fig. 6.6.1 Program FILTR1.

The phase shift is calculated by taking the negative of the arctangent of the ratio of the imaginary to the real part of the denominator in Eq. (6.6.1).

The calculation of the attenuation and phase shift as a function of the normalized frequency x is accomplished by the program of Fig. 6.6.1. In this program, the index of the DO loop is used to generate values of X. The IF statement on line 00060 limits the maximum value of X to an arbitrary value of 1.9.

The reader should note that in this program it would have been an easy matter to count the number of iterations needed to reach a value of 1.9. However, in some cases it would be inconvenient to calculate this number. Used in this manner, the IF statement is an easy way of terminating the calculations when the desired value of a particular variable, in this case X, has been reached. The segment of the program from lines 00110 to 00320, inclusive, represents the same procedure that was used in the program discussed in Sec. 4.12.

There is one difficulty with the program as presented in Fig. 6.6.1. If the trial run shown in Fig. 6.6.2 is examined, it will be noted that at X = 1.4 the phase shift is −358.833°. At X = 1.5 the phase shift is −6.374°. However, the phase shift of the filter actually increases monotonically with frequency. Therefore, a modification in the program is required to increase the magnitude of all values of phase shift by 360° for all values of X beyond 1.5. The revised program is shown in Fig. 6.6.3.

```
NØRM.FREQ.          DECIBELS          PHASE,DEG.
     0.00               -.000               0.000
     0.10               -.000             -22.957
     0.20               -.001             -46.154
     0.30               -.009             -69.851
     0.40               -.033             -94.296
     0.50               -.067            -119.745
     0.60               -.063            -146.835
     0.70               -.001            -177.577
     0.80               -.343            -215.851
     0.90              -2.594            -260.125
     1.00              -6.990            -296.565
     1.10             -11.844            -320.863
     1.20             -16.357            -337.396
     1.30             -20.429            -349.479
     1.40             -24.113            -358.833
     1.50             -27.476              -6.374
     1.60             -30.570             -12.630
     1.70             -33.438             -17.932
     1.80             -36.114             -22.499
     1.90             -38.623             -26.484
STØP
```

Fig. 6.6.2 Trial run for FILTR1.

```
C        FILTR2
C        THIS PRØGRAM CALCULATES THE ATTENUATIØN AND PHASE SHIFT
C        ØF A CØNSTANT K FILTER. THE FILTER IS TERMINATED BY
C        CØNSTANT RESISTØRS ØF VALUE R.
         PRINT 99
      99 FØRMAT("NØRM.FREQ.          DECIBELS          PHASE,DEG.")
         BETØLD=0.0
         DØ 14 I=1,200
         Z=I
         X=Z/10.-0.1
         IF(X-1.9) 97,97,98
      97 CØNTINUE
         DEC=-4.343*ALØG(1.+4.*X**6*(2.*(X**2)-1.)**2)
         B=4.*X-10.*X**3+4.*X**5
         A=1.-8.*X**2+8.*X**4
         IF(A) 2,3,4
       3 IF(B) 5,6,7
       5 BETA=270.
         IF(BETØLD-358.8) 1,1,41
       6 PRINT 20
      20 FØRMAT(22HDEGENERATE CASE:A=B=0.)
         STØP
       7 BETA=90.
         IF(BETØLD-358.8) 1,1,41
       2 IF(B) 8,9,10
       8 BETA=180.+57.296*ATAN(ABS(B/A))
         IF(BETØLD-358.8) 1,1,41
       9 BETA=180.
         IF(BETØLD-358.8) 1,1,41
      10 BETA=180.-57.296*ATAN(ABS(B/A))
         IF(BETØLD-358.8) 1,1,41
       4 IF(B) 11,12,13
      11 BETA=360.-57.296*ATAN(ABS(B/A))
         IF(BETØLD-358.8) 1,1,41
      12 BETA=0.0
         IF(BETØLD-358.8) 1,1,41
      13 BETA=57.296*ATAN(ABS(B/A))
         IF(BETØLD-358.8) 1,1,41
      41 BETA=BETA+360.
       1 BETØLD=BETA
         THETA=-BETA
      14 PRINT 33,X,DEC,THETA
      33 FØRMAT(3X,F5.2,9X,F8.3,8X,F10.3)
      98 STØP
         END
```

Fig. 6.6.3 Program FILTR2.

```
NØRM.FREQ.          DECIBELS          PHASE,DEG.
     0.00             -.000              0.000
     0.10             -.000            -22.957
     0.20             -.001            -46.154
     0.30             -.009            -69.851
     0.40             -.033            -94.296
     0.50             -.067           -119.745
     0.60             -.063           -146.835
     0.70             -.001           -177.577
     0.80             -.343           -215.851
     0.90            -2.594           -260.125
     1.00            -6.990           -296.565
     1.10           -11.844           -320.863
     1.20           -16.357           -337.396
     1.30           -20.429           -349.479
     1.40           -24.113           -358.833
     1.50           -27.476           -366.374
     1.60           -30.570           -372.630
     1.70           -33.438           -377.932
     1.80           -36.114           -382.499
     1.90           -38.623           -386.484
STØP
```

Fig. 6.6.4 Trial run for FILTR2.

In the revised program, the current value of BETA is stored in BETOLD (line 00360). During each passage through the loop BETOLD is tested to determine whether it exceeds 358.8. If it does, BETA is increased by 360° (line 00350). The test value 358.8 was chosen on the basis of the data found in the first trial run (Fig. 6.6.2). As noted above, the phase shift was found to be −358.833° at X = 1.4, and −6.374° at X = 1.5. In the trial run of the *revised program*, the phase shift at X = 1.5 is found to be −366.374°, as shown in Fig. 6.6.4., and it may be seen that the phase shift now varies monotonically over the frequency range considered in the program.

6.7 FIRST OVERSHOOT AND UNDERSHOOT OF A NETWORK

The time response of an electrical or mechanical network to a unit step function is of importance in many areas of engineering. After the time response has been calculated by a method such as the Laplace transformation, it is relatively easy to write a program which will evaluate the time response for a specified set of times. Similarly, it is relatively simple to write a program which evaluates a function as a parameter changes. The program discussed in Sec. 6.6 is of this type. It is more difficult, however, to write a program which locates only particular points of a function such as maxima and minima. In this section, we will consider the problem of calculating only the time and amplitude of the first overshoot and the first undershoot of the response of a network to a unit step function.

The network chosen for study is the underdamped shunt compensation network,[1] and the time-scale response is

$$e(t) = 1. - (\cos 0.975t + 0.216 \sin 0.975t)\, e^{-1.21t} \qquad (6.7.1)$$

The program for calculating the overshoot and undershoot is shown in Fig. 6.7.1.

In this program, values of time are generated by the index of the DO loop (lines 00030 and 00040). At the first pass through the DO loop, T = 0.02 s. The voltage $e(t)$ is calculated at T = 0.02 s, and the numerical value is stored in E. This value of E is

[1] J. Millman and H. Taub, "Pulse, Digital and Switching Waveforms," McGraw-Hill, New York, 1965, p. 149.

```
C     SHØØT1
C     THIS PRØGRAM CALCULATES THE TIME AND AMPLITUDE ØF THE
C     FIRST ØVERSHØØT AND THE FIRST UNDERSHØØT ØF A SIMPLE
C     SHUNT PEAKED NETWØRK.
      EØLD=0.0
      DØ 6 I=1,500
      Z=I
      T=Z/50.
      E=1.-(CØS(.975*T)+.216*SIN(.975*T))/EXP(1.21*T)
      IF(EØLD-E) 5,5,7
    5 EØLD=E
    6 TØLD=T
      STØP
    7 PRINT 3,TØLD,EØLD
    3 FØRMAT(13HØVERSHØØT AT F6.2,19H SEC. AMPLITUDE IS F10.7)
      PRINT 4,T,E
    4 FØRMAT(2HT=F6.2,3H E=F10.7)
      IØLD=I
      DØ 15 I=IØLD,500
      E=1.-(CØS(.975*T)+.216*SIN(.975*T))/EXP(1.21*T)
      Z=I
      T=Z/50.
      IF(E-EØLD) 8,8,9
    8 EØLD=E
   15 TØLD=T
      STØP
    9 PRINT 20,TØLD,EØLD
   20 FØRMAT(14HUNDERSHØØT AT F6.2,19H SEC. AMPLITUDE IS F10.7)
      PRINT 21,T,E
   21 FØRMAT(2HT=F6.2,3H E=F10.7//)
      STØP
      END
```

Fig. 6.7.1 Program SHOOT1.

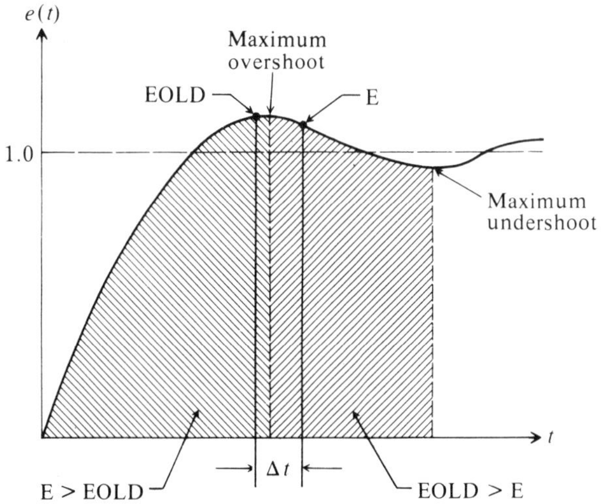

Fig. 6.7.2 Time response of a network.

```
ØVERSHØØT AT   2.52 SEC. AMPLITUDE IS  1.0302433
T=  2.54 E= 1.0302360
UNDERSHØØT AT   5.76 SEC. AMPLITUDE IS  0.9993871
T=  5.78 E= 0.9993872


STØP
```

Fig. 6.7.3 Trial run for SHOOT1.

compared with EOLD, which has been set at an initial value of zero (line 00010). E exceeds EOLD, so the program branches to statement 5, and the current value of E is

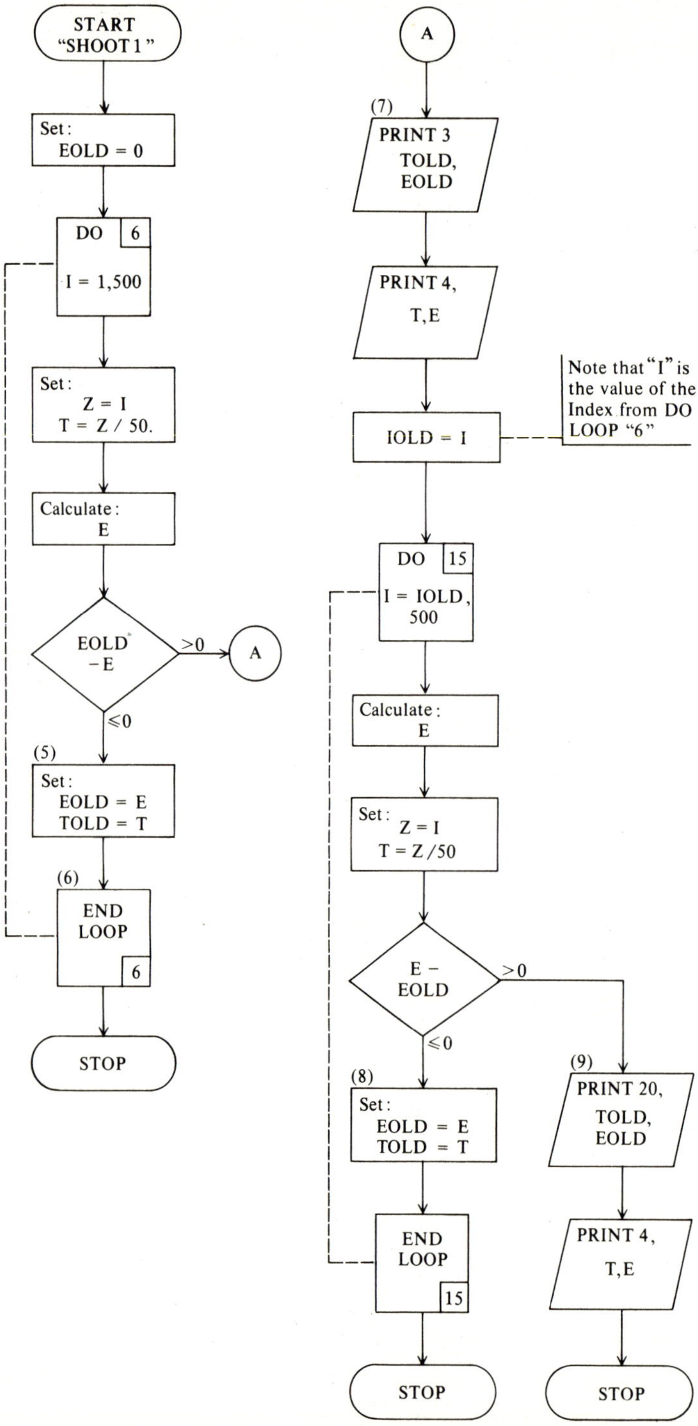

Fig. 6.7.4 Flow chart of program SHOOT1.

assigned to EOLD. Also, the current value of T, namely 0.02, is assigned to TOLD.

Iteration continues until EOLD exceeds E. At this point, the maximum overshoot has been reached or exceeded (see Fig. 6.7.2), and the program branches out of the DO loop. The values of TOLD, T, EOLD, and E, are then printed. The trial run in Fig. 6.7.3 shows that at the time of exit from the DO loop EOLD = 1.0302433 and E = 1.0302360. Thus, the amplitude of the first overshoot is 1.0302433 V, and occurs at 2.52 s. At 2.54 s the voltage decreases to 1.0302360 V.

Calculation of the approximate value of the undershoot is accomplished by the second DO loop (line 00150). Note that the initial value of the index of the second loop has been set equal to the final value of the index of the first DO loop. (See the discussion in Sec. 5.4 on the retention of the DO loop index when abnormal exit occurs. The reader is also urged to study carefully the flow chart shown in Fig. 6.7.4.) At this time EOLD exceeds E. Therefore, the IF statement on line 00190 causes the iteration to proceed until E exceeds EOLD. The smaller the value of the time increment used, the more closely will the true values of the maximum and minimum be approached. However, the use of smaller time increments increases computation time, and, therefore, judgment must be exercised as to whether the increased cost is justified by the increase of accuracy attained.

6.8 EFFECT OF FRICTION ON VELOCITY

If a body of constant mass is acted upon by a constant force, a constant acceleration is produced. However, if the mass is also acted on by a friction force (whose magnitude is a function of the velocity of the body), the velocity will tend to approach a limiting value for certain cases.[1] One case producing such a result occurs when the functional relationship between the coefficient of friction and the velocity is

$$\mu = 0.1758(\nu)^{0.09} \tag{6.8.1}$$

where μ = coefficient of kinetic friction
ν = velocity of body, cm/s

In this section we will examine a case of sliding friction in which Eq. (6.8.1) describes the coefficient of friction. A free-body diagram for the problem is shown in Fig. 6.8.1.

The equation governing the motion is

$$F - \mu N = ma \tag{6.8.2}$$

and since the normal force is

$$N = mg \tag{6.8.3}$$

we have

$$a = \frac{F}{m} - \mu g \tag{6.8.4}$$

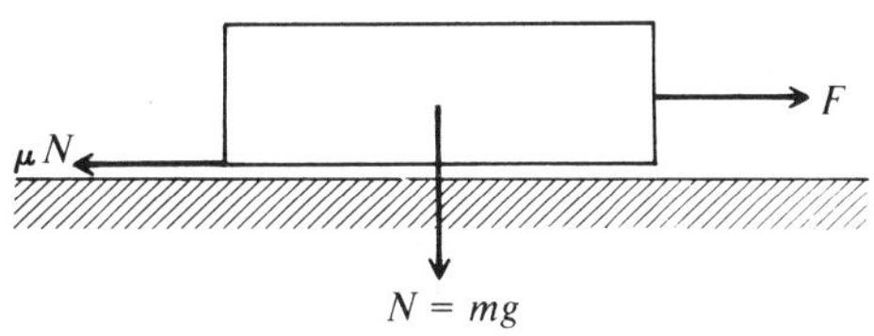

Fig. 6.8.1 Free-body diagram of friction effect.

[1] W. Arthur and S. K. Fenster, "Mechanics," Holt, New York, 1969, pp. 113–115.

We will solve the problem by assuming that the friction effect (and therefore the acceleration) remains constant over small increments of time, and then calculating the change in velocity as the product of the constant acceleration and the magnitude of the time increment. At the *end* of each time interval we will calculate the coefficient of friction to be used during the *next* time interval. For the *first* time interval we will assume the coefficient of friction is equal to the constant coefficient in Eq. (6.8.1), that is, 0.1758. For *all succeeding intervals*, the value of μ, obtained from Eq. (6.8.1) by using the velocity at the beginning of the interval, will be used in Eq. (6.8.4). Therefore, the velocity at the end of a time increment can be found from:

$$v_{\text{final}} = v_{\text{init}} + a\,\Delta t \tag{6.8.5}$$

where Δt is the magnitude of a time increment. The actual time at which a particular velocity occurs can be found by summing the time increments.

We will solve this problem using two different programs, entitled FRICT1 and FRICT2, respectively. Both programs will follow the calculation procedure outlined above, but different methods of producing the looping action and of summing the time intervals will be employed. The results produced by the two programs are identical. The authors do not categorically recommend one method over the other but have presented the two methods only to demonstrate the programming alternatives that are available to the reader at this point in the text. The variable names used in both these programs are defined as follows.

Symbol	*Name*	*Meaning*	*Units*
F	F	Applied force	dynes
M	XM	Mass of the body	g
v	V	Velocity	cm/s
Δt	TINC	Time increment	s
...	N	Number of printed lines of results	
g	G	Gravitational acceleration	cm/s^2
...	KPRINT	Counter controlling printing operations	
...	K	DO loop index (FRICT1)	
...	KOUNT	Cycle counter (FRICT2)	
...	T	Current time	s
μ	XMU	Coefficient of kinetic friction	
a	A	Acceleration of body	cm/s^2

The first program, FRICT1, is flow charted in Fig. 6.8.2 and uses a DO statement for producing the necessary loop. To permit the program user to control the duration of the calculation period, the loop was terminated by an abnormal exit from the DO after $2^{(n-1)}$ iterations, and therefore the test value of the DO was set at 50,000 to provide a *more than sufficient number of cycles.*

Because the velocity changes rapidly during the early stages of the motion and then changes slowly during the later stages, the results of the calculations were not printed at equal time intervals but at *geometrically increasing intervals.* The *first* result was printed after *one* time increment; the *second* result after *two* more time increments; the *third* result after *four* additional time intervals; and so on.

The program for FRICT1 is shown in Fig. 6.8.3. The DO loop that produces the cycling starts on line 00040 and ends on line 00170. Note that a CONTINUE statement was used to avoid having the arithmetic IF statement on line 00160 as the final statement in the range of the DO (see Sec. 5.5).

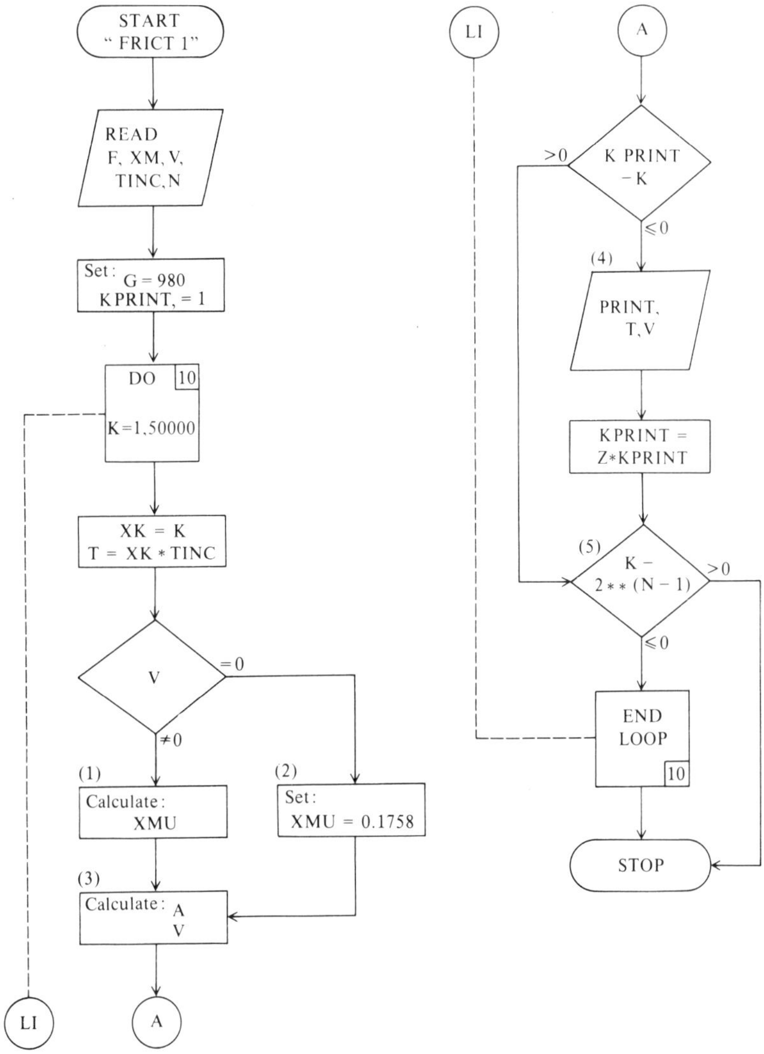

Fig. 6.8.2 Flow chart of program FRICT1.

In this program, the time at which the velocity is determined has been calculated (on line 00060) by multiplying the number of cycles by the time increment per cycle. The IF statement on line 00070 provides for initializing the coefficient of friction during the first cycle, when the velocity will be zero if the problem solution is started from rest. If the problem is started with the body in motion (that is, if an initial value other than zero is read-in for V), the statement on line 00100 will provide the initial value of the coefficient of friction. In either case, for all cycles beyond the first, the statement on line 00100 will be used to evaluate the coefficient.

The IF statement on line 00130 controls the printing operation and only prints the results of the calculations when K is equal to KPRINT. Each time the results are printed, the statement on line 00150 doubles the value of KPRINT and thereby increases the interval before the next set of results will be printed. The output of a sample run of FRICT1 is shown in Fig. 6.8.4, and the reader can see that a terminal velocity is approached.

```
00001C       FRICT1
00002C       PRØGRAM TØ CALCULATE THE VELØCITY AT TIME, T, RESULTING
00003C       FRØM A CØNSTANT FØRCE ACTING ØN A CØNSTANT MASS WITH THE
00004C       EFFECT ØF SLIDING FRICTIØN CØNSIDERED.
00010        READ,F,XM,V,TINC,N
00020        G=980.
00030        KPRINT=1
00040        DØ 10 K=1,50000
00050        XK=K
00060        T=XK*TINC
00070        IF(V)1,2,1
00080      2 XMU=0.1758
00090        GØ TØ 3
00100      1 XMU=0.1758*V**0.09
00110      3 A=F/XM-XMU*G
00120        V=V+A*TINC
00130        IF(KPRINT-K)4,4,5
00140      4 PRINT,T,V
00150        KPRINT=2*KPRINT
00160      5 IF(K-2**(N-1))10,10,6
00170     10 CØNTINUE
00180      6 STØP
00190        END
```

Fig. 6.8.3 Program FRICT1.

The basic changes in the alternative method of solution are best seen in the flow chart for FRICT2 shown in Fig. 6.8.5, and in the program listing in Fig. 6.8.6. The calculation scheme is the same as outlined earlier, but the cycle count and the current time are maintained by incrementing the appropriate variables within the iteration loop designed by the programmer. The reader should note that for certain values of TINC (values which cannot be represented exactly in the computer because of the conversion from decimal to binary, see Sec. 1.5) an appreciable error in the value of T can accumulate if the number of cycles is large. In the trial runs for these programs the time increment was taken as 0.125, a value that *can* be represented exactly, and no error was produced. While this type of error *may* occur in the *real* arithmetic statement on line 00150, no such problem presents itself for the statement on line 00160, where we are dealing with *integers*.

The IF statement on line 00170 will terminate the calculation procedure after 2^{n-1} cycles. For this program, it was not necessary (as it was for the program in Fig. 6.8.3) to *guess* the greatest number of PRINT statements any potential program user would request. The reader should note that if the programmer had guessed 10000

```
INPUT:00010
 ?2940.,10.,0.,0.125,15

          0.1250          15.2145
          0.2500          24.4502
          0.5000          39.7775
          1.0000          64.4986
          2.0000         102.5358
          4.0000         156.9712
          8.0000         226.4939
         16.0000         299.9420
         32.0000         355.1125
         64.0000         376.7454
        128.0000         379.2130
        256.0000         379.2411
        512.0000         379.2411
       1024.0000         379.2411
       2048.0000         379.2411
STØP
```

Fig. 6.8.4 Trial run for FRICT1.

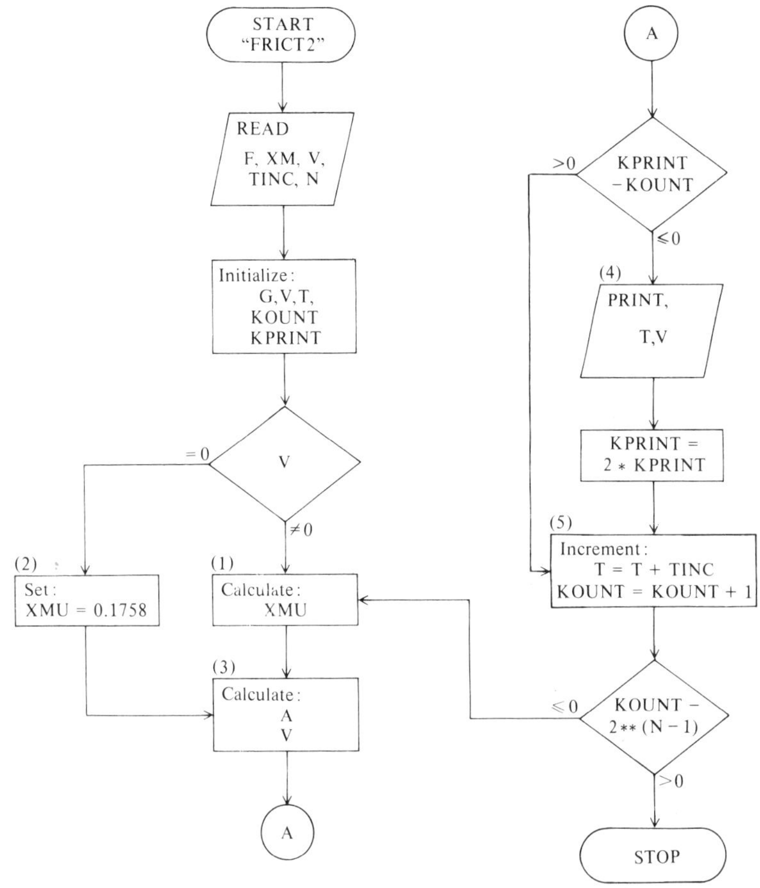

Fig. 6.8.5 Flow chart of program FRICT2.

```
C         FRICT2
C         PRØGRAM TØ CALCULATE THE VELØCITY AT TIME, T, RESULTING
C         FRØM A CØNSTANT FØRCE ACTING ØN A CØNSTANT MASS WITH THE
C         EFFECT ØF SLIDING FRICTIØN CØNSIDERED.
          READ,F,XM,V,TINC,N
          G=980.
          KØUNT=1
          KPRINT=1
          T=TINC
          IF(V)1,2,1
        2 XMU=0.1758
          GØ TØ 3
        1 XMU=0.1758*V**0.09
        3 A=F/XM-XMU*G
          V=V+A*TINC
          IF(KPRINT-KØUNT)4,4,5
        4 PRINT,T,V
          KPRINT=2*KPRINT
        5 T=T+TINC
          KØUNT=KØUNT+1
          IF(KØUNT-2**(N-1))1,1,6
        6 STØP
          END
```

Fig. 6.8.6 Program FRICT2.

```
INPUT:00010
 ?2940.,10.,0.,0.125,15

          0.1250            15.2145
          0.2500            24.4502
          0.5000            39.7775
          1.0000            64.4986
          2.0000           102.5358
          4.0000           156.9712
          8.0000           226.4939
         16.0000           299.9420
         32.0000           355.1125
         64.0000           376.7454
        128.0000           379.2130
        256.0000           379.2411
        512.0000           379.2411
       1024.0000           379.2411
       2048.0000           379.2411
STØP
```

Fig. 6.8.7 Trial run for FRICT2.

rather than 50000 as the maximum number of DO loop cycles for FRICT1, the last line of output in Fig. 6.8.4 would not have been printed. The 2048. s, reached in steps of 0.125 s, requires 16,384 cycles. With 10000 as the test value, the DO loop would have executed a normal exit and the last line of output would have been that for 1024. s or 8,192 cycles.

The results of a run of FRICT2, using the same data as before, are shown in Fig. 6.8.7, and, as would be expected, are the same as the earlier results for FRICT1.

PROBLEMS

6-1. Modify the program written as a solution to Prob. 5-3 so that an abnormal exit from the DO loop will occur if the absolute value of any term in the series is less than 0.000001.

The program should read-in values of X and N, but the output should show the actual number of terms used as well as the value of the sin x.

Sample data

Set	*X*	*N*
1	30°	20
2	47°	3
3	47°	15

6-2. The displacement along a line of a point having initial velocity v and undergoing acceleration a is given for any time t by the following formula:

$$s = vt + \frac{at^2}{2}$$

where s = distance, ft
v = initial velocity, ft/s
a = acceleration, ft/s^2
t = time, s

Write a FORTRAN program that will read-in values for v and a, and will printout the time and the displacement at that time, for 0.1 s intervals from 0 to 3 s. The problem should be programmed *without* using a DO loop.

6-3. The Fibonacci number sequence is a series in which the first two members are equal to unity:

$$X_1 = X_2 = 1$$

and all succeeding members are defined by the recursion expression:

$$X_{n+1} = X_n + X_{n-1}$$

Write a program, *without* using a DO loop, which calculates and prints the terms of the Fibonacci sequence and also the value of the ratio X_{n+1}/X_n. The calculation should be halted when the magnitude of the difference between the ratio and 1.6180 is equal to or less than 0.0001.

6-4. The harmonic mean is defined as

$$x_h = \frac{n}{\sum_i^n (1/x_i)}$$

Write a program which will calculate the harmonic mean of the first n integers and will result in a value as close as possible to 2.69. The program should print the value of n and the calculated value of the harmonic mean.

6-5. The formula for the sum of n terms of a geometric series is

$$S_n = \frac{a - ar^n}{1 - r}$$

where a is the first term of the series and r is the ratio of any term to the term immediately preceding it in the series.

Write a FORTRAN program which will calculate the sum of n terms of the geometric series shown below, and will determine the number of terms needed to achieve a sum as close as possible to 1.99. The program should print the value of n and the sum attained when iteration is halted.

The given series is

$$1 + \tfrac{1}{2} + \tfrac{1}{4} + \tfrac{1}{8} + \ldots$$

6-6. The geometric mean of the product of n terms in a series is

$$X_g = \sqrt[n]{X_1 X_2 X_3 \ldots X_n}$$

Given the sequence

1.10, 1.40, 1.70, 2.00, . . .

write a program which calculates the number of terms needed to attain a geometric mean as close as possible to 1.90. The program should print the number of terms needed, the value of the geometric mean attained when iteration is halted, and the last term included in the calculation of the geometric mean.

6-7. The magnitude of the transfer function of a low-pass Butterworth filter is

$$\left|\frac{V_o}{V_i}\right| = \frac{1}{\sqrt{1 + \omega^{2n}}}$$

where V_i = voltage at input terminals, V (rms)
V_o = voltage at output terminals, V (rms)
ω = radian frequency, rad/s
η = an integer, called order of filter

It is convenient to express the attenuation of the filter as follows:

$$\text{Attenuation} = 20 \log_{10} \left|\frac{V_o}{V_i}\right| = 20 \log_{10} \frac{1}{\sqrt{1 + \omega^{2n}}} \qquad \text{dB}$$

Write a program which will calculate the order of filter needed to achieve an attenuation of −40 dB at 2 rad/s. In writing the program, use a REAL name such as XN for the order of the filter. The user will then have the option of rounding the number which is computed. Also, remember that the formula for decibels must be expressed in terms of the common logarithm.

6-8. Write a program that will solve Prob. 6-5 without using the formula for the sum of the series. (Have the program generate the sum during execution.)

6-9. The normalized upper 3-dB frequency of a certain compensated amplifier may be found by calculating that value of α which satisfies the equality

$$\frac{1 - 0.12\alpha^2 + 0.02145\alpha^4}{1 - 0.12\alpha^2 + 0.02145\alpha^4 + 0.02145\alpha^6} = \frac{1}{2}$$

Write a progam which will calculate α by use of a DO loop and an IF statement. At each iteration α should be incremented by 0.01. In choosing the indices of the DO loop, take into account the fact that α is known to lie between 1 and 2.

6-10. In Prob. 5-8, equations were given for the velocity of an electron with and without relativistic correction. These velocities were denoted μ' and μ, respectively. Write a program which calculates the potential difference V for which $\mu'/\mu = 0.99$.

6-11. An electron is given enough energy to leave a positive electrode with an initial velocity of 8.4×10^5 m/s. It moves toward a negative electrode which is parallel to the positive electrode, as shown below:

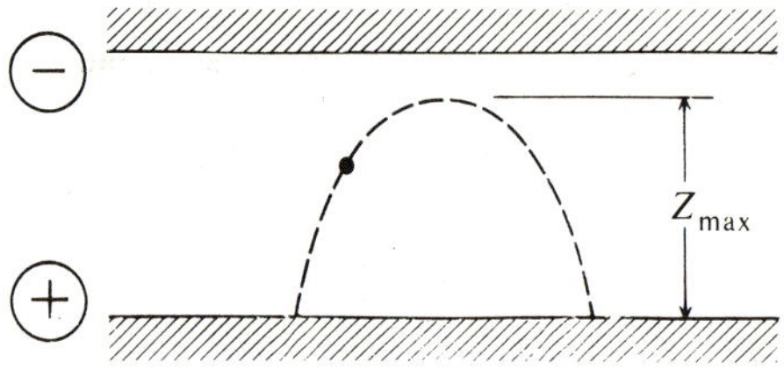

Because of the retarding field between the electrodes, the electron velocity decreases, eventually reaches zero, and then reverses its direction and returns to the positive electrode. At any time during the flight of the electron, its distance from the positive electrode is given by

$$Z = -100 \frac{e}{m} t^2 + (8.4 \times 10^5)t$$

where Z = distance of electron from positive plate, m
e/m = electron charge to mass ratio = 1.759×10^{11} C/kg
t = time, s

Approximate the maximum value of Z and the time at which it is attained by writing a program which calculates Z as a function of t as the latter is increased from zero in steps of 1×10^{-10} s.

6-12. The catalyzed reaction of ethylene and water vapor to form ethyl alcohol has the following equilibrium expression:

$$\frac{n_a(1.6 - n_a)}{(1 - n_a)(0.6 - n_a)} = 0.156$$

where n_a = number of moles of alcohol.

Write a computer program that will evaluate and print the value of n_a.

6-13. The Bessel function of the first kind, of integer order n, for an argument x, is

$$J_n(x) = \sum_{k=0}^{\infty} (-1)^k \left(\frac{x}{2}\right)^{n+2k} \frac{1}{k!(n+k)!} \tag{1}$$

where $n = 0, 1, 2, 3, \ldots$.

If n is negative, the following relation should be used in conjunction with Eq. (1):

$$J_{-n}(x) = (-1)^n J_n(x) \tag{2}$$

For example, if $n = -3$, calculate $J_3(x)$ by means of Eq. (1), and then $J_{-3}(x)$ by means of Eq. (2).

Write a program that will evaluate and print the value of the Bessel function given the order n and the argument x. Have the program terminate when the magnitude of any term in the series is equal to or less than 0.00001.

Sample data

Set	n	x
1	2	7
2	0	7
3	−1	7

6-14. Probs. 4-14 and 5-15 dealt with the combustion of propane gas with varying percentages of the theoretical air requirement. Modify the program written for Prob. 5-15 so it will determine the percentage of theoretical air required to reduce the CO_2 content in the combustion products to a value *less than* some specified value.

The input to the program should be the maximum tolerable percentage of CO_2. The output should show the minimum percent air needed as well as the composition of the combustion products.

Sample data

Set	% CO_2
1	5
2	2

6-15. In a traffic flow model, the merging delay $U(A)$ is given (in seconds) by one of the following equations, depending on the value of A:

(*a*) $A = 1$

$$U(1) = q^{-1}(e^{qT} - 1 - qT)$$

(*b*) $A = 2$

$$U(2) = \frac{e^{2qT} - 1 - 2qT - 2(qT)^2}{q(1 + 2qT)}$$

(*c*) $A = 3$

$$U(3) = \frac{e^{3qT} - 1 - 3qT - 4.5(qT)^2 - 4.5(qT)^3}{q[1 + 3qT + 4.5(qT)^2]}$$

where q = traffic flow, vehicles per second
T = critical gap, s

Write a FORTRAN program which will accept a value of A and will calculate and print the delay for various flow densities and critical gaps.

Sample data

q varies from 400 to 800 vehicles per *hour* in steps of 200
$T = 2, 3, 4, 5, 6$ (s)
$A = 2$

7
Computer Programs Using Subscripts

There are many situations in engineering analysis and design in which subscript notation is inherent. Some of these are illustrated in this chapter and in subsequent chapters. There are other cases in which the use of subscript notation is not essential, but proves to be very useful. This latter case is illustrated in Sec. 7.4.

7.1 SUBSCRIPTED VARIABLES

Any FORTRAN variable (integer, real, complex, or logical mode) can have one, two, or three subscripts.[1] That is, any variable name can represent a one-dimensional, two-dimensional, or three-dimensional *array.*

The general form of a subscripted variable or array is

name (n_1)

or

name (n_1, n_2)

or

name (n_1, n_2, n_3)

where *name* is the FORTRAN name used to designate the array and n_1,n_2,n_3 are acceptable subscript expressions.

In most versions of the language, the value of a subscript expression must be a *positive* integer, other than zero, and *not greater* than the *dimension* assigned to that particular subscript by a DIMENSION statement (see Sec. 7.2).

[1] Some FORTRAN compilers permit more than three subscripts.

The following subscript expressions are the only ones which are acceptable to *all* FORTRAN compilers:[1]

1. v
2. c
3. $v + c$
4. $v - c$
5. $c * v$
6. $c * v + k$
7. $c * v - k$

where v is a positive integer variable, and c and k are positive integer constants.

Some examples of subscripted variables or arrays are

```
A(I,6)
JOB(I+1)
SET(2*K,2*L+4,M)
XMAT(I,J)
POINT(12,6)
```

Note that in FORTRAN B(6) represents a subscripted variable while B6 represents an ordinary nonsubscripted variable.

7.2 DIMENSION STATEMENT

The DIMENSION statement is a nonexecutable specification statement of the form

DIMENSION *name*(n_1,n_2,n_3)

where *name* is the FORTRAN name given to the subscripted variable or array and n_1,n_2,n_3 are *integer constants* indicating the *maximum* values of the subscripts.

The DIMENSION statement reserves storage locations in the computer memory for each element of the array. For example, if the statement

```
DIMENSION SET(10,15),JOB(5)
```

were encountered in a program, the computer would set aside 150 (10×15) storage locations for the members of the array named SET and 5 storage locations for the members of the array named JOB. Because it serves to hold space in storage, the DIMENSION statement *must precede* the first mention of the array in the source program. The DIMENSION statement should specify the *maximum anticipated size* of each array to ensure the reservation of sufficient storage. Not all locations will be used every time a program is run; the number actually employed depends, in some cases, on the data supplied to the computer. Examples of DIMENSION statements are

```
DIMENSION A(50,50),M(15)
DIMENSION X(5),Y(5),Z(5),SUM(5)
DIMENSION MATRIX(10,10,10)
```

[1] For some compilers the rules are less restrictive, and any arithmetic operations with integer or real values (even subscripted variables) are permitted. The programmer should consult the manual for the system he is using.

7.3 IMPLIED LOOPS

The elements of an ARRAY may be handled individually just as all other data values are processed during I/O operations, or they may be handled with an *implied* DO loop. For example, the statement

```
READ 15, (X(I), I=1,10)
```

will cause values for the first 10 elements of the array named X to be read according to the FORMAT specified in statement 15.

If X only contained 10 elements and had been previously dimensioned, the following statements would also read in the first 10 elements according to the FORMAT of statement 15:

```
DIMENSION X(10)
READ 15, X
```

The same procedures may be used for output operations with arrays. That is, the statement

```
WRITE (3,60)(A(K),K=12,25)
```

would cause the values of 14 elements of the array named A (starting with the twelfth element and ending with the twenty-fifth) to be transmitted to output device 3 and transcribed according to the FORMAT specified in statement number 60. The statement

```
WRITE (3,60) A
```

would cause the values of *all* the elements of array A to be processed by output device 3.

It should be noted that for multidimensioned arrays the values of the elements are stored (or retrieved) according to a predefined pattern when only the name of the array is used in an I/O statement. The values of the elements are stored sequentially in the computer memory with the first subscript varying most rapidly, the second subscript next most rapidly, etc. To illustrate this let us consider a two-dimensional array named SET:

```
DIMENSION SET(4,2)
```

The eight elements of SET will be stored sequentially in the following order:[1]

```
SET(1,1)
SET(2,1)
SET(3,1)
SET(4,1)
SET(1,2)
SET(2,2)
SET(3,2)
SET(4,2)
```

During output operations, the commands

```
   PRINT 25,SET
25 FORMAT(F8.2)
```

[1]That is, the elements are stored column by column. However, the programmer has complete control over I/O operations and may, for example, print or read an array row by row.

would produce the values of the elements in the order shown. Conversely, a command to read values into SET during input operations would require that the values be presented in the indicated order.

7.4 PRINTING A TABLE OF SQUARE ROOTS

Writing a program for printing a table of square roots is a simple matter. If we wrote such a program in routine fashion using a DO Loop and if the table were printed, we would realize that this is a wasteful method. The first square root would appear on the first line, the second square root on the second line, and so forth. The results would occupy only a small part of each line.

The program shown in Fig. 7.4.1 prints a table of square roots in eight columns. The table is arbitrarily limited to square roots of numbers from 1 through 40. The first DO loop (beginning on line 00020) calculates the square roots of these numbers and stores them in the subscripted variable Z(I). Thus, the square root of 1 is stored in Z(1), the square root of 2 is stored in Z(2), and so on. Notice that the statement DIMENSION Z(40) precedes the first use of the variable Z. The second DO loop in conjunction with the PRINT statement on line 00110 produces an eight-column table. The variables K, L, and M in the PRINT statement are functions of I and are defined on lines 00080 through 00100. Thus, in the first pass through the DO loop, 1, Z(1), 11, Z(11), 21, Z(21), 31, and Z(31) are printed as shown in the trial run in Fig. 7.4.2. A flow chart for this program is shown in Fig. 7.4.3.

```
00001C |    RØØT
00002C |    THIS PRØGRAM CALCULATES AND PRINTS A TABLE ØF SQUARE RØØTS
00010  |    DIMENSIØN Z(40)
00020  |    DØ 1 I=1,40
00030  |    X=I
00040  |  1 Z(I)=SQRT(X)
00050  |    PRINT 4
00060  |  4 FØRMAT(4(7X1HN4X4HSQRT))
00070  |    DØ 9 I=1,10
00080  |    K=I+10
00090  |    L=I+20
00100  |    M=I+30
00110  |  9 PRINT 2,I,Z(I),K,Z(K),L,Z(L),M,Z(M)
00120  |  2 FØRMAT(4(I8,F8.2))
00130  |    STØP
00140  |    END
```

Fig. 7.4.1 Program ROOT.

N	SQRT	N	SQRT	N	SQRT	N	SQRT
1	1.00	11	3.32	21	4.58	31	5.57
2	1.41	12	3.46	22	4.69	32	5.66
3	1.73	13	3.61	23	4.80	33	5.74
4	2.00	14	3.74	24	4.90	34	5.83
5	2.24	15	3.87	25	5.00	35	5.92
6	2.45	16	4.00	26	5.10	36	6.00
7	2.65	17	4.12	27	5.20	37	6.08
8	2.83	18	4.24	28	5.29	38	6.16
9	3.00	19	4.36	29	5.39	39	6.24
10	3.16	20	4.47	30	5.48	40	6.32

STØP

Fig. 7.4.2 Trial run for ROOT.

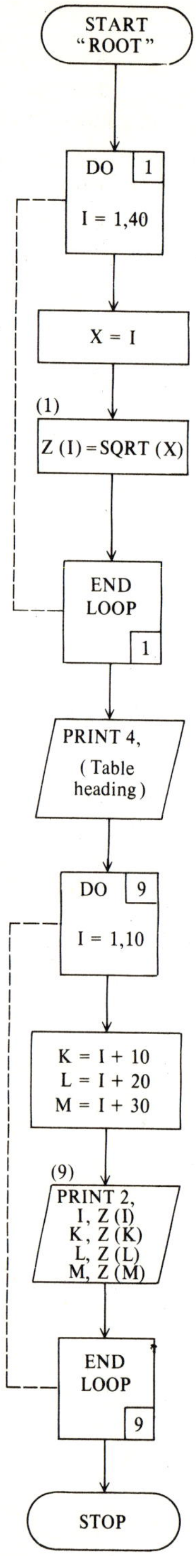

Fig. 7.4.3 Flow chart of program ROOT.

7.5 EQUIVALENT ANNUAL EXPENSE[1]

The engineer is frequently faced with the problem of making a comparison of the cost of manufacturing a product using alternative methods or processes. To make an intelligent comparison, he must reduce all the factors contributing to the operating costs for each alternative to a common basis. In most cases the actual expense of operating a piece of equipment will not be the same from year to year. Therefore, to make a comparison, it is necessary that the actual expenses over a period of years be reduced to a single *equivalent annual expense.* In determining this amount, the *time value* of money must be taken into account, and, therefore, a simple average will not suffice.

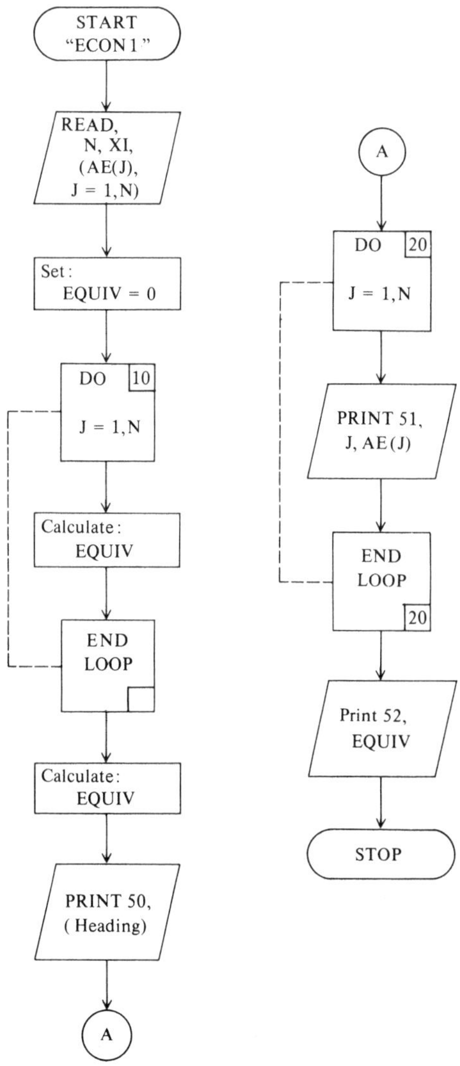

Fig. 7.5.1 Flow chart of program ECON1.

[1] H. G. Thuesen, "Engineering Economics," Prentice-Hall, Englewood Cliffs, N.J., 1957, p. 55.

```
C     ECØN1
C     PRØGRAM TØ CALCULATE THE EQUIVALENT UNIFØRM ANNUAL
C     EXPENSE ØF ØPERATING A PIECE ØF EQUIPMENT ØVER
C     A PERIØD ØF "N" YEARS
      DIMENSIØN AE(20)
      READ,N,XI,(AE(J),J=1,N)
      EQUIV=0.0
      DØ 10 J=1,N
   10 EQUIV=EQUIV+AE(J)*(1.+XI)**(N-J)
      EQUIV=EQUIV*XI/((1.+XI)**N-1.)
      PRINT 50
   50 FØRMAT(/4HYEAR5X7HEXPENSE/)
      DØ 20 J=1,N
   20 PRINT 51,J,AE(J)
   51 FØRMAT(I3,5XF7.0)
      PRINT 52,EQUIV
   52 FØRMAT(//27HEQUIVALENT ANNUAL EXPENSE =F9.1)
      STØP
      END
```

Fig. 7.5.2 Program ECON1.

To find the equivalent annual expense, the *present value* of the expenditures made over the years must be totaled. A calculation is then made to determine the *sinking-fund payment* that would have been required each year during the period of operation in order to have available, *now*, the present value of the expenditures. The amount of this sinking-fund payment is the *equivalent uniform annual expense* and can be calculated with the formula

$$(AE)_{\text{eq}} = \frac{i}{(1+i)^n - 1} \sum_{j=1}^{j=n} (AE)_j (1+i)^{n-j}$$

The symbols used have the following values:

Symbol	*Computer name*	*Meaning*	*Units*
$(AE)_{\text{eq}}$	EQUIV	Equivalent annual expense	\$
i	XI	Interest rate	% as a decimal
n	N	Number of years of operation	
$(AE)_j$	AE(J)	Annual expense during the jth year	\$

A flow chart of the procedure used in making the calculations is shown in Fig. 7.5.1, and the FORTRAN program is presented in Fig. 7.5.2. The expenses for each

```
INPUT:00020
?8,0.07,1630.,1540.,1720.,1205.,970.,865.,850.,725.


YEAR     EXPENSE

  1        1630.
  2        1540.
  3        1720.
  4        1205.
  5         970.
  6         865.
  7         850.
  8         725.


EQUIVALENT ANNUAL EXPENSE =    1241.1
STØP
```

Fig. 7.5.3 Trial run for ECON1.

year, starting with the first year of operation, are entered using the implied DO loop in the READ statement on line 00020. Note that the DIMENSION statement on line 00010 has made provision for storing up to 20 values in the array AE. The equivalent annual expense is ultimately stored in the computer location named EQUIV, but this same location is also used to hold the sum obtained in the DO loop on line 00050. When the DO has been satisfied, EQUIV will contain the *present value* of the actual expenditures. This value is then multiplied by the *sinking-fund factor*, on line 00060, and the result again stored in EQUIV.

The PRINT statement on line 00070 produces a table heading, and the DO loop, lines 00090 and 00100, echoes the problem data. Finally, the equivalent annual expense is printed by the statement on line 00120. A sample run is shown in Fig. 7.5.3.

7.6 LOWER 3-dB FREQUENCY OF CASCADED AMPLIFIERS

The frequencies at which the magnitude of the gain of an amplifier falls to 0.707 of its value at a reference frequency are called the upper and lower 3-dB frequencies. When several noninteracting, identical amplifiers are cascaded, the 3-dB frequencies of the cascade are less than the 3-dB frequencies of an individual stage. However, the 3-dB frequencies of the cascade can be calculated from formulas.[1]

```
00001C    LØCØRN
00002C    THIS PRØGRAM CALCULATES THE LØW FREQUENCY CØRNER ØF A
00003C    CASCADE ØF N RC CØUPLED STAGES NØT NECESSARILY ALIKE.
00010     DIMENSIØN FL(20),T(20)
00020     PRINT 4
00030   4 FØRMAT(16HNUMBER ØF STAGES)
00040     READ,N
00050     READ,(FL(I),I=1,N)
00060     C=ALØG(2.)
00070     DØ 1 J=1,1000
00080     Z=J
00090     F=Z/2.
00100     SUMLØG=0.0
00110     DØ 2 I=1,N
00120     T(I)=ALØG(1.+(FL(I))**2/F**2)
00130   2 SUMLØG=SUMLØG+T(I)
00140     IF(SUMLØG-C) 3,3,1
00150   1 CØNTINUE
00160     PRINT,"THE TEST FAILED"
00170     STØP
00180   3 PRINT 5,F
00190   5 FØRMAT(21HØVERALL LØWER CØRNER=F7.2//)
00200     STØP
00210     END
```

Fig. 7.6.1 Program LOCORN.

```
NUMBER ØF STAGES
INPUT:00040
 ?5

INPUT:00050
 ?54.9,12.7,8.8,25.4,61.1

ØVERALL LØWER CØRNER=  98.00


STØP
```

Fig. 7.6.2 Trial run for LOCORN.

[1] J. Millman and C. C. Halkias, "Electronic Devices and Circuits," McGraw-Hill, New York, 1967, p. 469.

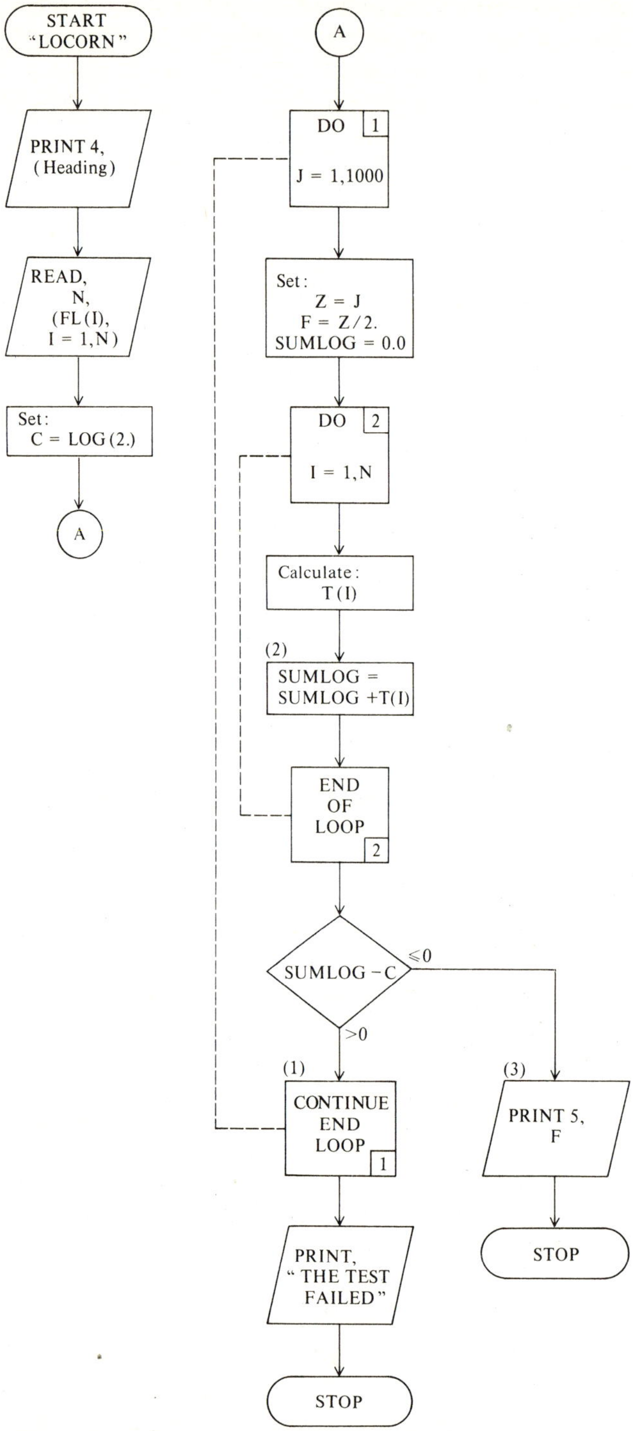

Fig. 7.6.3 Flow chart of program LOCORN.

If the cascade consists of noninteracting stages which are *different*, the upper and lower 3-dB frequencies can be found graphically. The program of Fig. 7.6.1 obviates the need for a graphical procedure. The transfer function of n low-pass noninteracting stages is

$$G(jf) = \prod_{i=1}^{n} \frac{1}{1 + j(\bar{f}_i/f)} \tag{7.6.1}$$

where $\bar{f}_i$ represents the lower 3-dB frequency of the ith stage. Now take the magnitude of the transfer function, set the frequency f equal to f_x, and equate the magnitude of the transfer function to $1/\sqrt{2}$. The result is Eq. (7.6.2). The unknown frequency f_x is that frequency at which the response has fallen the amount of 3-dB from the reference value (unity).

$$G(jf) = \prod_{i=1}^{n} \frac{1}{\sqrt{1 + (\bar{f}_i/f_x)^2}} = \frac{1}{\sqrt{2}} \tag{7.6.2}$$

Finally, we take the logarithm of the *reciprocal* of Eq. (7.6.2):

$$\sum_{i=1}^{n} \log \left[1 + \left(\frac{\bar{f}_i}{f_x}\right)^2\right] = 2 \log 1.414 = \log 2 \tag{7.6.3}$$

Equation (7.6.3) is in a form convenient for computer solution. In the program, Fig. 7.6.1, the terms on the left-hand side of the above equation are represented by the subscripted variable T(I). T(I) is evaluated at different frequencies, beginning at $F = \frac{1}{2}$ Hz and increasing in steps of $\frac{1}{2}$ Hz. The sum of the n logarithms is stored in the variable SUMLOG. The calculation of T(I) and SUMLOG is performed by the nested DO loop (lines 00110 through 00130). At each frequency SUMLOG is compared with the constant ALOG(2.). At low frequencies SUMLOG exceeds ALOG(2.), but as the frequency increases, a value of F is reached at which SUMLOG = ALOG(2.). At this point control passes out of the range of the DO loop to statement 3, and the overall lower corner frequency is printed. A flow chart for the program can be seen in Fig. 7.6.3.

In the trial run shown in Fig. 7.6.2, the lower 3-dB frequency of a cascade of five noninteracting amplifiers is calculated. The lower 3-dB frequencies of the individual stages are 54.9, 12.7, 8.8, 25.4, and 61.1 Hz. The overall lower 3-dB frequency is 98.0 Hz. The reader can check this answer by superimposing the frequency responses of the individual stages on a log-log plot.[1]

7.7 CHANGING THE INDEX PARAMETERS OF A DO LOOP

The index parameters of a DO loop may not be redefined *within its range.* There are occasions when it is desirable to use the same DO loop with a *new* set of index parameters, and one method of doing this is illustrated by the program of Fig. 7.7.1.

[1] J. Millman and C. C. Halkias, "Electronic Devices and Circuits," McGraw-Hill, New York, 1967, p. 470.

```
00001C      CYCLE
00002C      THIS PRØGRAM ILLUSTRATES ØNE METHØD BY WHICH THE INDEX
00003C      PARAMETERS ØF A DØ LØØP MAY BE CHANGED.
00010       PRINT 99
00020    99 FØRMAT(5H    J9H          W10H          MIN9H    KØUNT)
00030       MIN=1
00040       MAX=10
00050       INC=1
00060       KØUNT=0
00070       GØ TØ 40
00080    70 MIN=20
00090       MAX=100
00100       INC=10
00110       GØ TØ 40
00120    90 MIN=200
00130       MAX=1000
00140       INC=100
00150       GØ TØ 40
00160    40 DØ 1 J= MIN,MAX,INC
00170       W=J
00180       KØUNT=KØUNT+1
00190       PRINT 7,J,W,MIN,KØUNT
00200     7 FØRMAT(I5,4X,I5,5X,I4,5X,I3)
00210     1 CØNTINUE
00220       IF(KØUNT-10) 40,70,80
00230    80 IF(KØUNT-19) 40,90,50
00240    50 STØP
00250       END
```

Fig. 7.7.1 Program CYCLE.

The flow chart is shown in Fig. 7.7.2. In this program the index I of the DO loop is used to generate the real variable W which represents radian frequency.

When the ratio of the maximum value of W to the minimum value of W is large, it is convenient to vary the amount by which W is incremented. In the program, W is incremented in steps of one from 1 to 10, then in steps of 10 from 10 to 100, and finally in steps of 100 from 100 to 1000. This is accomplished by using variable index parameters.

The index parameters are given the descriptive names MIN, MAX, and INC. At the beginning of the program MIN, MAX, and INC are assigned the values 1, 10, and 1, respectively (lines 00030, 00040, 00050). Also, the variable KOUNT is initialized at a value of zero. The GO TO 40 statement on line 00070 causes the program to branch to statement 40 on line 00160. Then the DO loop is entered, and the statements within the range of the DO loop are executed until the index I (and also W) reaches its maximum value, namely 10. At this point the variable KOUNT also has attained a value of 10.

After exit from the DO loop, the first IF statement is executed. Because KOUNT is 10, the program transfers to statement 70, where MIN, MAX, and INC are given a new set of values. Then the program returns to the DO loop, and the statements in the range of the DO loop are executed until the index I (and also W) reaches its new maximum value, namely 100. At this point the variable KOUNT has attained a value of 19. After exit from the DO loop, the second IF statement transfers the program to statement 90, and a new set of values are assigned to the index parameters. Obviously, this cycle could be repeated as many times as desired. Another method for achieving the change of index parameters would be through the use of the computed GO TO statement (see Sec. 4.4).

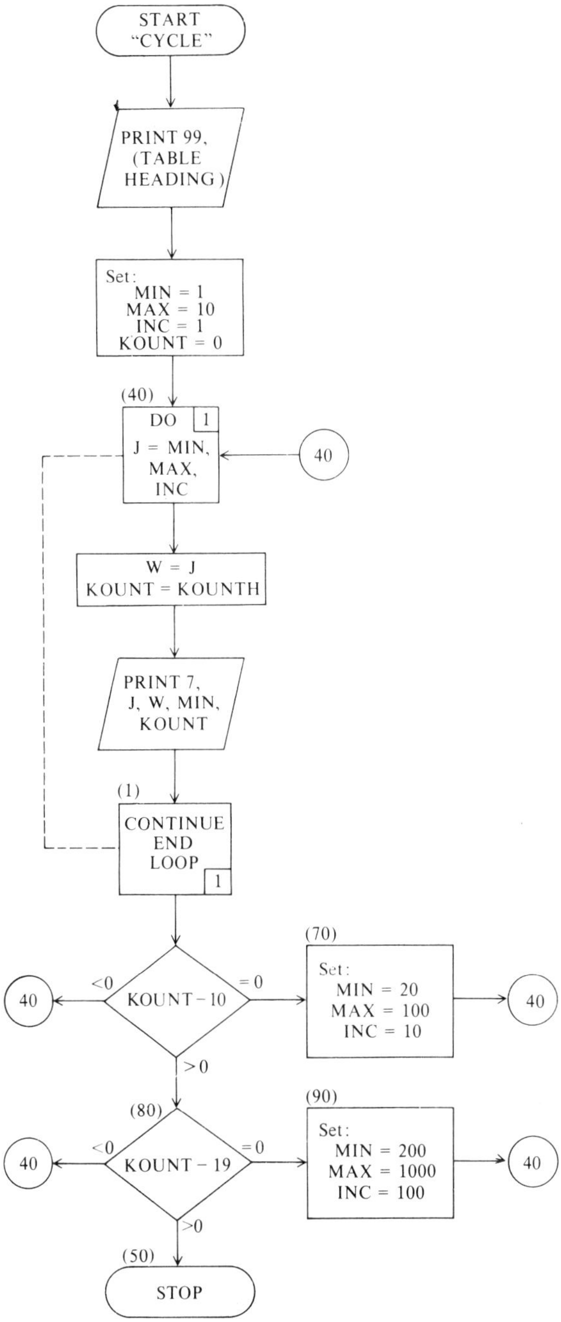

Fig. 7.7.2 Flow chart of program CYCLE.

J	W	MIN	KØUNT
1	1	1	1
2	2	1	2
3	3	1	3
4	4	1	4
5	5	1	5
6	6	1	6
7	7	1	7
8	8	1	8
9	9	1	9
10	10	1	10
20	20	20	11
30	30	20	12
40	40	20	13
50	50	20	14
60	60	20	15
70	70	20	16
80	80	20	17
90	90	20	18
100	100	20	19
200	200	200	20
300	300	200	21
400	400	200	22
500	500	200	23
600	600	200	24
700	700	200	25
800	800	200	26
900	900	200	27
1000	1000	200	28

STØP

Fig. 7.7.3 Trial run for CYCLE.

7.8 THE BODE PLOT[1]

The Bode plot, also called corner plot or logarithmic plot, is widely used in automatic control theory. It consists of a plot of the variation with frequency of the magnitude in decibels and phase in degrees of a transfer function. The plot is made on semilogarithmic paper. The most general form of the transfer function is

$$G(j\omega) = \frac{S \prod_{i=1}^{m} [1 + j(\omega/\omega_i)]}{(j\omega)^L \{(j\omega/\omega_n)^2 + (2\xi/\omega_n) j\omega + 1\} \prod_{k=1}^{n} [1 + j(\omega/\omega_k)]} \tag{7.8.1}$$

where S = scale factor (in this analysis assumed to be a real number)

ξ = damping ratio

ω_n = natural frequency

The term in braces is called the quadratic factor.

The magnitude of the transfer function in decibels is

$$\begin{aligned} 20 \log_{10} |G(j\omega)| = {} & 20 \log_{10} S + 20 \sum_{i=1}^{m} \log_{10} \left|1 + j\frac{\omega}{\omega_i}\right| \\ & -20L \log_{10} j\omega - 20 \sum_{k=1}^{n} \log_{10} \left|1 + j\frac{\omega}{\omega_k}\right| \\ & -20 \log_{10} \left|1 - \left(\frac{\omega}{\omega_n}\right)^2 + j2\xi\frac{\omega}{\omega_n}\right| \end{aligned} \tag{7.8.2}$$

[1] R. Saucedo and E. E. Schiring, "Introduction to Continuous and Digital Control Systems," Macmillan, New York, 1968, p. 393.

The phase angle of the transfer function is shown in Eq. (7.8.3). In this equation the symbol $\angle$ denotes "angle of."

$$\underline{/G(j\omega)} = \sum_{i=1}^{m} \underline{/1 + j\frac{\omega}{\omega_i}} - L\underline{/j\omega}$$

$$- \sum_{k=1}^{n} \underline{/1 + j\frac{\omega}{\omega_k}} - \underline{/1 - \left(\frac{\omega}{\omega_n}\right)^2 + j2\xi\frac{\omega}{\omega_n}} \tag{7.8.3}$$

The program of Fig. 7.8.1 evaluates Eqs. (7.8.2) and (7.8.3). The symbols used in these equations are defined as follows.

Formula symbol	*Computer name*	*Meaning*	*Units*
ω	W	Radian frequency	rad/s
ω_i	WNUM(I)	Numerator corner frequency	rad/s
ω_k	WDEN(K)	Denominator corner frequency	rad/s
S	SF	Scale factor	
ω_n	WN	Natural frequency	rad/s
ξ	DEL	Damping factor	
L	L	Order of the pole	
m	M	Number of numerator factors $(1 + j\omega/\omega_i)$	
n	N	Number of denominator factors $(1 + j\omega/\omega_k)$	

In the program, ω is generated by the procedure discussed in Sec. 7.7. For each value of ω, the phase and magnitude contribution of the quadratic factor (denoted by the names QUANG and QUADLG, respectively) is calculated by the segment of the program from lines 00610 through 00670. The phase angle of a quadratic factor monotonically increases from zero and approaches 180° as ω increases from 0 rad/s. When $\omega = \omega_n$, the phase angle is 90°.[1] If the quadratic term in Eq. (7.8.3) is examined, it is seen that when $\omega = \omega_n$, the *denominator* of the arctangent function $[1 - (\omega/\omega_n)^2]$ is zero, and the arctangent function is infinite. The IF statement on line 00620 places QUANG in the first quadrant when the quantity $[1 - (\omega/\omega_n)^2]$ is positive, in the second quadrant when $[1 - (\omega/\omega_n)^2]$ is negative, and sets QUANG equal to 90° when $[1 - (\omega/\omega_n)^2]$ is zero.

The phase and magnitude contribution of the pole at the origin (denoted by the names ANGPOL and POLEDB, respectively) are calculated on lines 00600 and 00590. The contributions of the other denominator terms of the transfer function are calculated and summed in the nested DO loop extending from lines 00530 through 00570, and the contributions of the numerator terms are calculated and summed in the nested DO loop extending from lines 00470 through 00510. The variables TOPLOG, TOTHET, BOTLOG, and BOTHET (lines 00500, 00510, 00560, 00570) are used to store the phase and decibel contributions of FNUM(I), THETN(I), FDEN(K), and THETD(K) for each value of radian frequency. Each time the program returns to the outer DO loop, these variables are initialized at a value of zero (lines 00400 through 00430).

[1] R. Saucedo and E. E. Schiring, "Introduction to Continuous and Digital Control Systems," Macmillan, New York, 1968, p. 400.

In practice, every transfer function does not have all of the types of factors indicated in Eq. (7.8.1). *If there is no pole at the origin,* a value of zero must be entered for L, the order of the pole at the origin. This eliminates the contributions of POLEDB and ANGPOL (see lines 00580, 00590, and 00600). *If there is no quadratic factor*, values of 1.E33 and 0. are entered for WN and DEL, respectively. This effectively eliminates the contributions of QUANG and QUADLG (see lines 00610 through 00670). *If there are no numerator factors of type* $(1 + j\omega/\omega_i)$, a value of zero must be entered for M. In this event, the variable WNUM(I) is also zero. In order to avoid division by zero, when M = 0, the IF statement on line 00460 causes the nested DO loop which begins on line 00470 to be bypassed. Finally, *if there are no*

```
00001C      BØDE
00002C      THIS PRØGRAM CALCULATES THE MAGNITUDE IN DECIBELS,AND
00003C      THE PHASE ANGLE IN DEGREES, ØF A TRANSFER FUNCTIØN AS A
00004C      FUNCTIØN ØF RADIAN FREQUENCY. IF THE TRANSFER FUNCTIØN
00005C      DØES NØT HAVE A PØLE AT THE ØRIGIN, ENTER A VALUE ØF
00006C      ZERØ FØR L,THE ØRDER ØF THE PØLE. IF THE TRANSFER FUNCTIØN
00007C      DØES NØT HAVE A QUADRATIC TERM, ENTER A VALUE ØF 1.E33
00008C      FØR WN, THE NATURAL FREQUENCY, AND A VALUE ØF 0 FØR DEL,
00009C      THE DAMPING RATIØ.
00010       DIMENSIØN WNUM(10),WDEN(10),FNUM(10),FDEN(10),THETN(10)
00020       DIMENSIØN THETD(10)
00030       PRINT 5
00040     5 FØRMAT("ENTER NØ. ØF NUMERATØR CØRNERS + THEIR VALUES")
00050       READ,M,(WNUM(I),I=1,M)
00060       PRINT 7
00070     7 FØRMAT("ENTER NØ. ØF DENØMINATØR CØRNERS + THEIR VALUES")
00080       READ,N,(WDEN(K),K=1,N)
00090       PRINT 9
00100     9 FØRMAT(22HENTER THE SCALE FACTØR)
00110       READ,SF
00120       PRINT 11
00130    11 FØRMAT("IF NØ QUADRATIC TERM ENTER WN=1.E33 +DEL=0")
00140       PRINT 13
00150    13 FØRMAT(26HENTER THE QUADRATIC CØRNER)
00160       READ,WN
00170       PRINT 15
00180    15 FØRMAT(23HENTER THE DAMPING RATIØ)
00190       READ,DEL
00200       SCALØG=8.686*ALØG(SF)
00210       PRINT 19
00220    19 FØRMAT(41HENTER THE ØRDER ØF THE PØLE AT THE ØRIGIN)
00230       READ,L
00240       PRINT 17
00250    17 FØRMAT("     W       DECIBELS      PHASE")
00260       MIN=1
00270       MAX=10
00280       INC=1
00290       KØUNT=0
00300       GØ TØ 40
00310    70 MIN=20
00320       MAX=100
00330       INC=10
00340       GØ TØ 40
00350    90 MIN=200
00360       MAX=1000
00370       INC=100
00380       GØ TØ 40
00390    40 DØ 1 J=MIN,MAX,INC
00400       TØPLØG=0.
00410       BØTLØG=0.
00420       TØTHET=0.
00430       BØTHET=0.
00440       W=J
00450       KØUNT=KØUNT+1
00460       IF(M) 71,71,72
00470    72 DØ 2 I=1,M
```

```
00480 |     FNUM(I)=4.343*ALØG(1.+(W**2)/(WNUM(I)**2))
00490 |     THETN(I)=ATAN(W/WNUM(I))
00500 |     TØPLØG=TØPLØG+FNUM(I)
00510 |   2 TØTHET=TØTHET+THETN(I)
00520 |  71 IF(N) 73,73,74
00530 |  74 DØ 3 K=1,N
00540 |     FDEN(K)=4.343*ALØG(1.+(W**2)/(WDEN(K)**2))
00550 |     THETD(K)=ATAN(W/WDEN(K))
00560 |     BØTLØG=BØTLØG+FDEN(K)
00570 |   3 BØTHET=BØTHET+THETD(K)
00580 |  73 A=L
00590 |     PØLEDB=A*8.686*ALØG(W)
00600 |     ANGPØL=A*90.
00610 |     QUADLG=4.343*ALØG((1.-(W/WN)**2)**2+(2.*DEL*W/WN)**2)
00620 |     IF(1.-(W/WN)**2) 22,23,24
00630 |  22 QUANG=3.1416-ATAN(ABS((2.*DEL*W/WN)/(1.-(W/WN)**2)))
00640 |     GØ TØ 29
00650 |  23 QUANG=1.5708
00660 |     GØ TØ 29
00670 |  24 QUANG=ATAN(ABS((2.*DEL*W/WN)/(1.-(W/WN)**2)))
00680 |  29 ANGTØT=57.3*(TØTHET-BØTHET-QUANG)-ANGPØL
00690 |     DBTØT=SCALØG+TØPLØG-BØTLØG-QUADLG-PØLEDB
00700 |     PRINT 21,W,DBTØT,ANGTØT
00710 |  21 FØRMAT(I5,5X,F6.1,5X,F6.1)
00720 |   1 CØNTINUE
00730 |     IF(KØUNT-10) 40,70,80
00740 |  80 IF(KØUNT-19) 40,90,50
00750 |  50 STØP
00760 |     END
```

Fig. 7.8.1 Program BODE.

denominator factors of type $(1 + j\omega/\omega_k)$, a value of zero must be entered for N. In this case, the variable WDEN(K) is zero. In order to avoid division by zero, when $N = 0$, the IF statement on line 00520 causes the nested DO loop which begins on line 00530 to be bypassed. A flow chart for the program is shown in Fig. 7.8.3.

Trial runs are shown in Fig. 7.8.2*a, b, c,* and *d*. The transfer functions which are evaluated are:

$$\frac{2}{1 + j\omega/10} \qquad (a)$$

$$\frac{10(1 + j\omega/10)}{(1 + j\omega/1)\,(1 + j\omega/100)} \qquad (b)$$

$$\frac{1}{(j\omega/10)^2 + 2(0.2)j\omega/10 + 1} \qquad (c)$$

$$\frac{5(1 + j\omega/1.67)}{(j\omega)^2\,(1 + j\omega/0.25)} \qquad (d)$$

```
ENTER NØ. ØF NUMERATØR CORNERS + THEIR VALUES
INPUT:00050
 ?0,0

ENTER NØ. ØF DENØMINATØR CØRNERS + THEIR VALUES
INPUT:00080
 ?1,10

ENTER THE SCALE FACTØR
INPUT:00110
 ?2.

IF NØ QUADRATIC TERM ENTER WN=1.E33 +DEL=0
ENTER THE QUADRATIC CØRNER
INPUT:00160
 ?1.E33

ENTER THE DAMPING RATIØ
INPUT:00190
 ?0.

ENTER THE ØRDER ØF THE PØLE AT THE ØRIGIN
INPUT:00230
 ?0

     W    DECIBELS     PHASE
     1       6.0        -5.7
     2       5.9       -11.3
     3       5.6       -16.7
     4       5.4       -21.8
     5       5.1       -26.6
     6       4.7       -31.0
     7       4.3       -35.0
     8       3.9       -38.7
     9       3.4       -42.0
    10       3.0       -45.0
    20      -1.0       -63.4
    30      -4.0       -71.6
    40      -6.3       -76.0
    50      -8.1       -78.7
    60      -9.7       -80.5
    70     -11.0       -81.9
    80     -12.1       -82.9
    90     -13.1       -83.7
   100     -14.0       -84.3
   200     -20.0       -87.1
   300     -23.5       -88.1
   400     -26.0       -88.6
   500     -28.0       -88.9
   600     -29.5       -89.1
   700     -30.9       -89.2
   800     -32.0       -89.3
   900     -33.1       -89.4
  1000     -34.0       -89.4
STØP
```

Fig. 7.8.2a Trial runs for BODE.

```
ENTER NØ. ØF NUMERATØR CØRNERS + THEIR VALUES
INPUT:00050
 ?1,10

ENTER NØ. ØF DENØMINATØR CØRNERS + THEIR VALUES
INPUT:00080
 ?2,1.,100.

ENTER THE SCALE FACTØR
INPUT:00110
 ?10.

IF NØ QUADRATIC TERM ENTER WN=1.E33 +DEL=0
ENTER THE QUADRATIC CØRNER
INPUT:00160
 ?1.E33

ENTER THE DAMPING RATIØ
INPUT:00190
 ?0.

ENTER THE ØRDER ØF THE PØLE AT THE ØRIGIN
INPUT:00230
 ?0

     W      DECIBELS     PHASE
     1        17.0        -39.9
     2        13.2        -53.3
     3        10.4        -56.6
     4         8.3        -56.5
     5         6.8        -55.0
     6         5.6        -53.0
     7         4.7        -50.9
     8         4.0        -48.8
     9         3.4        -46.8
    10         2.9        -45.0
    20         0.8        -35.0
    30         0.1        -33.2
    40         -.4        -34.4
    50         -.8        -36.7
    60        -1.2        -39.5
    70        -1.6        -42.3
    80        -2.1        -45.1
    90        -2.5        -47.7
   100        -3.0        -50.1
   200        -7.0        -66.0
   300       -10.0        -73.3
   400       -12.3        -77.3
   500       -14.1        -79.7
   600       -15.7        -81.4
   700       -17.0        -82.6
   800       -18.1        -83.5
   900       -19.1        -84.2
  1000       -20.0        -84.8
STØP
```

Fig. 7.8.2b Trial runs for BODE.

```
ENTER NØ. ØF NUMERATØR CØRNERS + THEIR VALUES
INPUT:00050
 ?0,0

ENTER NØ. ØF DENØMINATØR CØRNERS + THEIR VALUES
INPUT:00080
 ?0,0

ENTER THE SCALE FACTØR
INPUT:00110
 ?1.

IF NØ QUADRATIC TERM ENTER WN=1.E33 +DEL=0
ENTER THE QUADRATIC CØRNER
INPUT:00160
 ?10.

ENTER THE DAMPING RATIØ
INPUT:00190
 ?.2

ENTER THE ØRDER ØF THE PØLE AT THE ØRIGIN
INPUT:00230
 ?0

      W      DECIBELS      PHASE
      1          0.1          -2.3
      2          0.3          -4.8
      3          0.7          -7.5
      4          1.4         -10.8
      5          2.2         -14.9
      6          3.3         -20.6
      7          4.7         -28.8
      8          6.3         -41.6
      9          7.8         -62.2
     10          8.0         -90.0
     20         -9.8        -165.1
     30        -18.2        -171.5
     40        -23.6        -173.9
     50        -27.6        -175.2
     60        -30.9        -176.1
     70        -33.6        -176.7
     80        -36.0        -177.1
     90        -38.1        -177.4
    100        -39.9        -177.7
    200        -52.0        -178.9
    300        -59.1        -179.2
    400        -64.1        -179.4
    500        -68.0        -179.6
    600        -71.1        -179.6
    700        -73.8        -179.7
    800        -76.1        -179.7
    900        -78.2        -179.8
   1000        -80.0        -179.8
STØP
```

Fig. 7.8.2c Trial runs for BODE.

```
ENTER NØ. ØF NUMERATØR CØRNERS + THEIR VALUES
INPUT:00050
 ?1,1.67

ENTER NØ. ØF DENØMINATØR CØRNERS + THEIR VALUES
INPUT:00080
 ?1,.25

ENTER THE SCALE FACTØR
INPUT:00110
 ?5.

IF NØ QUADRATIC TERM ENTER WN=1.E33 +DEL=0
ENTER THE QUADRATIC CØRNER
INPUT:00160
 ?1.E33

ENTER THE DAMPING RATIØ
INPUT:00190
 ?0.

ENTER THE ØRDER ØF THE PØLE AT THE ØRIGIN
INPUT:00230
 ?2

     W     DECIBELS     PHASE
     1        3.0      -225.1
     2      -12.3      -212.7
     3      -20.5      -204.3
     4      -25.9      -199.1
     5      -30.0      -195.6
     6      -33.3      -193.2
     7      -36.1      -191.4
     8      -38.5      -190.0
     9      -40.5      -188.9
    10      -42.4      -188.0
    20      -54.5      -184.1
    30      -61.6      -182.7
    40      -66.6      -182.0
    50      -70.5      -181.6
    60      -73.6      -181.4
    70      -76.3      -181.2
    80      -78.6      -181.0
    90      -80.7      -180.9
   100      -82.5      -180.8
   200      -94.6      -180.4
   300     -101.6      -180.3
   400     -106.6      -180.2
   500     -110.5      -180.2
   600     -113.6      -180.1
   700     -116.3      -180.1
   800     -118.6      -180.1
   900     -120.7      -180.1
  1000     -122.5      -180.1
STØP
```

Fig. 7.8.2d Trial runs for BODE.

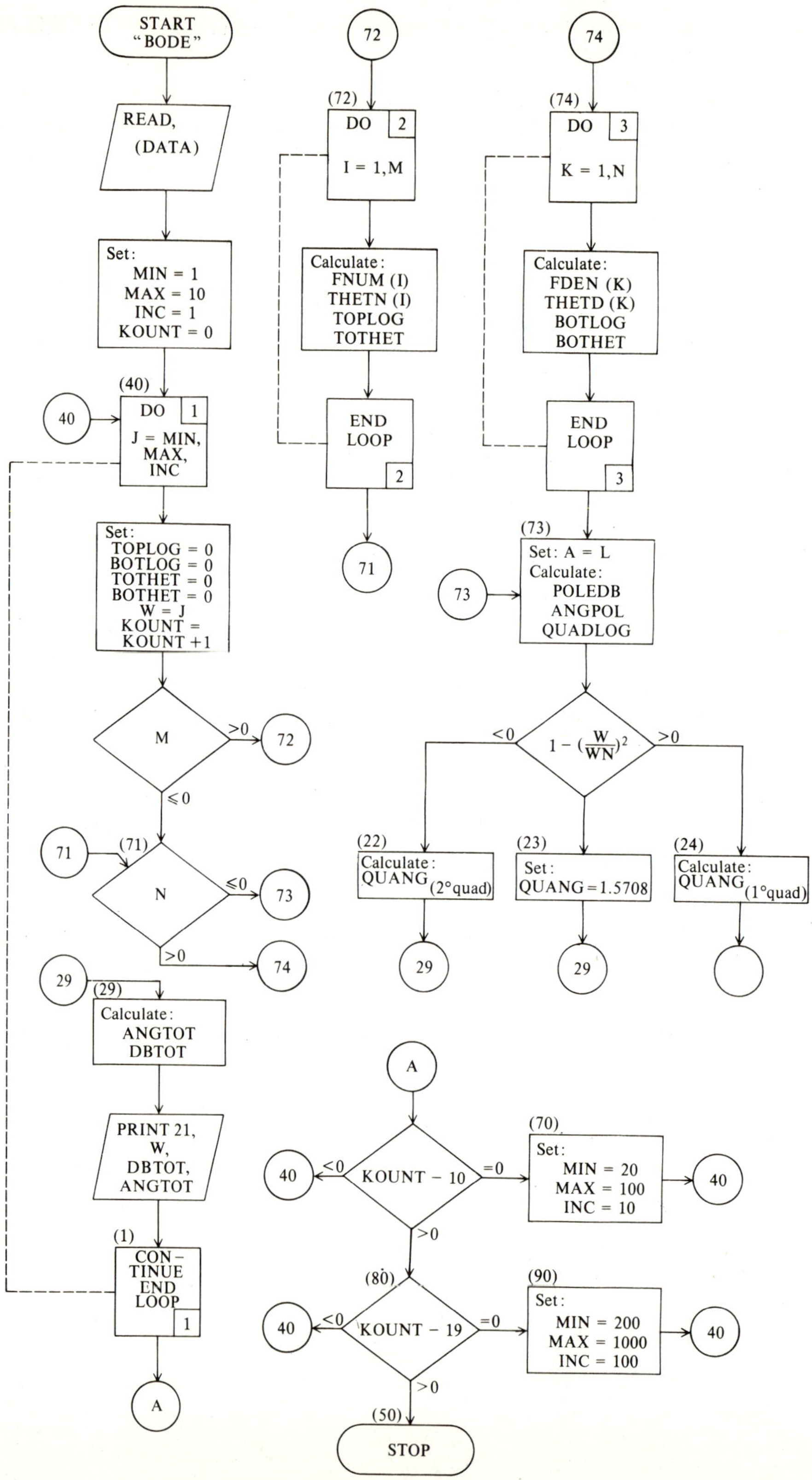

Fig. 7.8.3 Flow chart of program BODE.

7.9 THE PROBABILITY DENSITY OF THE SUM OF TWO RANDOM VARIABLES[1]

The probability density function $p(i)$ of the sum Y of two independent discrete random variables A and B which have only nonnegative integers for possible values is

$$p(i) = \sum_{j=1}^{i} a_j b_{i-j} \tag{7.9.1}$$

The right side of Eq. (7.9.1) is called a convolution.

Let the values which random variable A may assume be denoted by a and the values which random variable B may assume be denoted by b. As a specific example, suppose the following table of probabilities is available:

a	$P(A = a)$	b	$P(B = b)$
0	0.4	0	0.3
1	0.3	1	0.2
2	0.2	2	0.2
3	0.1	3	0.2
		4	0.1

If Y is the random variable which is the sum of random variables A and B, we can calculate the probabilities of the various values which Y may assume. For example, $P(Y = 0) = (0.4)(0.3) = 0.12$, and $P(Y = 1) = (0.4)(0.2) + (0.3)(0.3) = 0.17$. The calculation of other probabilities is similar.

The program of Fig. 7.9.1 carries out these calculations. The various probabilities of A and B are entered by means of unformatted input statements on lines 00040 and 00050. The name MAX (line 00070) represents the maximum number of values

```
00001C     CØNVØ
00002C     THIS PRØGRAM CALCULATES THE PRØBABILITY DENSITY FUNCTIØN
00003C     ØF THE SUM ØF TWØ INDEPENDENT DISCRETE RANDØM VARIABLES.
00010      DIMENSIØN A(20),B(20),P(20)
00020      READ,NA
00030      READ,NB
00040      READ,(A(K),K=1,NA)
00050      READ,(B(L),L=1,NB)
00060      SUM=0.0
00070      MAX=NA+NB-1
00080      DØ 9 I=1,MAX
00090      DØ 9 J=1,I
00100      M=I-J+1
00110    9 P(I)=P(I)+A(J)*B(M)
00120      PRINT 20,(P(I),I=1,MAX)
00130   20 FØRMAT(//11F6.2)
00140      DØ 11 I=1,MAX
00150   11 SUM=SUM+P(I)
00160      PRINT 90,SUM
00170   90 FØRMAT(F6.2//)
00180      STØP
00190      END
```

Fig. 7.9.1 Program CONVO.

[1] T. J. Healy, Convolution Revisited, *IEEE Spectrum,* April 1969. Also: J. R. McCord, III and R. M. Moroney, Jr., "Introduction to Probability Theory," Macmillan, New York, 1964, p. 171.

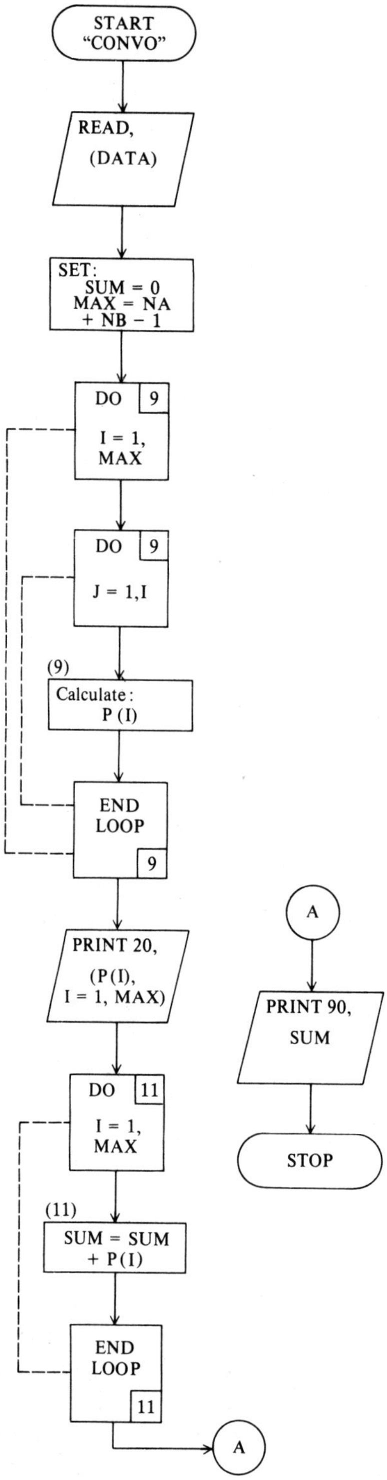

Fig. 7.9.2 Flow chart of program CONVO.

```
INPUT:00020
 ?4

INPUT:00030
 ?5

INPUT:00040
 ?.4,.3,.2,.1

INPUT:00050
 ?.3,.2,.2,.2,.1


  0.12  0.17  0.20  0.21  0.16  0.09  0.04  0.01
  1.00


STØP
```

Fig. 7.9.3 Trial run for CONVO.

which the random variable Y may assume, and the subscripted variable P(I) represents the probability of each of these values. For example, P(1) = 0.12 and P(2) = 0.17 in accordance with the calculations of the previous paragraph. Notice that the subscripts of the variable P do *not* represent the values which Y may assume. Thus, P(1) represents the probability that $Y = 0$, P(2) represents the probability that $Y = 1$, and so forth. As a check on the results, the DO loop DO 11 sums the values of P(I) to ascertain that they add to 1., as they should.

The operation of the program has been diagrammed in Fig. 7.9.2 and a trial run is shown in Fig. 7.9.3.

7.10 THE ADDITION OF MATRICES[1]

Two matrices $\bar{A}$ and $\bar{B}$ are said to be conformable for addition if they are of the same order. Let the sum of $\bar{A}$ and $\bar{B}$ be $\bar{C}$. The rule for calculating each element of $\bar{C}$, c_{ij}, is

$$c_{ij} = a_{ij} + b_{ij}$$

For example, $\bar{A}$ and $\bar{B}$ are added as follows:

$$\bar{A} = \begin{bmatrix} 1 & 2 & 3 \\ 0 & 1 & 0 \\ 5 & 1 & -1 \end{bmatrix} \qquad \bar{B} = \begin{bmatrix} 2 & 4 & 1 \\ -1 & 0 & 1 \\ 1 & 1 & 1 \end{bmatrix}$$

$$\bar{C} = \bar{A} + \bar{B} = \begin{bmatrix} 3 & 6 & 4 \\ -1 & 1 & 1 \\ 6 & 2 & 0 \end{bmatrix}$$

A program which performs matrix addition is shown in Fig. 7.10.1, and a trial run based upon the example given above is shown in Fig. 7.10.2. The values of the elements of matrices $\bar{A}$ and $\bar{B}$ are entered using unformatted input (lines 00030 and 00040). The calculation of matrix $\bar{C}$, which is the sum of $\bar{A}$ and $\bar{B}$, is performed by the pair of nested DO loops (lines 00050 through 00070). It may be noted that the results of the trial run are in agreement with the example given above. Matrix addition is used in the program discussed in Sec. 7.12.

[1] J. M. Gere and W. Weaver, Jr., "Matrix Algebra for Engineers," Van Nostrand, Princeton, N.J., 1965, p. 14.

```
C      ADDMA1
C      THIS PRØGRAM ADDS TWØ MATRICES ØF THE SAME ØRDER.
       DIMENSIØN A(10,10),B(10,10),C(10,10)
       READ,M,N
       READ,((A(I,J),J=1,N),I=1,M)
       READ,((B(I,J),J=1,N),I=1,M)
       DØ 1 I=1,M
       DØ 1 J=1,N
     1 C(I,J)=A(I,J)+B(I,J)
       DØ 17 I=1,M
    17 PRINT 27,(C(I,J),J=1,N)
    27 FØRMAT(/5E12.3)
       STØP
       END
```

Fig. 7.10.1 Program ADDMA1.

```
INPUT:00020
 ?3,3

INPUT:00030
 ?1,2,3,0,1,0,5,1,-1

INPUT:00040
 ?2,4,1,-1,0,1,1,1,1

     0.300E+01   0.600E+01   0.400E+01

     -.100E+01   0.100E+01   0.100E+01

     0.600E+01   0.200E+01   0.000E+00
STØP
```

Fig. 7.10.2 Trial run for ADDMA1.

7.11 TEMPERATURE DISTRIBUTION IN A PLATE

The FORTRAN program discussed in this section illustrates the use of two-dimensional arrays. The program has been designed to determine the temperatures at selected points in a thin plate after thermal equilibrium has been established. The

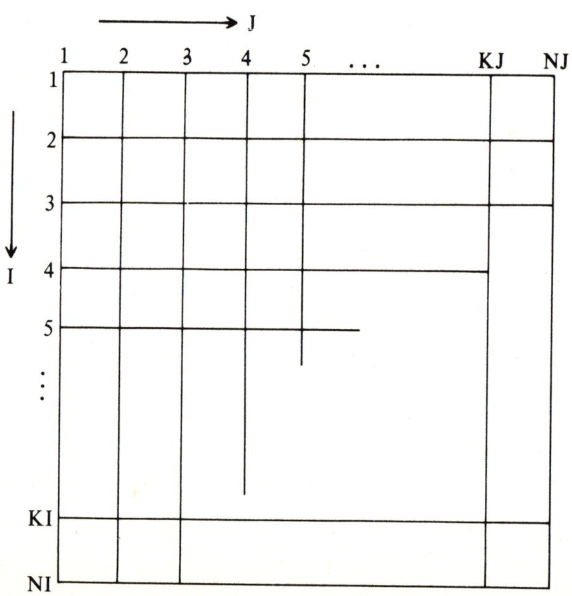

Fig. 7.11.1 Imaginary square grid on plate.

selected points are the *nodal points* determined by placing an imaginary square grid on the plate as shown in Fig. 7.11.1. On this grid, the *columns* are designed by j and the *rows* by i. In Fig. 7.11.1, and in the computer program, we have used the names J and I to designate the column and row, respectively. Note that the grid contains NI times NJ nodes and that KI and KJ are the next-to-last row and column, respectively.

Using finite difference methods, an equation can be derived[1] for evaluating the temperature of the material surrounding the node (shaded area in Fig. 7.11.2) for a typical *interior point* in the grid:[2]

$$T_o = \frac{T_1 + T_2 + T_3 + T_4}{4} \tag{7.11.1}$$

Writing this equation in terms of a general point in the grid of Fig. 7.11.1, we have

$$T_{ij} = \frac{T_{i,j+1} + T_{i+1,j} + T_{i,j-1} + T_{i-1,j}}{4} \tag{7.11.2}$$

We shall assume the four edges of the plate are *maintained at constant temperatures*, though not necessarily at the same temperature. In the program shown in Fig. 7.11.4, the names TT and TB have been used to store the temperatures of the *top* and *bottom* edges, respectively. The names TL and TR have been similarly used for the *left* and *right* edges of the plate.

The calculation procedure outlined in the flow chart of Fig. 7.11.3 is based on a numerical method called relaxation and is as follows:

1. Read-in the data and establish the size of the grid.
2. Initialize all the grid points (array T) at zero (arbitrarily).[3]
3. Set the temperatures on the isothermal edges.
4. Calculate the new temperatures (array TNEW) at all the interior grid points using Eq. (7.11.2).
5. Compare the new temperature with the old temperature at *each* grid point to determine if the difference is less than some arbitrarily chosen ϵ.

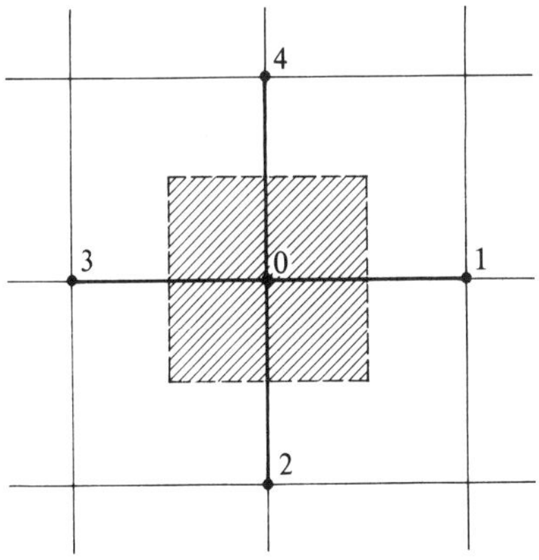

Fig. 7.11.2 Typical interior point in the grid.

[1] F. Kreith, "Principles of Heat Transfer," International Textbook. Scranton, Pa., 1958, p. 97.

[2] This same equation represents the solution for the potential field in a charge-free bounded region.

[3] Convergence is assured by this method regardless of the initial values chosen. However, the choice of initial value may influence the number of iterations required to achieve convergence.

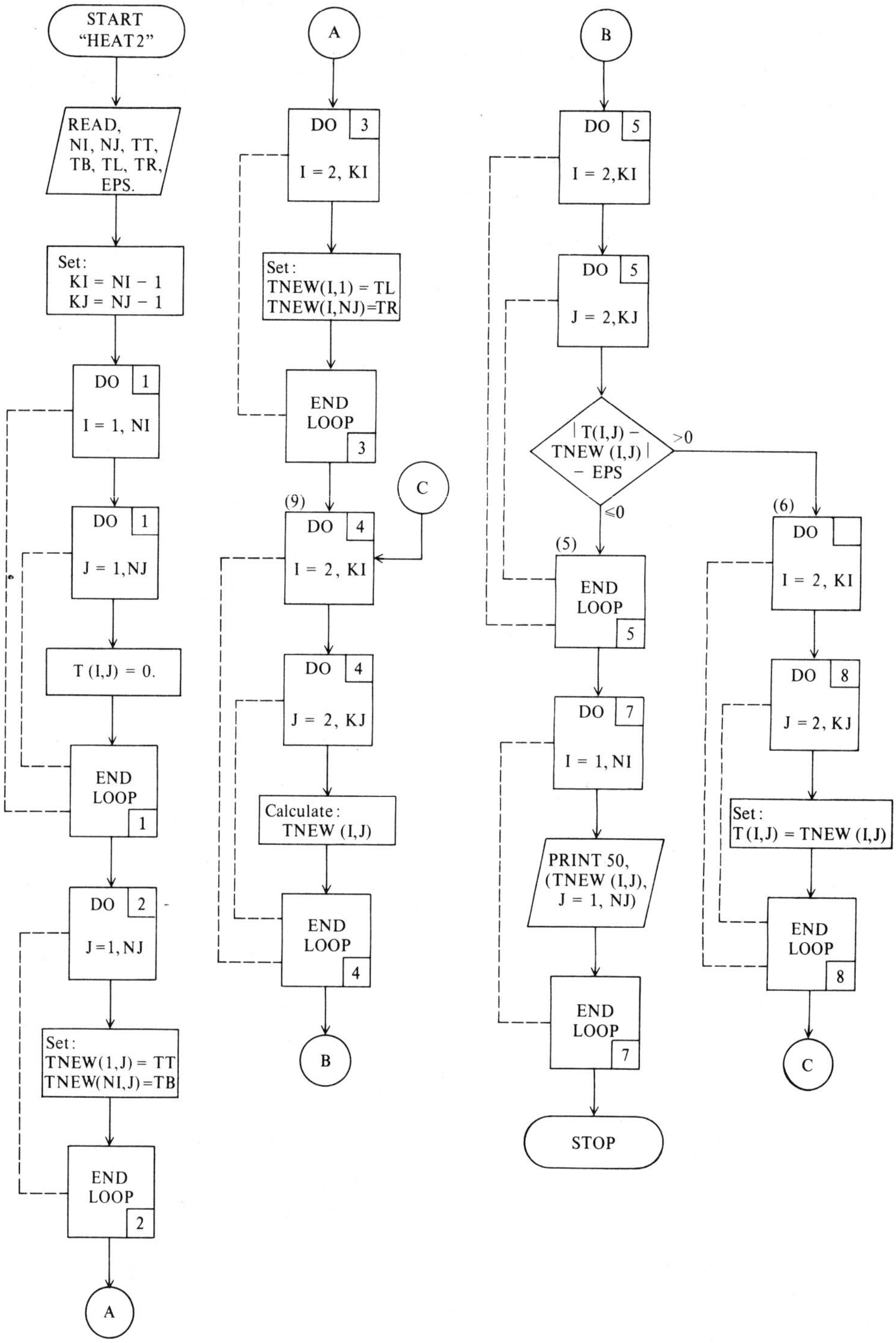

Fig. 7.11.3 Flow chart of program HEAT2.

```
C     HEAT2
C     PRØGRAM TØ CALCULATE THE STEADY STATE TEMPERATURE
C     DISTRIBUTIØN IN A PLATE
      DIMENSIØN T(20,20),TNEW(20,20)
      READ,NI,NJ,TT,TB,TR,TL,EPS
      KI=NI-1
      KJ=NJ-1
      DØ 1 I=1,NI
      DØ 1 J=1,NJ
    1 T(I,J)=0.
      DØ 2 J=1,NJ
      TNEW(1,J)=TT
      TNEW(NI,J)=TB
      T(1,J)=TT
    2 T(NI,J)=TB
      DØ 3 I=2,KI
      TNEW(I,1)=TL
      TNEW(I,NJ)=TR
      T(I,1)=TL
    3 T(I,NJ)=TR
    9 DØ 4 I=2,KI
      DØ 4 J=2,KJ
    4 TNEW(I,J)=(T(I+1,J)+T(I-1,J)+T(I,J+1)+T(I,J-1))/4.
      DØ 5 I=2,KI
      DØ 5 J=2,KJ
      IF(ABS(T(I,J)-TNEW(I,J))-EPS)5,5,6
    5 CØNTINUE
      DØ 7 I=1,NI
    7 PRINT 50,(TNEW(I,J),J=1,NJ)
   50 FØRMAT(/10F6.1)
      STØP
    6 DØ 8 I=2,KI
      DØ 8 J=2,KJ
    8 T(I,J)=TNEW(I,J)
      GØ TØ 9
      END
```

Fig. 7.11.4 Program HEAT2.

6. Continue the comparison, point by point, as long as the differences are less than ϵ.
7. If *any* pair has a difference greater than ϵ, cease the comparison procedure and immediately go to step 9 below.
8. If *all* the pairs have differences less than ϵ, the relaxation process is completed. Print the results in matrix form and stop.
9. Store all the values of TNEW for the interior points in the corresponding locations in array T and return to step 4. Note that only the interior points are "swapped." *The temperatures at the edges do not change.*

```
INPUT:00020
?8,10,250.,180.,100.,200.,0.2

 250.0 250.0 250.0 250.0 250.0 250.0 250.0 250.0 250.0 250.0

 200.0 222.3 229.4 230.9 229.7 226.1 219.1 205.3 175.4 100.0

 200.0 209.9 214.5 215.0 212.2 206.1 195.2 176.9 146.6 100.0

 200.0 203.2 204.3 202.9 198.9 191.4 179.4 160.9 134.3 100.0

 200.0 198.8 197.1 194.3 189.7 182.3 170.8 153.8 129.9 100.0

 200.0 195.1 191.7 188.4 184.2 178.0 168.6 154.0 131.8 100.0

 200.0 190.4 186.4 184.0 181.4 177.8 172.1 162.3 143.5 100.0

 180.0 180.0 180.0 180.0 180.0 180.0 180.0 180.0 180.0 180.0
STØP
```

Fig. 7.11.5 Trial run for HEAT2.

The program shown in Fig. 7.11.4 carries out the steps described above. The DIMENSION statement on line 00010 provides for a *maximum* size grid that is 20 X 20. In the READ statement on line 00020, the user enters the *actual* size of the grid to be used as NI and NJ. The FORTRAN statement on line 00200 represents the calculation discussed above in Eq. (7.11.2). The output from a sample run is shown in Fig. 7.11.5.

7.12 COST ANALYSIS

The problem programmed in this section is an example of one method of storing and handling data in triple-subscripted arrays. The problem treated is a simple cost analysis. Cost and usage records for two separate computer systems were kept for a period of one year. Both systems were available to six departments in one division of a company. The program is intended to:

1. Make the following calculations for *each* system:
   a. Calculate the total amount spent by all departments for each month
   b. Calculate the total amount spent by each department during the year
   c. Calculate the total amount spent by the division during the year
   d. Calculate the total time the system was used by all departments during each month
   e. Calculate the total time the system was used by each department during the year
   f. Calculate the total time the system was used by the division during the year
2. Take the above results for each system and combine them in all categories for *both* systems (a simple matrix addition operation; see Sec. 7.10).
3. Calculate the cost ratio (dollars spent per hour of use) for *each* department, *and* for the entire division, for *both* systems and for the combination of the two systems.
4. Produce a summary of the totals for the entire division.

With six departments operating for 12 months it is obvious that the money spent on *one* of the systems can be stored in an array having 6 columns and 12 rows. To store the monthly totals (amount spent by the division each month) a *seventh* column was added to the matrix. To obtain the annual expenditures of each department a *thirteenth* row was added. The annual total for the division was stored in the thirteenth row of the seventh column. The reader is referred to Fig. 7.12.4*a*, which is part of the output of a sample run made with this program and should serve to clarify the above description of the storage method employed. (While the reader is examining this figure it may be interesting to note that department 2 made negligible use of system A. Although a charge appears in the COST table on two occasions, the time interval involved was too small to round up to 0.01 h and therefore does not appear for the corresponding months in the HOURS table.)

An identical 7 X 13 matrix was used for the second system, and a third matrix was used for the sum of the two systems. Therefore, we actually have *three* 7 X 13 matrices which we have chosen to store in a *single* three-dimensional array named C. The dimensions of this triple-subscripted array are 3 X 7 X 13.

An alternative approach to this storage problem would have been to assign different names to each of the three 7 X 13 matrices, for example, A(7,13) for the costs incurred with system A, B(7,13) for the costs of system B, and C(7,13) for the combined costs. As a matter of fact, we actually did select this option when it came

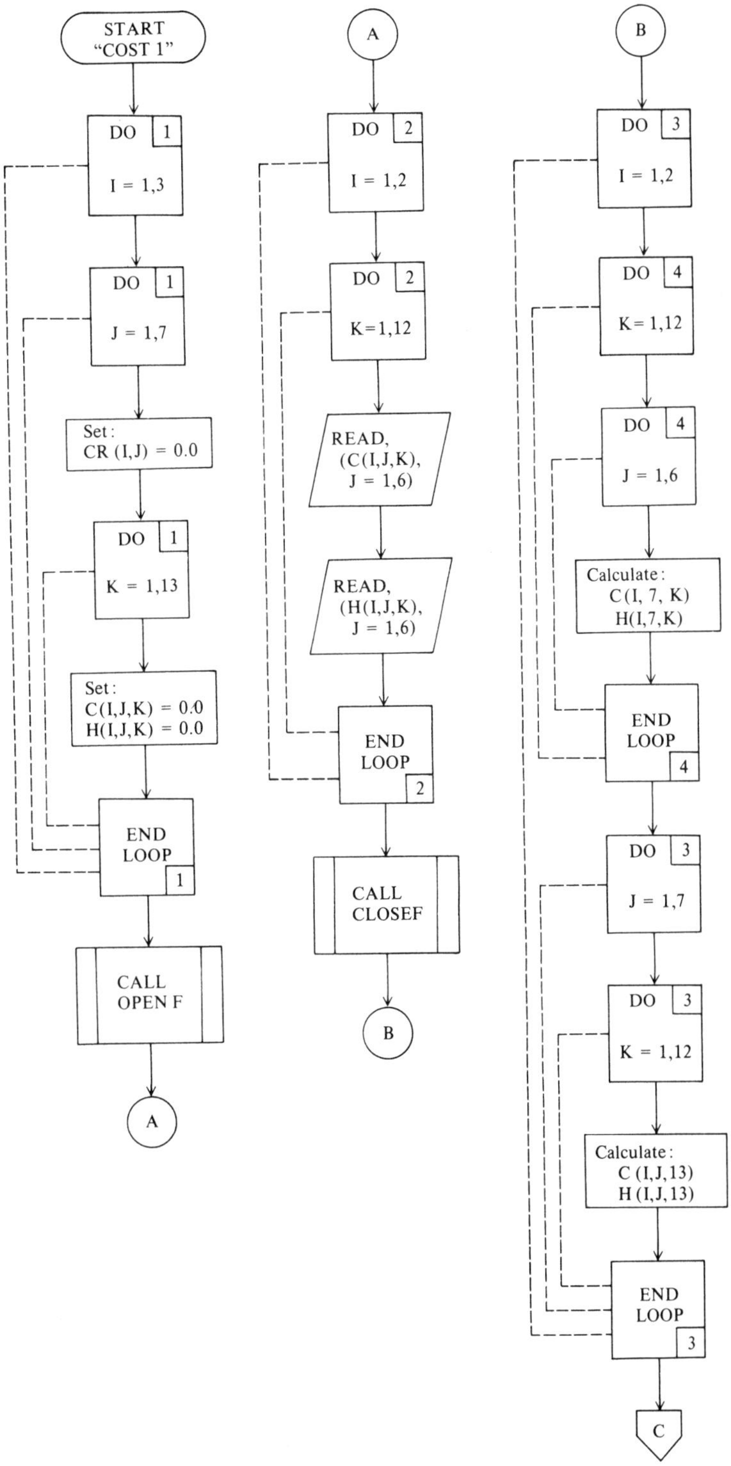

Fig. 7.12.1a Flow chart of program COST1.

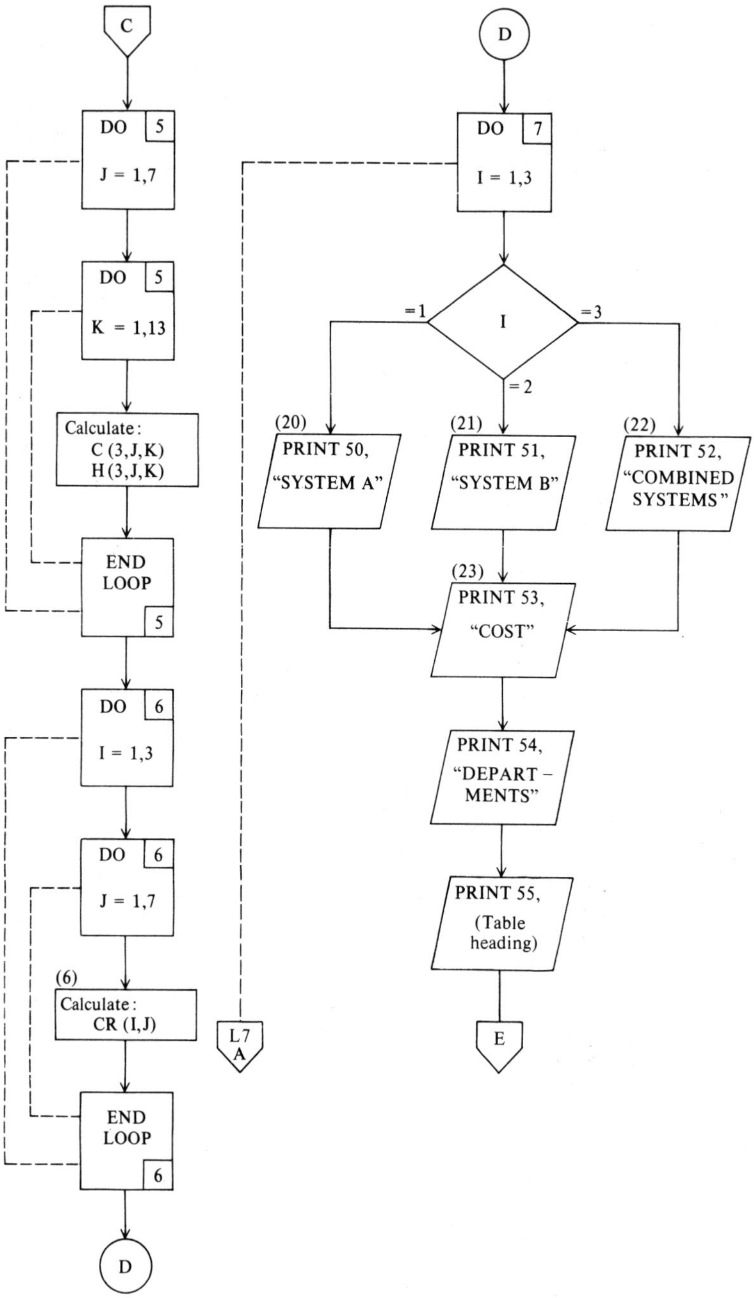

Fig. 7.12.1b Flow chart of program COST1.

time to also store the records of the usage time. A separate three-dimensional array, H(3,7,13), was used for storing the hours of use. Because *four*-dimensional arrays are *not standard* in the FORTRAN language, we could not store COST and HOURS in a single array dimensioned, for example, as X(2,3,7,13), where the 2 would create *two* three-dimensional arrays, one for COST and one for HOURS. (Some computer systems permit arrays with more than three dimensions, and in these systems, our hypothetical X array would be permitted.)

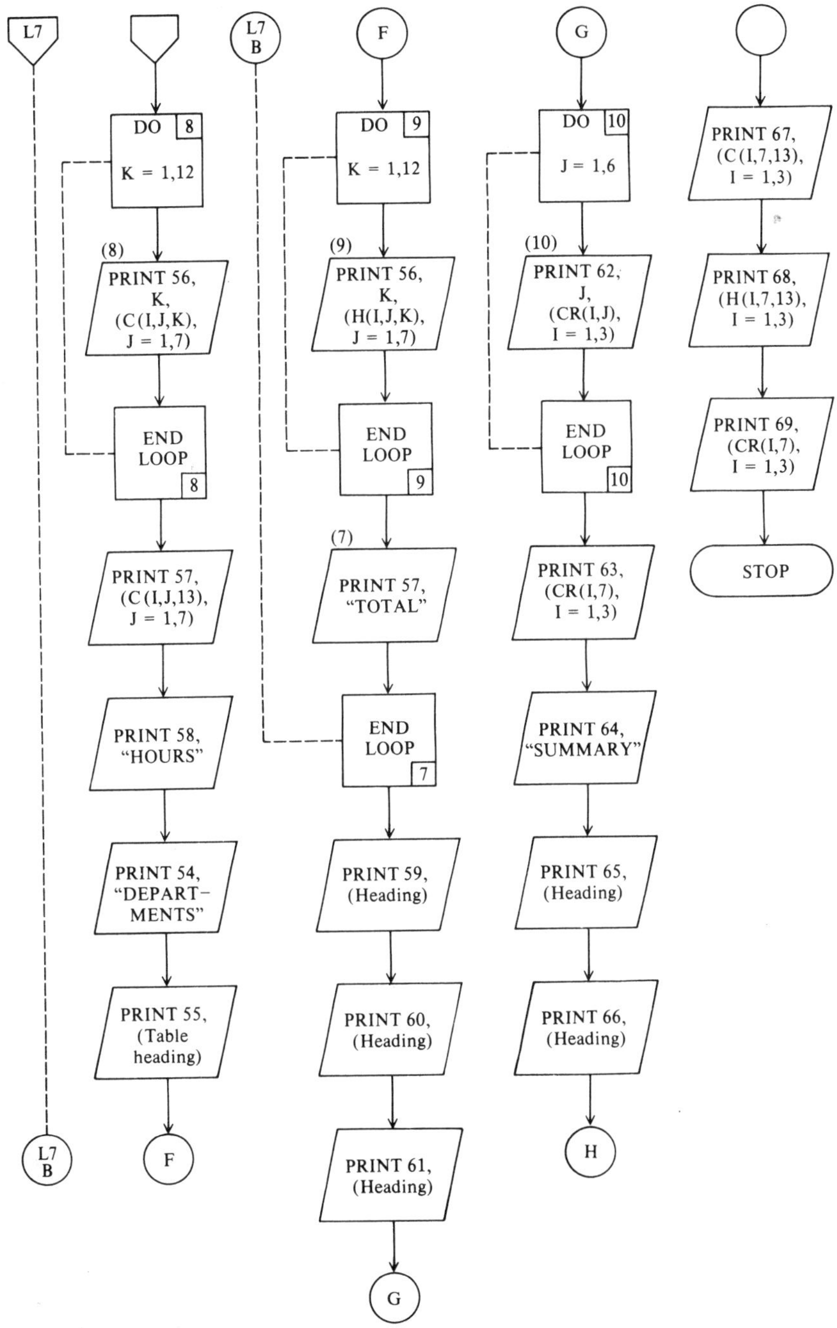

Fig. 7.12.1c Flow chart of program COST1.

Another approach to storing COST and HOURS in a *single* array would be to increase the *size* of one of the dimensions rather than the *number of dimensions.* That is, we could have stored both C(3,7,13) and H(3,7,13) in a single array, Y(6,7,13).

Figure 7.12.1 shows the flow chart for the calculation procedure outlined above. The symbols used are defined in Table 7.12.1.

Table 7.12.1

Name	*Meaning*	*Units*
I	Index for the two systems *and* the combination of the two (extent: 1 to 3)	
J	Index for the six departments *and* the monthly totals (extent: 1 to 7)	
K	Index for the 12 months *and* the annual totals (extent: 1 to 13)	
C(I,J,K)	Cost of operating	$
H(I,J,K)	Amount of time system was used	h
CR(I,J)	Cost ratio	$/h

```
C      CØST1
C      PRØGRAM TØ ANALYZE ØPERATIØNAL CØSTS FØR TWØ CØMPUTER
C      SYSTEMS USED BY SEVERAL DEPARTMENTS IN A DIVISIØN ØF
C      A CØMPANY
       DIMENSIØN C(3,7,13),H(3,7,13),CR(3,7)
       DØ 1 I=1,3
       DØ 1 J=1,7
       CR(I,J)=0.0
       DØ 1 K=1,13
       C(I,J,K)=0.0
     1 H(I,J,K)=0.0
       CALL ØPENF(1,"DATA")
       DØ 2 I=1,2
       DØ 2 K=1,12
       READ(1,)(C(I,J,K),J=1,6)
     2 READ(1,)(H(I,J,K),J=1,6)
       CALL CLØSEF(1,"DATA")
       DØ 3 I=1,2
       DØ 4 K=1,12
       DØ 4 J=1,6
       C(I,7,K)=C(I,7,K)+C(I,J,K)
     4 H(I,7,K)=H(I,7,K)+H(I,J,K)
       DØ 3 J=1,7
       DØ 3 K=1,12
       C(I,J,13)=C(I,J,13)+C(I,J,K)
     3 H(I,J,13)=H(I,J,13)+H(I,J,K)
       DØ 5 J=1,7
       DØ 5 K=1,13
       C(3,J,K)=C(1,J,K)+C(2,J,K)
     5 H(3,J,K)=H(1,J,K)+H(2,J,K)
       DØ 6 I=1,3
       DØ 6 J=1,7
     6 CR(I,J)=C(I,J,13)/H(I,J,13)
       DØ 7 I=1,3
       GØ TØ (20,21,22),I
    20 PRINT 50
       GØ TØ 23
    21 PRINT 51
       GØ TØ 23
    22 PRINT 52
    23 PRINT 53
       PRINT 54
       PRINT 55
       DØ 8 K=1,12
     8 PRINT 56,K,(C(I,J,K),J=1,7)
       PRINT 57,(C(I,J,13),J=1,7)
       PRINT 58
       PRINT 54
       PRINT 55
       DØ 9 K=1,12
     9 PRINT 56,K,(H(I,J,K),J=1,7)
     7 PRINT 57,(H(I,J,13),J=1,7)
       PRINT 59
       PRINT 60
       PRINT 61
       DØ 10 J=1,6
```

```
00530  10 PRINT 62,J,(CR(I,J),I=1,3)
00540     PRINT 63,(CR(I,7),I=1,3)
00550     PRINT 64
00560     PRINT 65
00570     PRINT 66
00580     PRINT 67,(C(I,7,13),I=1,3)
00590     PRINT 68,(H(I,7,13),I=1,3)
00600     PRINT 69,(CR(I,7),I=1,3)
00610  50 FØRMAT(///8HSYSTEM A/)
00620  51 FØRMAT(///8HSYSTEM B/)
00630  52 FØRMAT(///16HCØMBINED SYSTEMS/)
00640  53 FØRMAT(4HCØST)
00650  54 FØRMAT(22X11HDEPARTMENTS)
00660  55 FØRMAT(5HMØNTH4X1H18X1H28X1H38X1H48X1H58X1H66X5HTØTAL)
00670  56 FØRMAT(I3,2X7F9.2)
00680  57 FØRMAT(5HTØTAL7F9.2/)
00690  58 FØRMAT(5HHØURS)
00700  59 FØRMAT(///25HCØST RATIØ (DØLLARS/HØUR)/)
00710  60 FØRMAT(12X6HSYSTEM)
00720  61 FØRMAT(4HDEPT6X1HA9X1HB5X8HCØMBINED)
00730  62 FØRMAT(I3,2X3F10.3)
00740  63 FØRMAT(/5HTØTAL3F10.3/)
00750  64 FØRMAT(///7HSUMMARY/)
00760  65 FØRMAT(26X6HSYSTEM)
00770  66 FØRMAT(24X1HA9X1HB5X8HCØMBINED/)
00780  67 FØRMAT(10HTØTAL CØST8X3F10.3/)
00790  68 FØRMAT(11HTØTAL HØURS7X3F10.3/)
00800  69 FØRMAT(18HØVERALL CØST RATIØ3F10.3//)
00810     STØP
00820     END
```

Fig. 7.12.2 Program COST1.

The program used for solving this problem is shown in Fig. 7.12.2. DO loop 1, lines 00020 through 00070, initializes all the arrays at zero. On line 00080, a subroutine (see Sec. 10.6) was used to "open" a file containing the data for the problem. This special technique was used because of the large amount of data needed as input for this problem. The data were stored in the computer on a file named DATA (see Fig. 7.12.3). Entering all this data in the usual manner from the teletype, one line at a time, would have required establishing communication between the computer and the teletype on 54 occasions (the number of lines of data). Each occasion would consume several seconds, and the execution time for the program would be greatly increased.

The CALL statement on line 00080 assigned the number 1 as the *device code number* of the file named DATA. In DO loop 2, lines 00090 through 00120, the two READ statements direct the computer to read-in data from *device number* 1, that is, from the file named DATA (see Sec. 2.1, Part B). When all the data were entered, the subroutine on line 00130 "closed" the file and the *link* between the program and that file was broken.

DO loop 3 starts on line 00140 and selects *one* of the two systems, A or B. The nested loops extending from line 00150 to 00180 calculate the *monthly* costs (and hours) for each department for the system being examined. The DO loops starting on line 00190 and ending on 00220 calculate the *annual* costs (and hours) for each department for the system being considered. When these loops have been satisfied, the program returns to line 00140 and the entire calculation process is repeated for the *second* system. The reader should note that DO 3 . . . was used in three places in the program (lines 00140, 00190, and 00200). In each instance, the range of the DO in question was defined as ending with the statement on line 00220. For a better understanding of why it was convenient to write DO 3 on lines 00190 and 00200, the reader is referred to the flow chart in Fig. 7.12.1.

```
800.38,.03,.65,88.4,.03,11.97
92.8,0.,0.,12.5,0.,1.2
379.56,0.,.6,43.75,0.,11.21
39.,0.,0.,3.9,0.,1.2
336.07,0.,.6,118.3,0.,18.41
26.8,0.,0.,11.2,0.,1.9
462.35,0.,10.35,173.02,0.,339.98
41.7,0.,.6,16.5,0.,37.
671.94,0.,6.8,184.08,0.,575.47
69.6,0.,.4,19.,0.,64.
928.91,0.,1.04,187.45,97.15,181.01
83.1,0.,.1,17.,9.6,19.4
696.87,0.,28.91,107.25,348.9,126.26
54.3,0.,2.9,9.1,34.6,13.9
1126.83,.03,11.32,102.31,212.48,670.4
86.2,0.,1.2,7.1,21.2,76.4
1101.58,0.,30.09,158.6,77.4,473.7
107.4,0.,3.6,14.4,6.8,54.6
1292.59,0.,12.02,261.18,0.,473.12
125.,0.,1.2,26.8,1.5,56.3
438.9,0.,17.63,230.92,.05,335.11
46.2,0.,2.2,24.,0.,37.9
79.65,0.,.5,178.76,0.,0.
6.7,0.,0.,17.8,0.,0.
526.31,142.45,6.62,173.87,25.92,2.76
51.7,12.9,0.,7.1,0.,2.8
323.65,114.8,7.,224.75,27.25,18.73
24.8,5.7,0.,14.7,0.,.8
825.22,115.19,8.63,124.42,28.81,285.46
74.2,7.2,.2,2.4,.2,26.4
2643.2,208.68,3.3,248.44,38.5,310.95
264.,22.7,0.,13.,1.1,28.1
1642.95,300.86,3.5,189.5,74.48,88.47
171.7,30.7,0.,7.6,3.4,9.2
2132.35,515.32,4.46,326.53,34.68,70.58
234.7,57.6,.2,17.,.3,5.2
419.37,103.73,1.75,176.28,24.5,8.73
38.7,6.4,0.,4.1,0.,.2
1483.91,206.48,3.63,248.23,26.25,11.07
147.9,19.8,.2,10.5,0.,.4
1434.89,279.41,1.75,357.22,24.46,48.91
145.8,22.1,0.,21.,.7,4.7
1740.61,304.28,1.75,659.81,1.75,47.12
161.,20.2,0.,40.2,0.,2.6
583.2,271.38,1.75,169.72,1.75,60.45
49.6,12.2,0.,5.9,0.,3.7
175.58,143.37,1.75,50.65,1.75,7.
14.2,1.7,0.,.4,0.,0.
```

Fig. 7.12.3 Input data file for COST1.

DO loops 5 (lines 00230 through 00260) perform matrix addition on the matrices for the two systems. The sum is stored in a third matrix and represents the combined systems. The nested loops on lines 00270 through 00290 calculate the cost ratio for *each department* operating under *each system* by dividing the annual total cost by the annual total hours.

The remainder of the program (starting with line 00300) merely prints the results of the calculations, along with the data originally entered, in several formatted tables. The computed GO TO statement on line 00310, under control of the DO loop on line 00300, successively prints the headings specified in FORMAT statements 50, 51, and 52. The following program statements then complete the tables shown in Fig. 7.12.4 for each of the systems in turn. To follow the method used in constructing the tables, the reader should examine the FORMAT statements referenced by each of the PRINT statements, starting with the one on line 00370.

SYSTEM A

CØST

MØNTH	DEPARTMENTS 1	2	3	4	5	6	TØTAL
1	800.38	0.03	0.65	88.40	0.03	11.97	901.46
2	379.56	0.00	0.60	43.75	0.00	11.21	435.12
3	336.07	0.00	0.60	118.30	0.00	18.41	473.38
4	462.35	0.00	10.35	173.02	0.00	339.98	985.70
5	671.94	0.00	6.80	184.08	0.00	575.47	1438.29
6	928.91	0.00	1.04	187.45	97.15	181.01	1395.56
7	696.87	0.00	28.91	107.25	348.90	126.26	1308.19
8	1126.83	0.03	11.32	102.31	212.48	670.40	2123.37
9	1101.58	0.00	30.09	158.60	77.40	473.70	1841.37
10	1292.59	0.00	12.02	261.18	0.00	473.12	2038.91
11	438.90	0.00	17.63	230.92	0.05	335.11	1022.61
12	79.65	0.00	0.50	178.76	0.00	0.00	258.91
TØTAL	8315.63	0.06	120.51	1834.02	736.01	3216.64	14222.87

HØURS

MØNTH	DEPARTMENTS 1	2	3	4	5	6	TØTAL
1	92.80	0.00	0.00	12.50	0.00	1.20	106.50
2	39.00	0.00	0.00	3.90	0.00	1.20	44.10
3	26.80	0.00	0.00	11.20	0.00	1.90	39.90
4	41.70	0.00	0.60	16.50	0.00	37.00	95.80
5	69.60	0.00	0.40	19.00	0.00	64.00	153.00
6	83.10	0.00	0.10	17.00	9.60	19.40	129.20
7	54.30	0.00	2.90	9.10	34.60	13.90	114.80
8	86.20	0.00	1.20	7.10	21.20	76.40	192.10
9	107.40	0.00	3.60	14.40	6.80	54.60	186.80
10	125.00	0.00	1.20	26.80	1.50	56.30	210.80
11	46.20	0.00	2.20	24.00	0.00	37.90	110.30
12	6.70	0.00	0.00	17.80	0.00	0.00	24.50
TØTAL	778.80	0.00	12.20	179.30	73.70	363.80	1407.80

SYSTEM B

CØST

MØNTH	DEPARTMENTS 1	2	3	4	5	6	TØTAL
1	526.31	142.45	6.62	173.87	25.92	2.76	877.93
2	323.65	114.80	7.00	224.75	27.25	18.73	716.18
3	825.22	115.19	8.63	124.42	28.81	285.46	1387.73
4	2643.20	208.68	3.30	248.44	38.50	310.95	3453.07
5	1642.95	300.86	3.50	189.50	74.48	88.47	2299.76
6	2132.35	515.32	4.46	326.53	34.68	70.58	3083.92
7	419.37	103.73	1.75	176.28	24.50	8.73	734.36
8	1483.91	206.48	3.63	248.23	26.25	11.07	1979.57
9	1434.89	279.41	1.75	357.22	24.46	48.91	2146.64
10	1740.61	304.28	1.75	659.81	1.75	47.12	2755.32
11	583.20	271.38	1.75	169.72	1.75	60.45	1088.25
12	175.58	143.37	1.75	50.65	1.75	7.00	380.10
TØTAL	13931.24	2705.95	45.89	2949.42	310.10	960.23	20902.83

HØURS

MØNTH	DEPARTMENTS 1	2	3	4	5	6	TØTAL
1	51.70	12.90	0.00	7.10	0.00	2.80	74.50
2	24.80	5.70	0.00	14.70	0.00	0.80	46.00
3	74.20	7.20	0.20	2.40	0.20	26.40	110.60
4	264.00	22.70	0.00	13.00	1.10	28.10	328.90
5	171.70	30.70	0.00	7.60	3.40	9.20	222.60
6	234.70	57.60	0.20	17.00	0.30	5.20	315.00
7	38.70	6.40	0.00	4.10	0.00	0.20	49.40
8	147.90	19.80	0.20	10.50	0.00	0.40	178.80
9	145.80	22.10	0.00	21.00	0.70	4.70	194.30
10	161.00	20.20	0.00	40.20	0.00	2.60	224.00
11	49.60	12.20	0.00	5.90	0.00	3.70	71.40
12	14.20	1.70	0.00	0.40	0.00	0.00	16.30
TØTAL	1378.30	219.20	0.60	143.90	5.70	84.10	1831.80

Fig. 7.12.4a Trial run for COST1.

```
COMBINED SYSTEMS

COST
                              DEPARTMENTS
MONTH      1          2          3          4          5          6       TOTAL
  1     1326.69    142.48       7.27     262.27      25.95      14.73    1779.39
  2      703.21    114.80       7.60     268.50      27.25      29.94    1151.30
  3     1161.29    115.19       9.23     242.72      28.81     303.87    1861.11
  4     3105.55    208.68      13.65     421.46      38.50     650.93    4438.77
  5     2314.89    300.86      10.30     373.58      74.48     663.94    3738.05
  6     3061.26    515.32       5.50     513.98     131.83     251.59    4479.48
  7     1116.24    103.73      30.66     283.53     373.40     134.99    2042.55
  8     2610.74    206.51      14.95     350.54     238.73     681.47    4102.94
  9     2536.47    279.41      31.84     515.82     101.86     522.61    3988.01
 10     3033.20    304.28      13.77     920.99       1.75     520.24    4794.23
 11     1022.10    271.38      19.38     400.64       1.80     395.56    2110.86
 12      255.23    143.37       2.25     229.41       1.75       7.00     639.01
TOTAL  22246.87   2706.01     166.40    4783.44    1046.11    4176.87   35125.70

HOURS
                              DEPARTMENTS
MONTH      1          2          3          4          5          6       TOTAL
  1      144.50     12.90       0.00      19.60       0.00       4.00     181.00
  2       63.80      5.70       0.00      18.60       0.00       2.00      90.10
  3      101.00      7.20       0.20      13.60       0.20      28.30     150.50
  4      305.70     22.70       0.60      29.50       1.10      65.10     424.70
  5      241.30     30.70       0.40      26.60       3.40      73.20     375.60
  6      317.80     57.60       0.30      34.00       9.90      24.60     444.20
  7       93.00      6.40       2.90      13.20      34.60      14.10     164.20
  8      234.10     19.80       1.40      17.60      21.20      76.80     370.90
  9      253.20     22.10       3.60      35.40       7.50      59.30     381.10
 10      286.00     20.20       1.20      67.00       1.50      58.90     434.80
 11       95.80     12.20       2.20      29.90       0.00      41.60     181.70
 12       20.90      1.70       0.00      18.20       0.00       0.00      40.80
TOTAL   2157.10    219.20      12.80     323.20      79.40     447.90    3239.60

COST RATIO (DOLLARS/HOUR)

              SYSTEM
DEPT        A          B       COMBINED
  1      10.677     10.108     10.313
  2       0.000     12.345     12.345
  3       9.878     76.483     13.000
  4      10.229     20.496     14.800
  5       9.987     54.404     13.175
  6       8.842     11.418      9.325

TOTAL    10.103     11.411     10.843


SUMMARY

                                    SYSTEM
                              A            B         COMBINED

TOTAL COST               14222.870    20902.830    35125.700

TOTAL HOURS               1407.800     1831.800     3239.600

OVERALL COST RATIO          10.103       11.411       10.843

STOP
```

Fig. 7.12.4b Trial run for COST1.

PROBLEMS

7-1. It is sometimes necessary to calculate the effective or rms value of a current which consists of a number of sinusoidal currents of different frequencies. The effective or rms value is

$$I_e = \sqrt{I_{e1}{}^2 + I_{e2}{}^2 + \ldots + I_{en}{}^2}$$

For example, if a current in a particular circuit consists of the three component currents: 2 A, dc, zero Hz; 0.9 A, rms, 120 Hz; and 0.5 A, rms, 240 Hz, then:

$$I_e = \sqrt{(2)^2 + (0.9)^2 + (0.5)^2}$$

$$= \sqrt{5.06}$$

$$= 2.25 \text{ A, rms}$$

Write a program that will calculate the rms value of a current which may have up to 10 components. Denote the components by the subscripted variable XI(I).

Sample data

Set	*XI(1)*	*XI(2)*	*XI(3)*	*XI(4)*
1	2.0	0.9	0.5	
2	3.7	2.8	1.4	0.6

7-2. Write a program which will print a table of the exponential function e^y for values of y from −0.95 to 1.00 increasing in steps of 0.05. Use a tabular form similar to that used in the program of Sec. 7.4.

7-3. Write a program which will find the largest element of a matrix with NRO rows and MCOL columns, where NRO ⩽ 10 and MCOL ⩽ 10. The program should read-in the matrix and print the value of the largest element.

Sample data

Set 1:

$$\begin{bmatrix} 4 & -7 & 9 & 8 & 3 \\ 1 & 6 & 2 & -2 & 2 \\ 7 & -4 & 2 & 0 & 1 \end{bmatrix}$$

Set 2:

$$\begin{bmatrix} 10 & -37 & 68 \\ 42 & 84 & 13 \\ 63 & 75 & -12 \\ 4 & 92 & 61 \end{bmatrix}$$

Set 3:

$$\begin{bmatrix} 16 & 8 & -9 \\ 22 & 96 & 48 \\ 96 & 51 & 8 \end{bmatrix}$$

Set 4:

$$\begin{bmatrix} 157.63 & 97.14 \\ 85.92 & 148.90 \\ 287.95 & 165.40 \\ 49.00 & -86.90 \\ -367.92 & 287.96 \end{bmatrix}$$

7-4. Modify the program written for Prob. 7-3 so that the program will find the element of largest absolute value. Test the program with the sample data given for Prob. 7.3.

7-5. Write a program which will calculate the mean, variance, and standard deviation of a set of data denoted by $X(I)$. The required formulas are

$$\text{Mean} = \frac{\sum_1^n X(I)}{n}$$

$$\text{Variance} = \frac{\sum_1^n [X(I) - \text{mean}]^2}{n}$$

$$\text{Standard deviation} = (\text{variance})^{0.5}$$

Sample data

Set	*N*	X_1	X_2	X_3	X_4	X_5	X_6
1	5	4	7	9	8	3	–
2	6	10	68	42	13	63	12

7-6. Given a matrix A, the matrix obtained from A by interchanging its rows and columns is called the transpose of A and is denoted by A^t. For example:

$$A = \begin{bmatrix} 5 & 9 & 4 \\ 2 & 7 & 3 \\ 8 & 1 & 6 \end{bmatrix} \qquad A^t = \begin{bmatrix} 5 & 2 & 8 \\ 9 & 7 & 1 \\ 4 & 3 & 6 \end{bmatrix}$$

Write a program which will determine and print the transpose of any matrix which has a maximum of 10 columns and 10 rows.

Use the sample data given with Prob. 7-3 to test the program.

7-7. Any square matrix in which all the diagonal elements are 1 and all other elements are 0 is called an identity matrix. For a matrix of order 3 the identity matrix is

$$\begin{bmatrix} 1 & 0 & 0 \\ 0 & 1 & 0 \\ 0 & 0 & 1 \end{bmatrix}$$

Any square matrix of order n can be *augmented* by having appended to it an identity matrix of the same order. For example, if the given matrix is

$$\begin{bmatrix} 1 & 3 \\ 5 & 7 \end{bmatrix}$$

and the identity matrix of the same order is

$$\begin{bmatrix} 1 & 0 \\ 0 & 1 \end{bmatrix}$$

then the augmented matrix is

$$\begin{bmatrix} 1 & 3 & 1 & 0 \\ 5 & 7 & 0 & 1 \end{bmatrix}$$

Write a FORTRAN program that will read-in values for the elements in an n-order matrix and will print out the augmented matrix.

Sample data

Set 1:

$$\begin{bmatrix} 5 & 9 & 4 \\ 2 & 7 & 3 \\ 8 & 1 & 6 \end{bmatrix}$$

Set 2:

$$\begin{bmatrix} 65 & 84 & 16 & 23 \\ -78 & 1 & -14 & -14 \\ 28 & 29 & 31 & 90 \\ 6 & -48 & 39 & 16 \end{bmatrix}$$

7-8. Using the transpose of a matrix, it is possible to express *any* square matrix as the sum of a symmetric matrix and an antisymmetric matrix:

$$A = \underbrace{\frac{A + A^t}{2}}_{\text{Symmetric}} + \underbrace{\frac{A - A^t}{2}}_{\text{Antisymmetric}}$$

Write a computer program that will determine and print the symmetric and antisymmetric parts of any square matrix.

Sample data

Set 1:

$$\begin{bmatrix} 1 & 4 & 6 \\ 9 & -2 & 3 \\ 7 & 4 & 8 \end{bmatrix}$$

Set 2:

$$\begin{bmatrix} 1 & 7 & 8 & 6 \\ 2 & 9 & 5 & 2 \\ 4 & 4 & 1 & 3 \\ 3 & 6 & 9 & 2 \end{bmatrix}$$

7-9. In a pneumatic control line the flow factor is defined as

$$F = \frac{\text{flow}}{P_u} \qquad \text{when } \frac{P_d}{P_u} = 0.5$$

where flow = volumetric flow rate, standard ft^3/min
P_d = downstream pressure, lb/in^2 absolute
P_u = upstream pressure, lb/in^2 absolute
F = flow factor

If there are n components connected in series, the equivalent flow factor is given by

$$F_{eq} = \frac{F_1F_2F_3F_4\cdots F_n}{(F_1F_2F_3\cdots F_n/F_1)^2 + (F_1F_2F_3\cdots F_n/F_2)^2 + \cdots}$$

or

$$F_{eq} = \frac{\prod_{i=1}^{n} F_i}{\sum_{i=1}^{n}\left[\left(\prod_{i=1}^{n} F_i\right)/F_i\right]^2}$$

Write a program that will accept N flow factors and will calculate and print the value of the equivalent flow factor for the series.

Sample data

$N = 5$ $F_2 = 1.3$ $F_4 = 1.0$
$F_1 = 1.0$ $F_3 = 1.15$ $F_5 = 0.9$

7-10. Given a rectangular plate that is simply supported at all of its edges and loaded with a single concentrated force P at the following specified location:

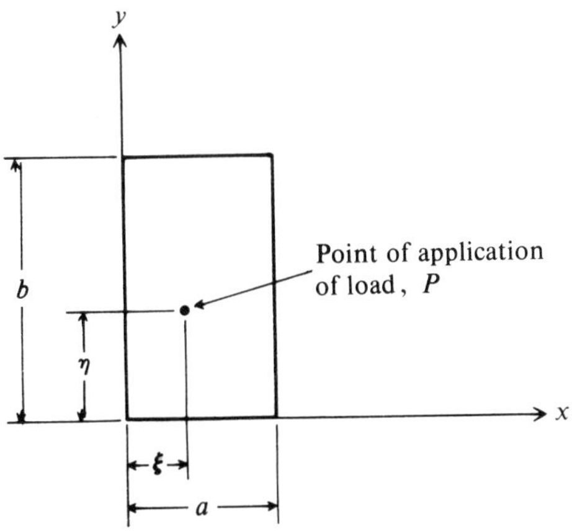

the deflection w at any point on the plate is given by

$$w = \frac{4P}{\pi^4 abD} \sum_{m=1}^{\infty} \sum_{n=1}^{\infty} \frac{1}{(m^2/a^2 + n^2/b^2)^2} \sin\frac{m\pi\xi}{a} \sin\frac{n\pi\eta}{b} \sin\frac{m\pi x}{a} \sin\frac{n\pi y}{b}$$

where

$$D = \frac{Eh^3}{12(1-\nu^2)}$$

and D = flexural rigidity of the plate, lb-in
E = modulus of elasticity, lb/in^2
ν = Poisson's ratio
h = plate thickness, in
w = deflection, in
P = load, lb
a,b = dimensions of the plate, in
ξ,η = location of the applied force, in
x,y = location of the deflected point, in

Divide the plate into a rectangular grid having 25 interior nodal points:

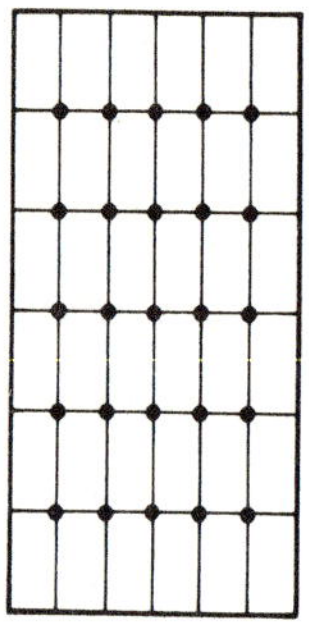

and write a FORTRAN program that will calculate the location of the interior nodal points and also calculate and print the value of the deflection at each of these points.

The input to the program should be the dimensions of the plate (a,b,h), the magnitude and location of the applied load (P,ξ,η), and the physical properties of the plate material (E,ν).

Sample data

$a = 14.$	$P = 15{,}000.$	$E = 3 \times 10^7$
$b = 28.$	$\xi = 5.$	$\nu = 0.3$
$h = 0.25$	$\eta = 10.$	

7-11. A periodic waveform $y = f(\theta)$ can be represented by the Fourier series

$$y = f(\theta) = A_0/2 + \sum_{n=1}^{\infty} (A_n \cos n\theta + B_n \sin n\theta)$$

When the various parts of the waveform can be defined by analytic expressions, the coefficients A_0, A_n, and B_n can be calculated by integration. There are occasions when the waveform cannot be defined by analytic expressions but it can be viewed on an oscilloscope. If a period (360°) of the waveform is observed and the amplitude is recorded at 15° intervals, the *approximate* values of the Fourier coefficients are

$$\frac{A_0}{2} = \frac{1}{24} \sum_{i=1}^{24} y_i$$

$$A_n = \frac{1}{12} \sum_{i=1}^{24} y_i \cos \frac{ni\pi}{12}$$

$$B_n = \frac{1}{12} \sum_{i=1}^{24} y_i \sin \frac{ni\pi}{12}$$

Write a program which will calculate the approximate Fourier coefficients for a periodic waveform. Test the program by calculating $A_0/2$, A_n, and B_n for $n = 1, 2, 3, 4$, for the half-wave rectified waveform shown.

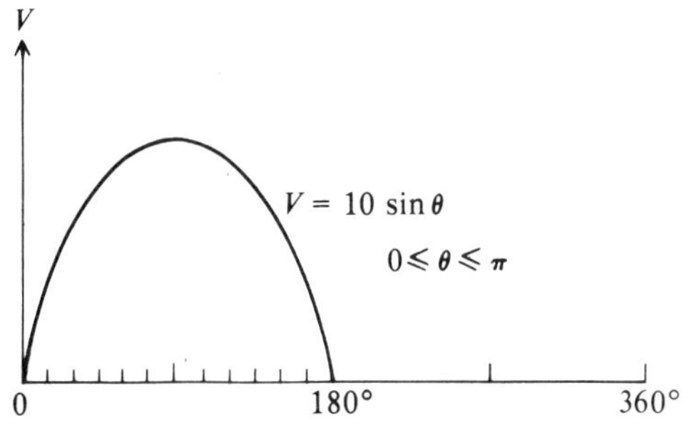

Sample data

θ	y	θ	y	
15	2.59	105	9.66	
30	5.00	120	8.66	
45	7.07	135	7.07	
60	8.66	150	5.00	
75	9.66	165	2.59	
90	10.00	180	0.0	
		180 to 360	0.0	(12 points)

7-12. Write a FORTRAN program that will accept up to 40 coordinate points (x,y) and will calculate and print the approximate perimeter of a smooth closed curve. Test the program by having it calculate the perimeter of a circle.

7-13. The state of stress at a point in an elastic body can be represented by the symmetric matrix

$$[T] = \begin{bmatrix} X & XY & ZX \\ XY & Y & YZ \\ ZX & YZ & Z \end{bmatrix}$$

where X, Y, Z = normal stresses, lb/in^2
XY, YZ, ZX = shear stresses, lb/in^2

If we define the *mean normal stress* S as

$$S = \frac{X + Y + Z}{3} \text{ lb/in}^2$$

the general stress matrix can be decomposed into two matrices representing the volumetric and deviatoric components of stress:

$$[T] = \underset{\text{Volumetric component}}{\begin{bmatrix} S & 0 & 0 \\ 0 & S & 0 \\ 0 & 0 & S \end{bmatrix}} + \underset{\text{Deviatoric component}}{\begin{bmatrix} (X-S) & XY & ZX \\ XY & (Y-S) & YZ \\ ZX & YZ & (Z-S) \end{bmatrix}}$$

The volumetric component produces dilatation of the stressed element; the deviatoric component produces distortion.

Write a FORTRAN program that will read-in the six components of stress at a point and will calculate and print the matrices representing the volumetric and deviatoric stress components at that point.

Sample data

Set	*X*	*Y*	*Z*	*XY*	*YZ*	*ZX*
1	10	8	6	2	3	4
2	12	8	10	7	6	5

7-14. The distance d between successive parallel planes in a crystal is specified by the Miller indices, h, k, and l. For a crystal consisting of cubic cells, the distance is given by

$$d_{hkl} = \frac{a}{\sqrt{h^2 + k^2 + l^2}}$$

where a is the length of the side of a cubic cell in Angstrom units.

Write a FORTRAN program which calculates the values of d_{hkl} and stores them in a variable with three subscripts, as h, k, and l each assume values of 0, 1, and 2. The value of a is to be 1 Å. The program should print each set of Miller indices and the corresponding value of d_{hkl}.

8
Double Precision and Complex Arithmetic

8.1 TYPE DECLARATION STATEMENTS

The type declaration statement is a nonexecutable specification statement that is used in the FORTRAN language to *explicitly* define the mode of the data being stored under a particular variable name.

There are five[1] general forms of the type declaration statement:

1. INTEGER $\nu_1, \nu_2, \nu_3, \ldots$
2. REAL $\nu_1, \nu_2, \nu_3, \ldots$
3. DOUBLE PRECISION $\nu_1, \nu_2, \nu_3, \ldots$
4. COMPLEX $\nu_1, \nu_2, \nu_3, \ldots$
5. LOGICAL $\nu_1, \nu_2, \nu_3, \ldots$

where $\nu_1, \nu_2, \nu_3, \ldots$ is a list of variable names or *dimensioned* array names.[2]

These declaration statements prepare the storage locations named in the list to receive a particular type of data. They are somewhat similar in function to the DIMENSION statement (see Sec. 7.2) in that they reserve storage locations and must therefore appear in the source program *before* any of the variables named in the list are used.

[1] An additional type declaration, the EXTERNAL statement, is used in connection with subprograms and is discussed separately in Sec. 10.10.

[2] If an array is *not* dimensioned in a type declaration statement, it *must* be dimensioned later in a DIMENSION statement. If it *is* dimensioned in a type statement, it *must not* be dimensioned elsewhere.

The predefined convention of implicitly indicating data mode by the initial letter of the variable name can be overridden by a type declaration statement. For example, the statement

```
REAL ABLE,I,J(10)
```

would indicate that locations I and all 10 elements in array J, which ordinarily would imply *integer* mode, are to be used to store *real* values. Of course, location ABLE would be understood to contain a real value even without being named in the declaration statement. Because the array J was dimensioned in the type declaration statement, no separate DIMENSION statement is required. Note that if the name IOTA were used in the program, it would imply integer mode as usual. That is, only the names specifically listed in the type statement are modified by it.

As double precision, complex, and logical variables have no predefined convention for implying mode, they must always be explicitly defined in a type declaration statement.

8.2 DOUBLE PRECISION CONSTANTS

As the name implies, a DOUBLE PRECISION number is one having twice the precision of an ordinary number, that is, provision for twice the number of significant digits. All double precision constants are real numbers and, as in the case of real constants, can be expressed either with or without an exponent. *Without an exponent* the value of the constant is expressed literally, using from 10 to 17 decimal digits with a decimal point in front of the first digit, or in back of the last digit, or between any two digits. Note that the mere presence of more digits than permitted for representing a real constant automatically establishes the number as being in the double precision mode, provided that the particular computer system being used permits double precision.

Double precision numbers are always written with a decimal point and may be either positive or negative. If no algebraic sign is present, the number is assumed to be positive. *With an exponent* the value of the constant is represented in the same manner as a real constant with an exponent except that the letter D replaces the letter E between the constant and the exponent. The constant part of the number can be either a double precision constant without an exponent or a real constant without an exponent. Note that this permits the use of *less* than 10 digits as long as the letter D is used. The exponent can be any positive or negative one- or two-digit decimal number, including zero. Some examples of double precision constants *without* an exponent are

```
247826.591562
.00022064219
3.1415926536
```

Some examples of double precision constants, *with* an exponent are:

```
27182818285.D-10
        30.D+6
      5.522D+0
```

8.3 DOUBLE PRECISION VARIABLES

All double precision variables are represented as names which refer to storage locations in the computer. A particular location in memory is designated as containing a double precision value by *explicitly stating* in the program that the designated location is in double precision mode. This is accomplished with a type declaration statement.

8.4 D SPECIFICATION

The D type specification is a *numeric* specification for DOUBLE PRECISION constants *with exponents* and has the form

D$w.d$

where w is an integer constant specifying the field width, and d is an integer constant indicating the number of digits to the right of the decimal point.

The same general rules that apply to the E specification (see Sec. 2.7) are also applicable to the D specification. The only difference in the two conversions is that under the D specification twice as many significant digits can be stored (or printed) as under the E specification.

8.5 DOUBLE PRECISION EXPRESSIONS

Double precision expressions are essentially the same as arithmetic expressions except that the value of the expression contains *twice* the number of significant digits.

Mixed-mode multiplication, division, addition, and subtraction is permitted between double precision and *real* numbers. The result will be in the double precision mode.

Exponentiation of double precision numbers is possible with exponents in the integer, real, and double precision modes. In all cases the result will be in double precision mode.

If A and B are real variables and D is a double precision variable, some examples of double precision expressions are

A + D	A*D+B	D**B
A − 7.25D + 04	A**D	3.E+6/D

8.6 COMPLEX CONSTANTS

In FORTRAN, a complex constant is an ordered pair of real constants separated by a comma and enclosed in parentheses. The first number of the pair is the real part of a complex number. The second number in the parentheses is the imaginary portion of the complex number. Some examples of complex constants are:

FORTRAN notation	*Algebraic notation*
(−6.2,1.E+3)	$-6.2 + 1000i$
(15.7,−.5E−3)	$15.7 - 0.0005i$
(3.E+4,−90.)	$30{,}000. - 90.i$

8.7 COMPLEX VARIABLES

All complex variables are represented by names in the FORTRAN language. A type declaration statement is used to explicitly define a particular name as being the address of a storage location containing both the real and imaginary parts of a complex number.

8.8 COMPLEX ARITHMETIC OPERATORS

All the arithmetic operators that are available for integer and real number arithmetic are also used for carrying out complex number arithmetic.

If A and C represent two complex numbers:

$$A = a + bi$$
$$C = c + di$$

the complex arithmetic operations are defined as shown in the following algebraic expressions:

1. *Addition:*
   $$A+C = (a + bi) + (c + di) = (a + c) + (b + d)i$$
2. *Subtraction:*
   $$A-C = (a + bi) - (c + di) = (a - c) + (b - d)i$$
3. *Multiplication:*
   $$A*C = (a + bi) * (c + di) = (ac - bd) + (ad + bc)i$$
4. *Division* (for $C \neq 0$):
   $$A/C = \frac{a + bi}{c + di} \frac{(ac + bd)}{c^2 + d^2} + \frac{(bc - ad)}{c^2 + d^2} i$$
5. *Exponentiation:*
   $$A{*}{*}N = (a + bi)^n = [(a^2 + b^2)^{1/2} (\cos \theta + i \sin \theta)]^n$$
   $$= (a^2 + b^2)^{n/2} (\cos n\theta + i \sin n\theta)$$

where $\theta = \arctan \dfrac{b}{a}$

n = an integer

In the FORTRAN exponentiation process, complex numbers may have *only* positive integer exponents and may never be used themselves as exponents.

An array (see Sec. 7.1) consisting of complex numbers can be defined and dimensioned entirely by a type declaration statement or with a combination of a type declaration statement followed by a DIMENSION statement (see Sec. 7.2). The subscripts for these complex number arrays must be integers.

8.9 COMPLEX EXPRESSIONS

In the FORTRAN language, a complex expression may consist of:

1. A single complex constant
2. A single complex variable
3. *Integer* powers of complex constants and complex variables
4. Arithmetic combinations of the above with real constants and real variables

Assuming A and B are real variables and C and D are complex variables, some examples of complex expressions are:

```
A+C              A*C+D             B+C**N
A-(2.6,-5.25)          3.7/D
```

8.10 CALCULATION OF π

The calculation of π to an increasing number of significant figures has long been of interest to mathematicians.[1] One well-known method uses the relation

$$\pi = 4[4 \tan^{-1}\tfrac{1}{5} - \tan^{-1} \tfrac{1}{239}] \qquad (8.10.1)$$

The arctangent can be evaluated by the infinite series

$$\tan^{-1} x = x - \frac{x^3}{3} + \frac{x^5}{5} - \frac{x^7}{7} + \ldots \qquad |x| \leq 1 \qquad (8.10.2)$$

This series can be expressed in compact form as

$$\tan^{-1} x = \sum \frac{(-1)^{n-1}}{(2n-1)} (x)^{2n-1} \qquad n = 1, 2, 3, \ldots \qquad (8.10.3)$$

The program shown in Fig. 8.10.1 calculates π to 16 significant figures. To attain this number of significant figures, A, B, TEMP, and PI are declared to be DOUBLE PRECISION variables. In the first DO loop, the arctangent series is evaluated, with X = 1./5., term by term in accordance with Eq. (8.10.3). The numerical value of 16 times the product of each term of the series [see Eq. (8.10.1)] is assigned to the variable TEMP, and the sum of these terms is stored in the variable A. For example, when I = 1, TEMP = 1./5. and A = 0. + 1./5.; when I = 2, TEMP = $-(1./3.)(1./5.)^3$ and A = $(1./5.) - (1./3.)(1./5.)^3$. Iteration continues until $|\text{TEMP}| \leqslant 10^{-25}$. At this point, the program transfers to statement 2, where TEMP is initialized at a value of zero.

The second DO loop is identical to the first except that X = 1./239. and the sum of the terms is stored in the variable B. When $|\text{TEMP}| \leqslant 10^{-25}$, the program transfers to statement 4, where PI is calculated as A—B. The value of PI is printed according to the format D24.16. The reader should note that the constants (1./5.) and (1./239.) are written as (1./5.D0) and (1./239.D0) so that when the division indicated *within* each parenthesis is performed, each number is regarded as a double precision number.

```
C     THIS PROGRAM ILLUSTRATES THE USE OF DOUBLE PRECISION
C     VARIABLES TO COMPUTE PI.
C
      DOUBLE PRECISION A,B,TEMP,PI
      A=0.0
      B=0.0
      TEMP=0.0
      DO 1 I=1,1000
      Z=(2*I-1)
      TEMP=16.*((-1)**(I-1)*(1./5.D0)**(2*I-1))/Z
      IF(DABS(TEMP)-1.D-25)2,2,1
    1 A=A+TEMP
    2 TEMP=0.0
      DO 3 I=1,1000
      Z=(2*I-1)
      TEMP=4.*((-1)**(I-1)*(1./239.D0)**(2*I-1))/Z
      IF(DABS(TEMP)-1.D-25)4,4,3
    3 B=B+TEMP
    4 PI=A-B
      WRITE(3,10)PI
   10 FORMAT(4H PI=,D24.16)
      STOP
      END
```

```
PI=    0.3141592653589791D 01
```

Fig. 8.10.1 Program PI and trial run.

[1] G. B. Thomas, Jr., "Calculus," 2d ed., Addison-Wesley, Reading, Mass., 1961, p. 684.

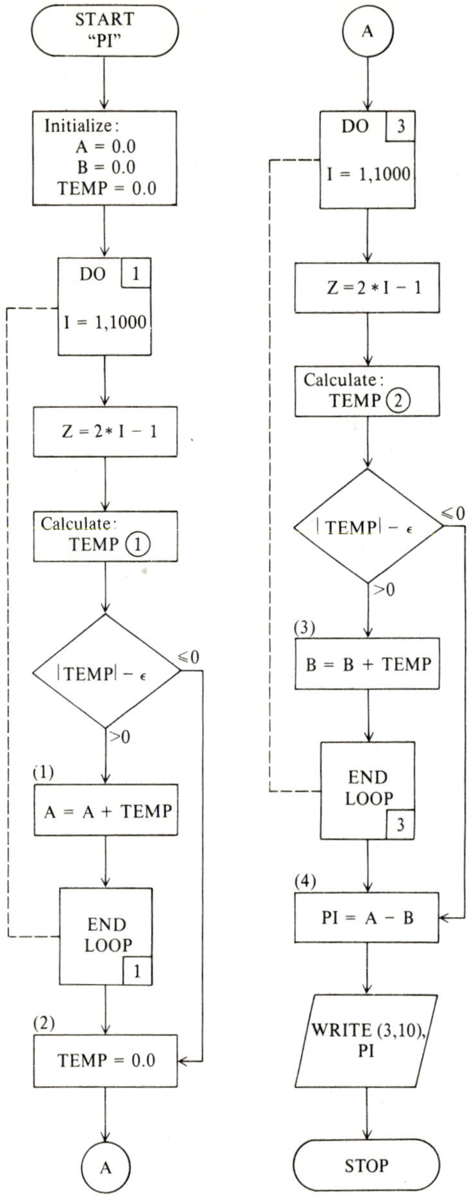

Fig. 8.10.2 Flow chart of program PI.

Similarly, the compiler-supplied function DABS is used to return a double precision value for use in the IF statement.

The procedure followed by the program is outlined in the flow chart of Fig. 8.10.2.

8.11 CHANGING COMPLEX NUMBERS FROM RECTANGULAR TO POLAR FORM

In Sec. 4.12 we discussed a program which transforms complex numbers from rectangular to polar form. The knowledge gained in this chapter makes it possible for

```
C      PØLAR2
C      THIS PRØGRAM CØNVERTS CØMPLEX NUMBERS FRØM RECTANGULAR
C      TØ PØLAR FØRM.
       CØMPLEX Z
     1 READ,A,B
       IF(A) 20,21,20
    21 IF(B) 20,22,20
    22 PRINT 5
     5 FØRMAT("DEGENERATE CASE:A=B=0."//)
       GØ TØ 1
    20 Z=CMPLX(A,B)
       RAD=CABS(Z)
       THETA=ATAN2(B,A)
       BETA=57.29578*THETA
       PRINT 15,RAD,BETA
    15 FØRMAT("RADIUS VECTØR=",F8.3,"    ANGLE=",F8.3," DEGREES"//)
       GØ TØ 1
       END
```

Fig. 8.11.1 Program POLAR2.

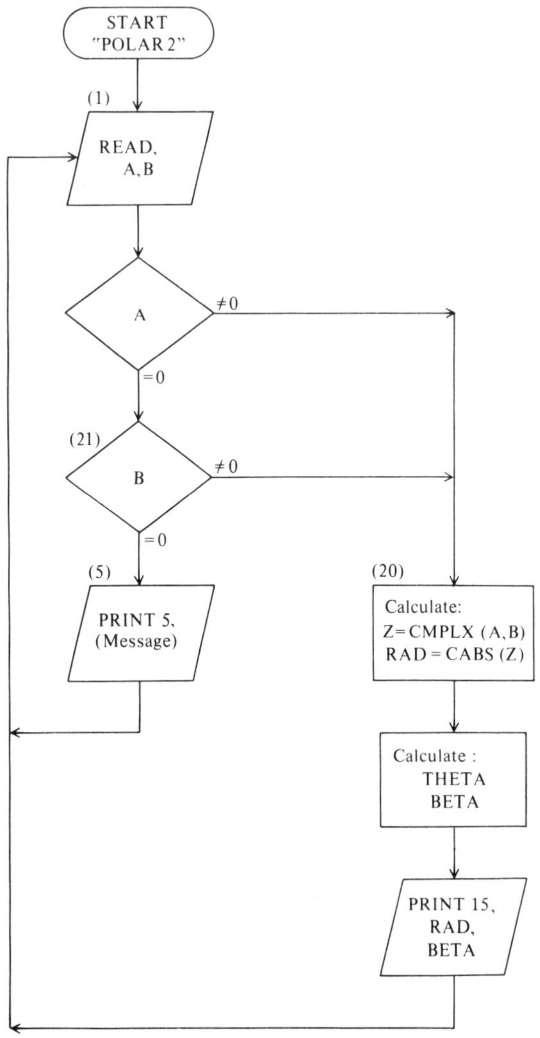

Fig. 8.11.2 Flow chart of program POLAR2.

us to write a more compact version of that program. This version is shown in Fig. 8.11.1; a flow chart is shown in Fig. 8.11.2.

The complex number which is to be transformed is designated Z. The program begins with the type declaration COMPLEX Z. Then the real and imaginary parts of Z, designated A and B, respectively, are entered. A and B are tested to determine if they are zero by the IF statements (lines 00030 and 00040). If both A and B are zero, the warning message DEGENERATE CASE: A = B = 0. is printed and the program branches back to statement 1 on line 00020. This gives the user the option of entering a new set of data.

If *both* A and B are *not* equal to zero, the program branches to statement 20 on line 00080. There the values of A and B are assigned to Z. Then the magnitude of Z is calculated and assigned to RAD. Next, the angle of Z is calculated by means of the compiler-supplied function ATAN2(B,A). This function calculates the arctangent of B/A in radians. The value calculated is assigned to THETA. It should be noted that the function ATAN2 returns values only in the range from 0 to $\pm\pi$ radians. For example, if A = 1. and B = −1., the value returned will be $-\pi/4$ rad. If the equivalent form $7\pi/4$ is desired, the program must be modified appropriately. Finally, on line 00110, the angle is converted from radians to degrees.

```
INPUT:00020
 ?0.,0.

DEGENERATE CASE:A=B=0.


INPUT:00020
 ?2.1,0.

RADIUS VECTØR=   2.100   ANGLE=    0.000 DEGREES


INPUT:00020
 ?0.,2.1

RADIUS VECTØR=   2.100   ANGLE=   90.000 DEGREES


INPUT:00020
 ?-2.1,0.

RADIUS VECTØR=   2.100   ANGLE=-180.000 DEGREES


INPUT:00020
 ?0.,-2.1

RADIUS VECTØR=   2.100   ANGLE= -90.000 DEGREES


INPUT:00020
 ?-5.,2.

RADIUS VECTØR=   5.385   ANGLE= 158.199 DEGREES


INPUT:00020
 ?4.,-3

RADIUS VECTØR=   5.000   ANGLE= -36.870 DEGREES


INPUT:00020
 ?STØP
```

Fig. 8.11.3 Trial runs for POLAR2.

Trial runs are shown in Fig. 8.11.3. These trial runs are the same as those used to test the program discussed in Sec. 4.12. If the reader compares the results of these trial runs, he will note two differences:

1. In the third trial run, the program of Fig. 4.12.3 calculated a value of $+180°$; the program of this section calculates a value of $-180°$.
2. In the fourth trial run, the program of Fig. 4.12.3 calculated a value of $+270°$; the program of this section calculates a value of $-90°$.

These results are mathematically equivalent. However, as noted above, if the positive form of the angle is desired, the program must be modified to attain this result.

8.12 A BILINEAR TRANSFORMATION[1]

As a second illustration of complex arithmetic, let us briefly consider a topic from complex variable theory. Suppose we have two complex variables Z and W that are related as follows:

$$W = \frac{j - Z}{j + Z} \tag{8.12.1}$$

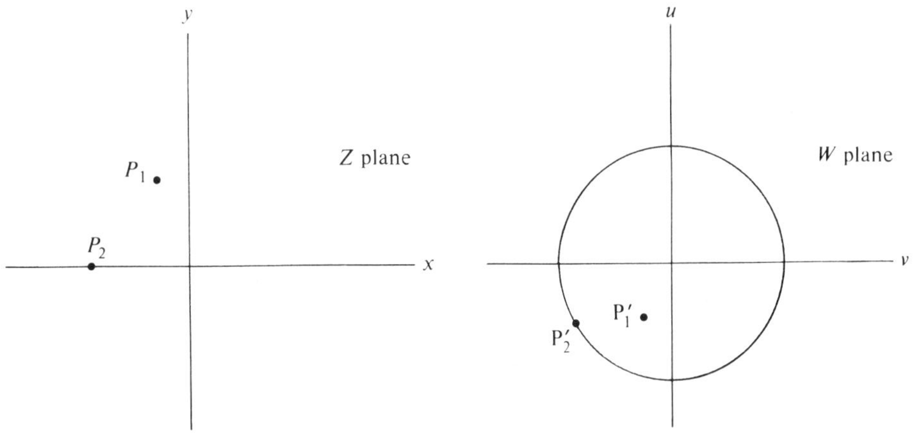

Fig. 8.12.1 Z and W planes for bilinear transformation.

```
C     BILIN1
C     THIS PRØGRAM DEMØNSTRATES THE BILINEAR TRANSFØRMATIØN
C     DISCUSSED IN SECTIØN 8.12
      CØMPLEX Z,W
      X=-4.
      PRINT 20
   20 FØRMAT(3X,"X",4X,"Y",14X,"U",11X,"V",9X,"WMAG")
      DØ 10 I=1,7
      X=X+1.
      Y=-1.
      DØ 10 J=1,3
      Y=Y+1.
      Z=CMPLX(X,Y)
      W=((0.,1.)-Z)/((0.,1.)+Z)
      WMAG=CABS(W)
   10 PRINT 30,Z,W,WMAG
   30 FØRMAT(2F5.0,4X,3F12.4)
      STØP
      END
```

Fig. 8.12.2 Program BILIN1.

[1] R. V. Churchill, "Complex Variables and Applications," 2d ed, McGraw-Hill, New York, 1960, p. 76.

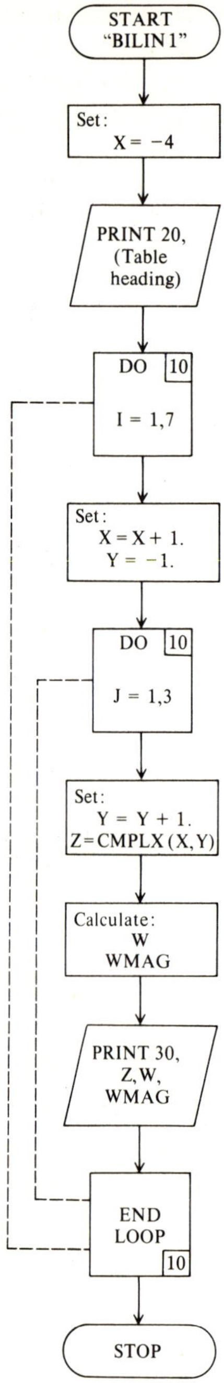

Fig. 8.12.3 Flow chart of program BILIN1.

(This relation is known as a bilinear transformation.) For each value of the variable *Z*, we can calculate a value of *W*. We will show the various values of *Z* in the *Z* plane and the corresponding values of *W* on the *W* plane (see Fig. 8.12.1). Let us now examine two of the interesting properties of bilinear transformations:

1. Points on the upper half of the Z plane are transformed into points in the interior of a circle of unit radius, centered at the origin, on the W plane.
2. Points on the abscissa of the Z plane are transformed into points on the perimeter of a circle of unit radius, centered at the origin, on the W plane.

Figure 8.12.2 shows a program which demonstrates these properties. The corresponding flow chart is in Fig. 8.12.3. The program begins with the declaration COMPLEX Z, W. X, the real part of Z, is arbitrarily initialized at −4, and the indices I and J are selected to suit this choice. The nested DO loops which begin on lines 00050 and 00080 vary X and Y. For each value of Z, a value of W is calculated according to the bilinear transformation on line 00110. Also, the absolute value of W is calculated (line 00120) and stored in WMAG.

Let us examine a typical calculation. When I = 1, X = −3. Y, the real part of Z, is initialized at a value of −1. (line 00070). Then the nested DO loop which begins on line 00080 is executed three times. At the first pass through the nested DO loop, Y = 0. At this point, Z = −1. + j0. A value of W corresponding to this value of Z is calculated. Also, the magnitude of W is calculated. Then the following quantities are printed:

1. The real and imaginary part of Z
2. The real and imaginary part of W
3. The magnitude of W

Note that there are five FORMAT specifications on line 00140 because a separate specification is required for each part of the complex number. After the nested loop has been satisfied, the program branches back to the beginning of the outer DO loop. Then I = 2, X = −2., Y is again initialized at a value of −1., and the nested DO loop is again executed three times.

A trial run is shown in Fig. 8.12.4. Note that the real and imaginary parts of W have been designated U and V, respectively. The reader can observe that all the calculated points on the W plane are either within the unit circle or else on the perimeter of the unit circle.

X	Y	U	V	WMAG
-3.	0.	-.8000	-.6000	1.0000
-3.	1.	-.6923	-.4615	0.8321
-3.	2.	-.6667	-.3333	0.7454
-2.	0.	-.6000	-.8000	1.0000
-2.	1.	-.5000	-.5000	0.7071
-2.	2.	-.5385	-.3077	0.6202
-1.	0.	0.0000	-1.0000	1.0000
-1.	1.	-.2000	-.4000	0.4472
-1.	2.	-.4000	-.2000	0.4472
0.	0.	1.0000	0.0000	1.0000
0.	1.	0.0000	0.0000	0.0000
0.	2.	-.3333	0.0000	0.3333
1.	0.	0.0000	1.0000	1.0000
1.	1.	-.2000	0.4000	0.4472
1.	2.	-.4000	0.2000	0.4472
2.	0.	-.6000	0.8000	1.0000
2.	1.	-.5000	0.5000	0.7071
2.	2.	-.5385	0.3077	0.6202
3.	0.	-.8000	0.6000	1.0000
3.	1.	-.6923	0.4615	0.8321
3.	2.	-.6667	0.3333	0.7454

STØP

Fig. 8.12.4 Trial run for BILIN1.

8.13 NETWORK CALCULATIONS[1]

The importance of complex numbers in the steady-state analysis of electric circuits was mentioned in Sec. 4.12. In this section we will discuss a computer program which illustrates the utility of FORTRAN in solving network problems involving complex quantities.[2]

Let us consider a network of the form shown in Fig. 8.13.1, driven by a sinusoidal voltage generator. In the most general case, each of the impedances Z_1, Z_2, and Z_3 may consist of resistance, and inductive and capacitive reactances. Let us assume that we are interested in calculating the following quantities:

1. The current delivered by the generator
2. The voltage across the parallel impedances Z_2 and Z_3
3. The average power delivered by the generator

We can calculate the phasor current I by use of Ohm's law. To carry out these calculations, we begin by combining the parallel impedances Z_2 and Z_3 and denote the result by Z_{eq}.

$$Z_{eq} = \frac{Z_2 Z_3}{Z_2 + Z_3} \tag{8.13.1}$$

Then we add Z_1 to Z_{eq} and use Ohm's law:

$$I = \frac{V}{Z_1 + Z_{eq}} \tag{8.13.2}$$

The phasor voltage V_p across Z_{eq} is

$$V_p = IZ_{eq} \tag{8.13.3}$$

As mentioned previously, in the most general case, each impedance may consist of resistance, inductance, and capacitance and may therefore be expressed as follows:

$$Z = R + j(\omega L - \frac{1}{\omega C}) \tag{8.13.4}$$

where R = resistance, ohms
L = inductance, H
C = capacitance, F
ω = radian frequency, rad/s

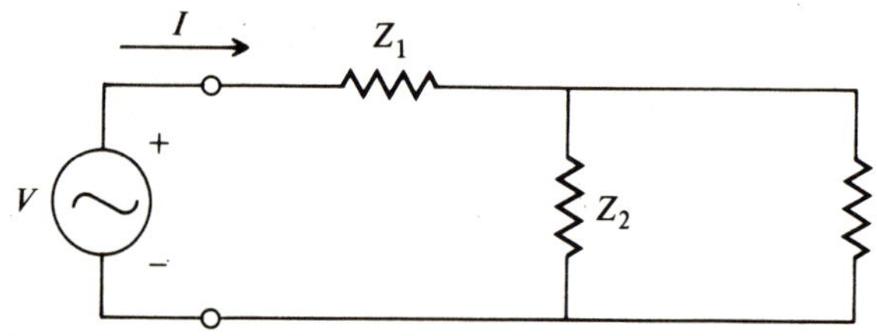

Fig. 8.13.1 Impedance network.

[1] W. H. Hayt, Jr., and G. W. Hughes, "Introduction to Electrical Engineering," McGraw-Hill, New York, 1968, p. 268.

[2] A voltage or current represented by a complex number is usually called a phasor. Thus, the expressions "phasor voltage" and "phasor current" denote complex representations of voltage and current, respectively.

The average power delivered by the generator is:

$$P_{av} = \tfrac{1}{2}\,\mathrm{Re}\,(VI^*) \tag{8.13.5}$$

where P_{av} = average power, W
Re = "real part of"
V = generator voltage, V
I^* = complex conjugate of current, A

A program for carrying out these calculations is shown in Fig. 8.13.2. The following table gives the program names for the symbols used in the equations above.

Equation symbol	*Program name*
Z_1, Z_2, Z_3	Z1, Z2, Z3
Z_{eq}	ZEQ
I	I
V	VG
V_p	VP
R	R
L_1, L_2, L_3	H1, H2, H3
C_1, C_2, C_3	C1, C2, C3
ω	W
P_{av}	PAV

The first statement of the program is a COMPLEX type declaration (line 00010). The reader should note that I, which is normally an integer name, becomes a complex variable because it is included in the COMPLEX type declaration. The type declaration is followed by READ statements which permit the following data to be entered: the frequency of the generator in radians per second, the generator voltage (a complex quantity), and the values of the resistance, inductance, and capacitance, which

```
00001C       CIRC1
00002C       THIS PRØGRAM SØLVES NETWØRKS ØF THE TYPE SHØWN
00003C       IN FIGURE 8.13.1.
00010        CØMPLEX Z1,Z2,Z3,I,VG,VP,ZEQ
00020        READ,W,VG,R1,R2,R3,H1,H2,H3,C1,C2,C3
00030        IF(C1) 1,2,1
00040      2 Z1=CMPLX(R1,W*H1)
00050        GØ TØ 3
00060      1 Z1=CMPLX(R1,W*H1-1./(W*C1))
00070      3 IF(C2) 4,5,4
00080      5 Z2=CMPLX(R2,W*H2)
00090        GØ TØ 6
00100      4 Z2=CMPLX(R2,W*H2-1./(W*C2))
00110      6 IF(C3) 7,8,7
00120      8 Z3=CMPLX(R3,W*H3)
00130        GØ TØ 20
00140      7 Z3=CMPLX(R3,W*H3-1./(W*C3))
00150     20 ZEQ=(Z2*Z3)/(Z2+Z3)
00160        I=VG/(Z1+ZEQ)
00170        VP=I*ZEQ
00180        PAV=0.5*REAL(VG*CØNJG(I))
00190        PRINT 10,I,ZEQ
00200     10 FØRMAT(2F8.4,4X,2F8.3)
00210        PRINT,"VP=",VP
00220        PRINT,"PAV=",PAV
00230        STØP
00240        END
```

Fig. 8.13.2 Program CIRC1.

```
INPUT:00020
?50000,14.49,3.885,10,10,0

INPUT:00030
?0,4.E-04,0

INPUT:00040
?0,0,2.E-06

  0.3535  0.6125          5.000 -15.000
VP=      10.9550         -2.2400
PAV=         3.7509
STØP
```

Fig. 8.13.3 Trial run for CIRC1.

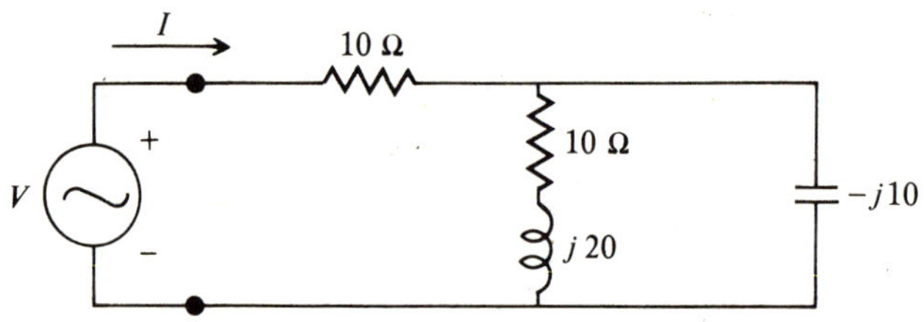

Fig. 8.13.4 Typical values for impedance network.

describe each of the impedances Z1, Z2, and Z3. On lines 00050, 00060, and 00070 the complex quantities Z1, Z2, and Z3 are evaluated in accordance with Eq. (8.13.4). Then ZEQ, I, VP, and PAV are calculated in accordance with Eqs. (8.13.1) to (8.13.3), and (8.13.5).

A trial run is shown in Fig. 8.13.3, in which the network shown in Fig. 8.13.4 is solved. It should be noted that the values of the complex quantities I, ZEQ, and VP are *always* printed in rectangular form. Therefore, FORMAT statements must provide *two* FORMAT specifications for each variable which is to be printed. This is illustrated by the statement on line 00130.

8.14 TRANSMISSION-LINE CALCULATIONS

In Sec. 4.13 we discussed a program which calculates the characteristic impedance Z_0 of a transmission line. In this section we will discuss a program which calculates the characteristic impedance Z_0 and also the complex propagation constant Γ of a transmission line.[1] Complex arithmetic will be used for several of the calculations.

The reader may recall that Z_0 is given by

$$Z_0 = \sqrt{Z/Y} \tag{8.14.1}$$

where

$$Z = R + j\omega L \tag{8.14.2}$$
$$Y = G + j\omega C \tag{8.14.3}$$

We can express Z_0 in rectangular form as

$$Z_0 = A + jB \tag{8.14.4}$$

[1] Charles A. Holt, "Introduction to Electromagnetic Fields and Waves," Wiley, New York, 1963, p. 487.

where

$$A = \text{Re}\sqrt{\frac{Z}{Y}} \tag{8.14.5}$$

$$B = \text{Im}\sqrt{\frac{Z}{Y}} \tag{8.14.6}$$

We can express Z_0 in polar form as

$$Z_0 = \text{ZOMAG } \underline{/\text{ANGLE}} \tag{8.14.7}$$

where

$$\text{ZOMAG} = \sqrt{A^2 + B^2} \tag{8.14.8}$$

$$\text{ANGLE} = \tan^{-1}(B/A) \tag{8.14.9}$$

The parameters R, L, G, and C have the same significance and units here as in Sec. 4.13. Another quantity of importance in the analysis of transmission lines is Γ, the complex propagation constant defined by

$$\Gamma = \sqrt{ZY} = \alpha + j\delta \tag{8.14.10}$$

where Z and Y are defined by Eqs. (8.14.2) and (8.14.3)

α = attenuation constant, nepers/mi

δ = phase constant, rad/mi

```
C      TRANS2
C      THIS PRØGRAM CALCULATES THE CHARACTERISTIC IMPEDANCE,
C      ATTENUATIØN CØNSTANT,AND THE PHASE CØNSTANT ØF A
C      TRANSMISSIØN LINE.
       CØMPLEX Z,Y,ZØ
       READ,ØM,XL,G,C,W
       Z=CMPLX(ØM,W*XL)
       Y=CMPLX(G,W*C)
       ZØ=CSQRT(Z/Y)
       ALPHA=REAL(CSQRT(Z*Y))
       DELTA=AIMAG(CSQRT(Z*Y))
       A=REAL(ZØ)
       B=AIMAG(ZØ)
       ZØMAG=SQRT(CABS(Z/Y))
       IF(A) 2,3,4
     3 IF(B) 5,6,7
     5 BETA=270.
       GØ TØ 1
     6 PRINT 20
    20 FØRMAT(22HDEGENERATE CASE:A=B=0.)
       GØ TØ 25
     7 BETA=90.
       GØ TØ 1
     2 IF(B) 8,9,10
     8 BETA=180.+57.296*ATAN(ABS(B/A))
       GØ TØ 1
     9 BETA=180.
       GØ TØ 1
    10 BETA=180.-57.296*ATAN(ABS(B/A))
       GØ TØ 1
     4 IF(B) 11,12,13
    11 BETA=360.-57.296*ATAN(ABS(B/A))
       GØ TØ 1
    12 BETA=0.
       GØ TØ 1
    13 BETA=57.296*ATAN(ABS(B/A))
     1 ANGLE=BETA
       PRINT,"ZØMAG=",ZØMAG,"    ANGLE=",ANGLE
       PRINT,"ZØ IN RECTANGULAR FØRM=",ZØ
       PRINT,"ALPHA IN NEPERS PER MILE=",ALPHA
       PRINT,"DELTA IN RADIANS PER MILE=",DELTA
    25 STØP
       END
```

Fig. 8.14.1 Program TRANS2.

The program shown in Fig. 8.14.1 calculates Z_0 in polar and in rectangular form, and Γ in rectangular form. In transmission-line analysis, Γ is needed in rectangular form because the real part of Γ, α, determines the attenuation per unit length along the line, and the imaginary part of Γ, δ, determines the phase shift per unit length along the line.

As in the program discussed in Sec. 4.13, the quantities R, L, G, C, and ω are entered into the program. The same names are used for these quantities as were used in Sec. 4.13. Z and Y are defined on lines 00030 and 00040 in accordance with Eqs. (8.14.2) and (8.14.3) by use of the CMPLX function. The reader should note that Z and Y have been previously declared to be complex variables by a complex type declaration statement.

Lines 00050 to 00100 illustrate a variety of calculations that are specific to complex arithmetic. These are tabulated as follows.

Line sequence number	*Calculation*	*Equation*
00050	ZO is the complex square root of (Z/Y). The answer is stored in rectangular form.	(8.14.1)
00060	The complex square root of (ZY) is calculated and stored in rectangular form. The REAL function selects the real part of the complex number, and thus calculates ALPHA, the attenuation constant. REAL is used here in the mathematical sense, and signifies the real part of a complex number.	(8.14.10)
00070	The complex square root of (ZY) is calculated and stored in rectangular form. The AIMAG function selects the imaginary part of the complex number, and thus calculates DELTA, the phase constant.	(8.14.10)
00080 00090	The complex square root of (Z/Y) is calculated and stored in rectangular form. The real and imaginary parts of the complex number are stored, by means of REAL and AIMAG functions, in the addresses A and B, respectively.	(8.14.4) (8.14.5) (8.14.6)
00100	The square root of the magnitude of the complex number (Z/Y) is calculated and stored in the address ZOMAG.	(8.14.8)
00110 to 00330	The previously calculated values of A and B are used to calculate the phase angle of ZO. The value of the phase angle is stored in the location ANGLE.	(8.14.9)

A trial run is shown in Fig. 8.14.2. The same data were used as in the trial run in Fig. 4.13.2. The value of Z_0 to two decimal places was

$$Z_0 = 451.47 - j56.103 = 454.94 \underline{/352.92^\circ}$$

```
INPUT:00020
 ?4,.003,1.E-06,1.5E-08,5000

ZØMAG=        454.9402    ANGLE=        352.9162
ZØ IN RECTANGULAR FØRM=          451.4676        -56.1033
ALPHA IN NEPERS PER MILE=          0.0047
DELTA IN RADIANS PER MILE=          0.0338
STØP
```

Fig. 8.14.2 Trial run for TRANS2.

It will be seen that the value of Z_0 in polar form is the same as previously calculated by the program of Fig. 4.13.1. The attenuation constant was 0.0047 nepers/mi, and the phase constant was 0.0338 rad/mi.

PROBLEMS

8-1. Write a FORTRAN program that will evaluate the following function:

$$y = \sqrt{a + jb} + \sqrt{c + jd}$$

The program should read-in values for a, b, c, and d and printout the value of the function.

Sample data

Set	a	b	c	d
1	1	1.732	1	−1.732
2	2	8	−4	6

8-2. Write a program to evaluate the function

$$y = ze^{jz}$$

where z is a complex number with real and imaginary parts (a and b) as the input values for the program.

Sample data

Set	a	b
1	+4	−3
2	+4	+3
3	−4	+3

8-3. Problem 4-4 required the preparation of a program to solve a system of two simultaneous equations using Cramer's rule. Revise that program to permit complex coefficients.

Sample data

Set 1:

$a_1 = 2 + j1$ $\quad a_2 = 2 + j0$
$b_1 = 5 - j2$ $\quad b_2 = 4 - j2$
$c_1 = 8 + j3$ $\quad c_2 = 8 - j1$

Set 2:

$a_1 = 5 - j3$ $\quad a_2 = 7 + j2$
$b_1 = 4 - j2$ $\quad b_2 = 3 + j3$
$c_1 = 6 + j5$ $\quad c_2 = 9 + j1$

8-4. Write a program which transforms a delta network to an equivalent wye network.

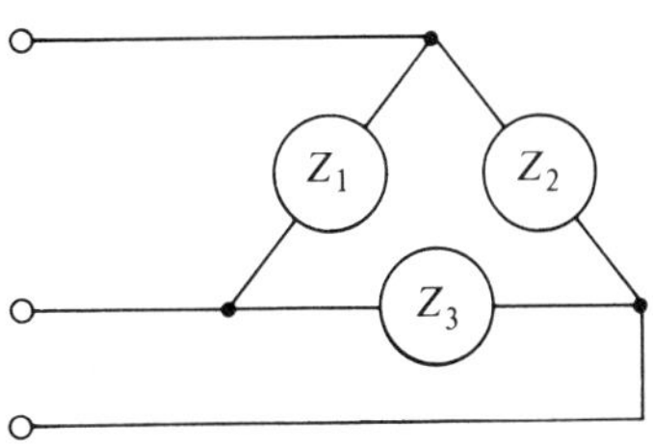

"Delta" network

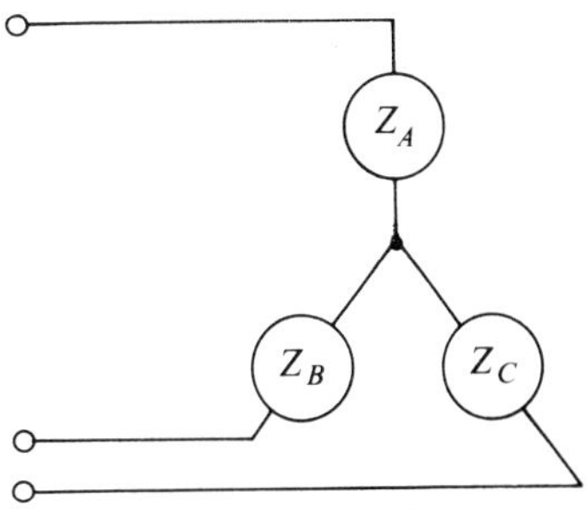

Equivalent "wye"

The conversion formulas are:

$$Z_A = \frac{Z_1 Z_2}{Z_1 + Z_2 + Z_3}$$

$$Z_B = \frac{Z_2 Z_3}{Z_1 + Z_2 + Z_3}$$

$$Z_C = \frac{Z_1 Z_3}{Z_1 + Z_2 + Z_3}$$

The program should read-in the values for the delta impedances and print the values for the wye impedances.

Sample data

$Z_1 = 1 + j10$
$Z_2 = 2 - j8$
$Z_3 = 4 + j0$

8-5. The transfer impedance of the following pi network

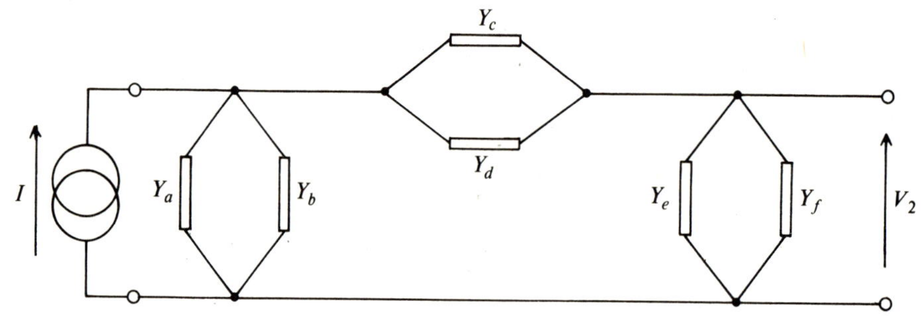

when driven by a sinusoidal current generator is

$$Z_{tr} = \frac{Z_1 Z_3}{Z_1 + Z_2 + Z_3} = \frac{V_2}{I}$$

where $Z_1 = \dfrac{1}{Y_a + Y_b}$

$$Z_2 = \frac{1}{Y_c + Y_d}$$

$$Z_3 = \frac{1}{Y_e + Y_f}$$

Write a program that will evaluate the transfer impedance.

Sample data

$Y_a = 1 \times 10^{-4} + j0$ $\quad$ $Y_d = 0$
$Y_b = 0.2 \times 10^{-4} + j0$ $\quad$ $Y_e = 0.2 \times 10^{-4} + j0$
$Y_c = 0 + j0.188 \times 10^{-4}$ $\quad$ $Y_f = 0$

8-6. The normalized transfer function V_2/V_1 of a twin-T network is

$$\frac{V_2}{V_1} = \frac{1-(f/f_0)^2}{[1-(f/f_0)^2] + j2(\sqrt{n} + 1/\sqrt{n})(f/f_0)}$$

where f = frequency, Hz
f_0 = frequency at which the magnitude of the transfer function is minimum, Hz
n = dimensionless parameter
V_1 = input voltage, V
V_2 = output voltage, V

Write a program which calculates the transfer function as f/f_0 varies from 0.4 to 1.6 in steps of 0.2 for three values of n: $n = 1$, 3, and 9. For each case the program should calculate the transfer function in rectangular form and also the magnitude of the transfer function.

8-7. The current I_r at the receiving end of a long transmission line of length l is

$$I_r = \frac{V_g}{(Z_g + Z_r)\cosh \Gamma l + (Z_0 + Z_g Z_r/Z_0)\sinh \Gamma l}$$

and

$\Gamma = \alpha + j\beta$
where V_g = generator voltage, V
Z_g = generator impedance, ohms
Z_r = load impedance, ohms
Z_0 = characteristic impedance, ohms
Γ = propagation constant per mile
α = attenuation constant = 0.0079 neper/mi
β = phase constant = 0.029 rad/mi
l = length of line, mi

Write a program to calculate and print the receiving current in amperes if

$V_g = 2 + j0$ V
$Z_g = 600 + j0$ ohms
$Z_r = 355 + j355$ ohms
$Z_0 = 690 - j180$ ohms
l = 100 mi

9
Program Control with FORTRAN Logic

9.1 LOGICAL CONSTANTS

In FORTRAN the two *logical constants* that are available are represented as

.TRUE. and .FALSE.

Note that the strings of symbols representing these constants are always written within a pair of decimal points.

9.2 LOGICAL VARIABLES

Logical variables are names of storage locations that have been explicitly declared to contain values expressed in the logical mode. A logical variable can assume one of only two possible values: true or false. A variable name is defined as being of logical mode by a type declaration statement (see Sec. 8.1).

9.3 L SPECIFICATION

The L specification is a *logical* specification used for the conversion of *logical* data and has the form

L*w*

where *w* is an integer constant indicating the field width.

For input operations

If the first nonblank character in the designated field is a T, a value of .TRUE. is assigned to the variable. If the first nonblank character in the field is an F, a value of .FALSE. is assigned to the variable. If any nonblank character other than T or F leads the field, an error message will be generated by the compiler. Some examples are:

Input	*Specification*	*Value stored*
TRUEbb	L6	.TRUE.
FINALb	L6	.FALSE.
bFb	L3	.FALSE.
bbF3	L4	.FALSE.
bb3F	L4	(ERROR)

For output operations

If the stored value is .TRUE., a T is printed preceded by $(w-1)$ blanks, i.e., right-justified in the field of w spaces. If the stored value is .FALSE., an F is printed preceded by $(w-1)$ blanks. Some examples are:

Stored value	*Specification*	*Output*
.TRUE.	L4	bbbT
.FALSE.	L1	F

9.4 RELATIONAL OPERATORS

The six binary relational operators used for making comparisons between the numeric values of two arithmetic expressions are as follows.

Meaning	*Algebraic notation*	*FORTRAN symbol*
Equal to	$=$	.EQ.
Not equal to	$\neq$	.NE.
Less than	$<$	.LT.
Less than or equal to	$\leq$	.LE.
Greater than	$>$	.GT.
Greater than or equal to	$\geq$	.GE.

Note that in FORTRAN notation the relational operators must be set off by a pair of decimal points. All the relational operators have the *same operational precedence.*

9.5 RELATIONAL EXPRESSIONS

If *exp*1 and *exp*2 are two arithmetic expressions having the same mode (see below), a relational expression is formed by placing any one of the relational operators between the two expressions. For example, *exp*1.LE.*exp*2 means: "The numeric value of *exp*1 is less than or equal to the numeric value of *exp*2." If this relation actually exists, the relational expression has the value .TRUE.. If the relation does *not* exist, the relational expression has the value .FALSE..

The two expressions being compared may be:

1. Both in integer mode
2. Both in real mode
3. Both in double precision mode
4. One in real mode and one in double precision mode

In all cases the *value* of a relational expression is a *logical value*, either .TRUE. or .FALSE.; therefore a *relational expression is equivalent to a logical variable.*

Some examples of relational expressions are

26.548D+06.GT.XL

J.LE.M

9.6 LOGICAL OPERATORS

The FORTRAN language has three logical operators. These operators and their meanings are as follows.

Operation	*Meaning*	*Boolean symbol* *	*FORTRAN notation*
1. Negation	not	$\neg$	.NOT.
2. Conjunction (logical product)	and	$\wedge$	.AND.
3. Disjunction (logical sum)	or	$\vee$	.OR.

* The symbols given here are widely used, but several variations are to be found in the literature. For example, the disjunction of X and Y is sometimes denoted by $X + Y$.

The first operator (.NOT.) is a unary operator and is applied to the single logical value that immediately follows it.[1] The other two operators (.AND. and .OR.) are binary operators and form compound statements with the logical values that are positioned before and after the operator.

Note that the string of alphabetic characters representing these logical operators is always enclosed within a pair of decimal points.

9.7 LOGICAL EXPRESSIONS

In a FORTRAN program, a logical expression may be a single logical constant, a single logical variable, or a single relational expression, or it may be formed from combinations of the above with the logical operators. The *value* of a logical expression is either .TRUE. or .FALSE..

If A, B, and C are *logical* variables and X and Y are *real* variables, some examples of *logical expressions* and their meanings are as follows:

1. .NOT.A — If A has the value .TRUE., the value of the expression is .FALSE. If A has the value .FALSE., the value of the expression is .TRUE.

[1] If the logical value being operated on is the result of a relational expression (Sec. 9.5) or of a logical expression (Sec. 9.7), it must be enclosed in parentheses to avoid ambiguity.

2. B.AND.C	If *both* B and C have the value .TRUE., the value of the expression is .TRUE. If *either* B or C has the value .FALSE., the value of the expression is .FALSE. If *both* B and C have the value .FALSE., the expression value is .FALSE.
3. A.OR.C	If *both* A and C have the value .TRUE., the value of the expression is .TRUE. If *either* A or C has the value .TRUE., the value of the expression is .TRUE. If *both* A and C have the value .FALSE., the expression value is .FALSE.
4. B.AND.X.LT.Y	This logical expression is the same as item 2, with the relational expression (X.LT.Y) replacing the logical variable C.

9.8 HIERARCHY OF LOGICAL OPERATIONS

When a logical expression involves arithmetic expressions, relational expressions, logical values (constants and/or variables), and logical operators, the *normal order* in which the operations are performed (regardless of their position in the expression) is according to the following *precedence order*:

1. Arithmetic operations
2. Relational operations
3. .NOT.
4. .AND.
5. .OR.

If there is more than one operation having the same precedence rating, those operations are performed in sequence from left to right. Just as with arithmetic operations, the normal order can be modified by the use of parentheses.

Some examples illustrating the *normal precedence* of execution and the *use of parentheses* follow. (A, B, and C are again assumed to be *logical* variables; X and Y are *real* variables.)

The expression

A.AND.B.OR.C.AND.X.LT.Y

will be evaluated according to the *normal precedence*, which is as follows:

1. The logical value of (X.LT.Y) will be determined (*highest* precedence).
2. The logical value of (A.AND.B) will be determined (*first* .AND. operation in the expression).
3. The logical value of C.AND. (the logical value determined in step 1) will be evaluated (*next* .AND. operation in the expression).
4. The logical value of (result of step 2 .OR. result of step 3) will be determined (*first* .OR. operation in the expression).

To summarize, the expression would be evaluated as if it were actually divided as follows:

(A.AND.B).OR.(C.AND.(X.LT.Y))

Now let us consider the evaluation of the expression

A.AND.(B.OR.C).AND.X.LT.Y

1. The logical value of (X.LT.Y) will be determined (*highest* precedence).
2. The logical value of (B.OR.C) will be determined (*innermost* parentheses).
3. The logical value of (A.AND. result of 2) will be determined (*first* .AND. operation).
4. The logical value of (result of step 3 .AND. result of step 1) will be determined (*next* .AND. operator in the expression).

Again, in summary, the expression would be evaluated as if it were actually divided as:

(A.AND.(B.OR.C)).AND.(X.LT.Y)

9.9 LOGICAL ASSIGNMENT STATEMENT

The *logical assignment statement* has the general form

v = *exp*

where *v* is a logical variable and *exp* is any logical expression.

This statement can be used to set the value of a logical variable or to evaluate and store the value of a logical expression. Some examples of these operations are:

```
A = .TRUE.
B = A.OR.X.LT.Y
C = J.LE.M
```

9.10 LOGICAL IF STATEMENT

The *logical IF statement* is an executable statement that is always used as a *pair* of statements in the form

IF (*exp*) *stat 1*
stat 2

where *exp* is a logical expression and *stat 1* and *stat 2* are any *executable* FORTRAN statements, with the exception that *stat 1* may *not* be either another logical IF statement or a DO statement.

The action produced by this statement is as follows:

1. The *logical value* of the expression is determined.
2. If the value of the expression is .TRUE., control passes to *stat 1* for execution.
   a. If *stat 1* is a *transfer* statement, control will be shifted according to the instructions given by *stat 1*.
   b. If *stat 1* is *not* a transfer statement, it will be executed and control will pass to *stat 2* for execution.
3. If the expression value is .FALSE., control passes *immediately* to *stat 2* for execution, and *stat 1* is bypassed.

Some examples of the logical IF statement follow.

1. IF(X.LT.Y) Z = SQRT(Y–X)
   P = A + Z
2. IF(A.OR.B) GO TO 25
   STOP

9.11 PROBABILITY DISTRIBUTION FUNCTIONS

In many engineering applications, the joint probability of two continuous random variables X and Y is of importance. A typical probability calculation involving two random variables is "Find the probability that $X \leq x$ and $Y \leq y$." This may be expressed by the following notation:

$$P(X \leq x \text{ and } Y \leq y)$$

A probability of this type can be calculated from the joint distribution function $F(x,y)$. That is:

$$P(X \leq x \text{ and } Y \leq y) = F(x,y)$$

Let us consider a particular joint distribution function:

$$F(x,y) = \begin{cases} 0 & x < 0 \text{ or } y < 0 \\ (1 - e^{-x})y^3 & 0 \leq x, 0 \leq y \leq 1 \\ (1 - e^{-x}) & 0 \leq x, 1 < y \end{cases}$$

Notice that the distribution function assumes *different values* as the *range* of the random variables changes. This is illustrated by the following examples:

$$\begin{aligned} P(X \leq 0.1 \text{ and } Y \leq -0.1) &= F(0.1,-0.1) \\ &= 0 \end{aligned}$$

$$\begin{aligned} P(X \leq -0.1 \text{ and } Y \leq 0.1) &= F(-0.1,0.1) \\ &= 0 \end{aligned}$$

$$\begin{aligned} P(0 < X \leq 1 \text{ and } 0 < Y \leq 0.5) &= F(1,0.5) \\ &= (1 - e^{-1})(0.5)^3 \\ &= 0.0790 \end{aligned}$$

$$\begin{aligned} P(0 < X \leq 1 \text{ and } 1 < Y) &= F(1, Y > 1) \\ &= (1 - e^{-1}) \\ &= 0.632 \end{aligned}$$

A program which carries out probability calculations of the type discussed above is shown in Fig. 9.11.1, and a flow chart is shown in Fig. 9.11.2. The three logical IF statements in the program determine in which of three ranges the random variables lie, and, based upon this determination, choose the required form of the distribution function. For example, if *either* X or Y is negative, the IF statement on line 00020 causes the program to branch to line 00080, and, as a result, FXY = 0. However, if X is positive and Y is positive but not greater than 1, the IF statement on line 00030 causes the program to branch to line 00060. FXY is calculated in accordance with the

```
00001C     CDF1
00002C     THIS PRØGRAM CALCULATES JØINT PRØBABILITIES FRØM
00003C     A PRØBABILITY DISTRIBUTIØN FUNCTIØN.
00010      READ,X,Y
00020      IF(X.LT.0..ØR.Y.LT.0.) GØ TØ 1
00030      IF(X.GE.0..AND.(Y.GE.0..AND.Y.LE.1.)) GØ TØ 2
00040      FXY=(1.-EXP(-X))
00050    5 PRINT 9,FXY
00060      STØP
00070    2 FXY=(1.-EXP(-X))*Y**3
00080      GØ TØ 5
00090    1 FXY=0.
00100      GØ TØ 5
00110    9 FØRMAT("THE JØINT PRØBABILITY =",F8.4//)
00120      END
```

Fig. 9.11.1 Program CDF1.

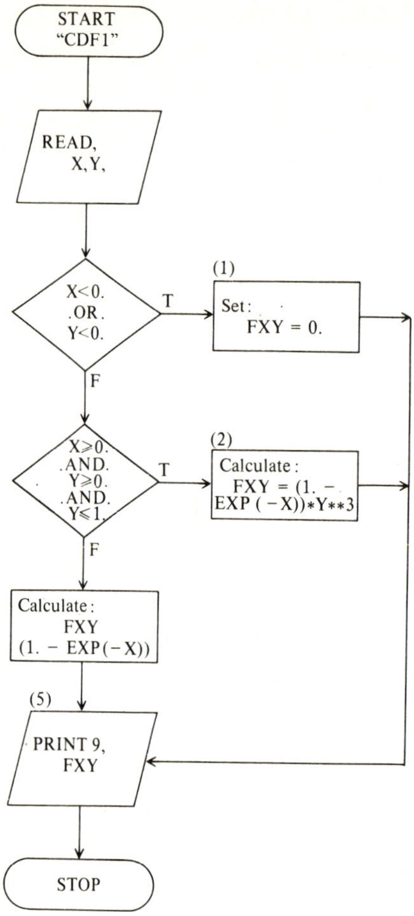

Fig. 9.11.2 Flow chart of program CDF1.

function on line 00060, and the GO TO statement on the next line causes the program to branch to statement 5 on line 00090. A trial run is shown in Fig. 9.11.3. In this trial, the four illustrative examples discussed above have been calculated.

```
INPUT:00010
?.1,-.1

THE JØINT PRØBABILITY =   0.0000


INPUT:00010
?-.1,.1

THE JØINT PRØBABILITY =   0.0000


INPUT:00010
?1.,.5

THE JØINT PRØBABILITY =   0.0790


INPUT:00010
?1.,2.

THE JØINT PRØBABILITY =   0.6321
```

Fig. 9.11.3 Trial runs for CDF1.

9.12 A RESISTIVE LADDER NETWORK

In this section we will analyze a program which calculates the branch currents and node voltages of a resistive network. The program is applicable to a network of the type shown in Fig. 9.12.1, generally called a ladder network. It should be noted that the subscript of each branch current is the same as the subscript of the resistance through which it flows. The subscript of each unknown node voltage is always the same as that of the shunt resistor at that node. The network is driven by a voltage source.

The analysis of the network is carried out by a step by step procedure which is easily applied to a ladder network. Beginning at the right, resistances are combined in groups of three by adding the two resistors which are in series and then calculating a resistance equivalent to the parallel resistances. In the program, the equivalent resistances are designated by the subscripted variable RE(J). The first calculation is

$$RE_{n-2} = \frac{(R_{n-2})(R_{n-1}+R_n)}{R_n + R_{n-1} + R_{n-2}} \tag{9.12.1}$$

The calculation indicated is carried out $(n/2 - 1)$ times, where n is the total number of resistors. (For a network of the type shown n is an even number.) The last equivalent resistance which is calculated is denoted RE_2. The current delivered by the voltage source is:

$$I_1 = \frac{V_1}{R_1 + RE_2} \tag{9.12.2}$$

where V_1 is the source voltage. V_2 and then I_2 and I_3 can be calculated by Eqs. (9.12.3), (9.12.4), and (9.12.5):

$$V_2 = V_1 - I_1R_1 \tag{9.12.3}$$

$$I_2 = \frac{V_2}{R_2} \tag{9.12.4}$$

$$I_3 = I_1 - I_2 \tag{9.12.5}$$

This procedure is repeated until all node voltages and branch currents have been found.

The program for performing these calculations is shown in Fig. 9.12.2. There is a REAL declaration (line 00010) so that the conventional name for current, I, can be used in real mode. The source voltage, the number of resistors, and the values of the resistors are entered by unformatted input (lines 00030 and 00040). The purpose of the DO loop on line 00050 is to assign the values stored in R(J) to RE(J). This does not, of course, affect the values stored in R(J).

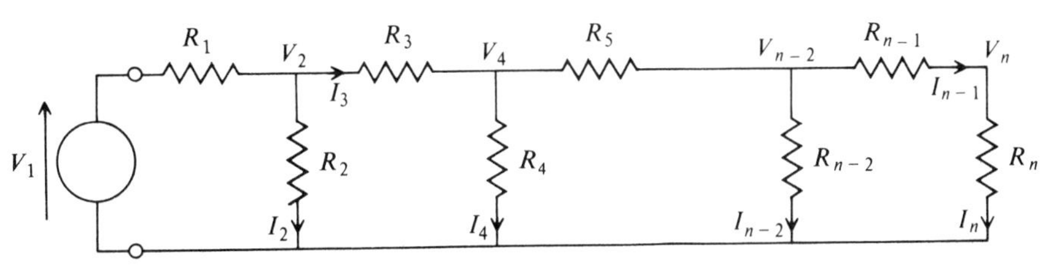

Fig. 9.12.1 A resistive ladder network.

```
00001C       LADD1
00002C       THIS PRØGRAM CALCULATES THE NØDE VØLTAGES AND BRANCH
00003C       CURRENTS ØF A NETWØRK ØF THE TYPE SHØWN IN
00004C       FIGURE 9.12.1 THE VALUES ØF THE RESISTANCES
00005C       MUST BE ENTERED IN THE ØRDER SHØWN IN THAT FIGURE.
00010        REAL I
00020        DIMENSIØN R(29),RE(29),I(29),V(20)
00030        READ,V(1)
00040        READ,N,(R(J),J=1,N)
00050        DØ 2 J=1,N
00060      2 RE(J)=R(J)
00070        MAX=N-2
00080        DØ 4 J=2,MAX,2
00090        K=N-J+1
00100        L=N-J+2
00110        M=N-J
00120      4 RE(M)=RE(M)*(RE(K)+RE(L))/(RE(M)+RE(K)+RE(L))
00130        RTØT=R(1)+RE(2)
00140        I(1)=V(1)/RTØT
00150        V(2)=V(1)-I(1)*R(1)
00160        I(2)=V(2)/R(2)
00170        DØ 6 J=4,N,2
00180        I(J-1)=I(J-3)-I(J-2)
00190        V(J)=V(J-2)-I(J-1)*R(J-1)
00200      6 I(J)=V(J)/R(J)
00210        PRINT 7
00220      7 FØRMAT("BRANCH CURRENTS")
00230        PRINT,(I(J),J=1,N)
00240        PRINT 8
00250      8 FØRMAT("NØDE VØLTAGES")
00260        PRINT,(V(J),J=2,N,2)
00270        STØP
00280        END
```

Fig. 9.12.2 Program LADD1.

Calculation of the equivalent resistances is carried out by the DO loop extending from line 00080 to line 00120. The reader is referred to Fig. 9.12.3 for an illustration of the reduction of a network consisting of six resistors. Note that program names are used to represent the network elements and that the resistors are represented by the array, RE(J), because the DO loop on line 00050 has assigned the values stored in R(J) to RE(J). Assume that the resistors R(1) to R(6) have the following values:[1] 15., 4.5, 6., 4.5, 3., and 6., respectively (see Fig. 9.12.5). As previously mentioned, after the DO loop which begins on line 00050 has been executed N times, the variables RE(J) have also been assigned these values. For example, RE(1) = 15., RE(2) = 4.5, and so forth. The program now passes to the DO loop which begins on line 00080. After one cycle of this DO loop, RE(4) has been assigned a new value (see Fig. 9.12.3*a*):

$$\text{RE}(4) = \frac{4.5(3.\ +\ 6.)}{4.5 + 3. + 6.} = 3.$$

Thus, the value stored in RE(4) changes from 4.5 to 3. After the second cycle of the DO loop, RE(2) has been assigned a new value (see Fig. 9.12.3*b*):

$$\text{RE}(2) = \frac{(4.5)(6. + 3.)}{4.5 + 6. + 3.} = 3.$$

The network, as viewed from the generator end, has now been reduced to two resistors as shown in Fig. 9.12.3*c*. The current I(1) is calculated as follows:

$$\text{I}(1) = \text{V}(1)/\text{RTOT} = 18./(15. + 3.) = 1.\ \text{amp}$$

[1] The same values of resistance have been used in the trial run (Fig. 9.12.4).

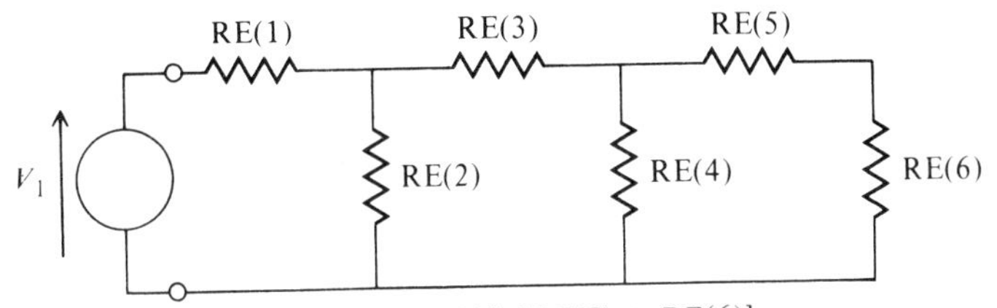

$$\text{RE}(4) = \frac{\text{RE}(4)\,[\text{RE}(5) + \text{RE}(6)]}{\text{RE}(4) + \text{RE}(5) + \text{RE}(6)}$$

(*a*)

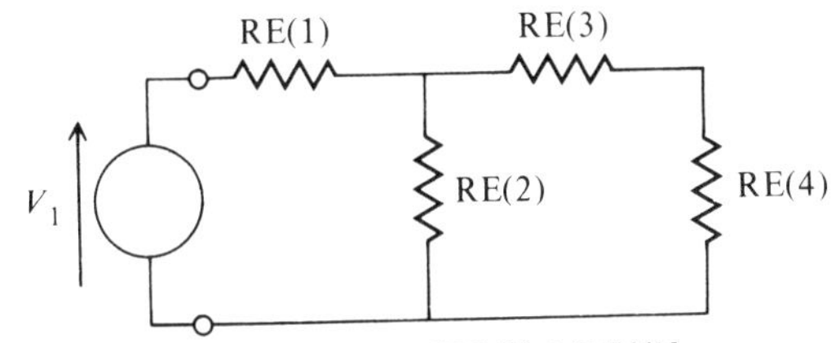

$$\text{RE}(2) = \frac{\text{RE}(2)[\text{RE}(3) + \text{RE}(4)]}{\text{RE}(2) + \text{RE}(3) + \text{RE}(4)}$$

(*b*)

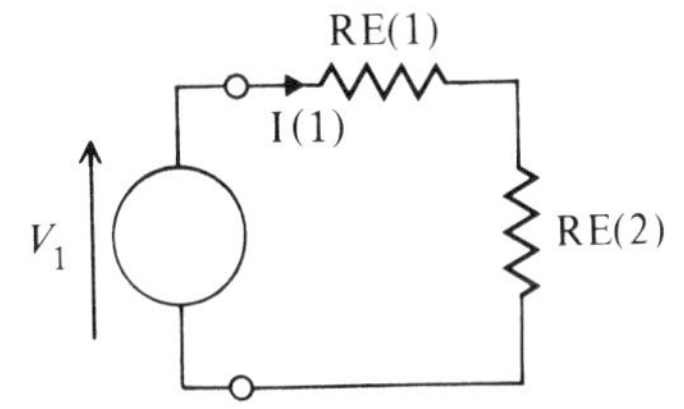

RTOT = RE(1) + RE(2)

I(1) = V(1) / RTOT

(*c*)

Fig. 9.12.3 Procedure for calculating equivalent resistances in a ladder network.

After I(1) has been calculated, the resistances RE(J) are not used. Following the calculation of I(1), I(2) and V(2) are calculated in accordance with Eqs. (9.12.3) and (9.12.4). Finally, I(3), I(4), and the other currents and voltages are calculated by

```
INPUT:00030
 ?18.

INPUT:00040
 ?6,15.,4.5,6.,4.5,3.,6.

BRANCH CURRENTS
        1.0000          0.6667          0.3333          0.2222          0.1111
        0.1111
NØDE VØLTAGES
        3.0000          1.0000          0.6667
STØP
```

Fig. 9.12.4 Trial run for LADD1.

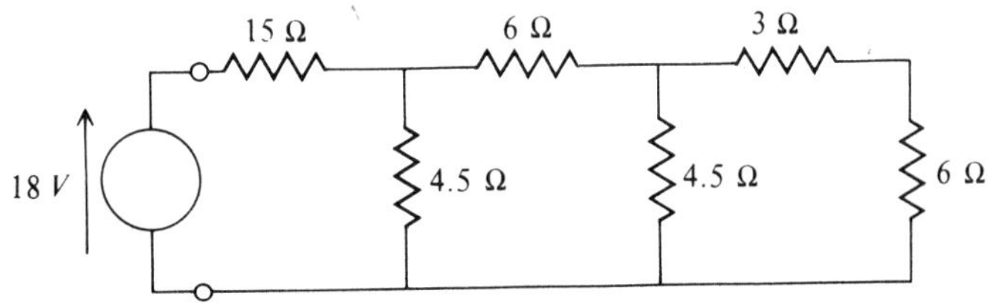

Fig. 9.12.5 Illustrative ladder network.

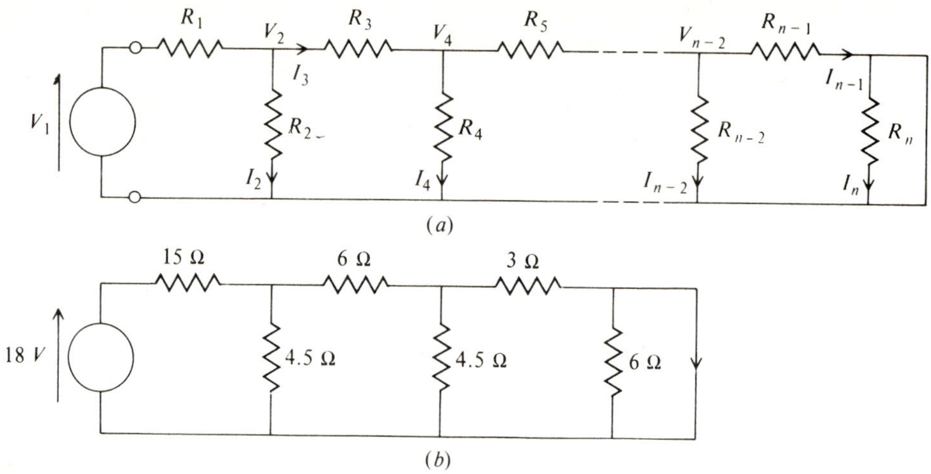

Fig. 9.12.6 (*a*) Short-circuited network; (*b*) typical resistances in a short-circuited network.

repeated use of Eqs. (9.12.3) to (9.12.5) in the DO loop beginning on line 00170. The branch currents and node voltages of the network shown in Fig. 9.12.5 have been calculated in the trial run of Fig. 9.12.4. Let us assume that we are interested in the branch currents and node voltages when the network is short-circuited. *Physically* we

```
00001C       LADD2
00002C       THIS PRØGRAM CALCULATES THE NØDE VØLTAGES AND BRANCH
00003C       CURRENTS ØF A NETWØRK ØF THE TYPE SHØWN IN
00004C       FIGURE 9.12.1 THE VALUES ØF THE RESISTANCES
00005C       MUST BE ENTERED IN THE ØRDER SHØWN IN THAT FIGURE.
00010        REAL I
00020        DIMENSIØN R(29),RE(29),I(29),V(20)
00030        READ,V(1)
00040        READ,N,(R(J),J=1,N)
00050        DØ 2 J=1,N
00060      2 RE(J)=R(J)
00070        MAX=N-2
00080        DØ 4 J=2,MAX,2
00090        K=N-J+1
00100        L=N-J+2
00110        M=N-J
00120      4 RE(M)=RE(M)*(RE(K)+RE(L))/(RE(M)+RE(K)+RE(L))
00130        RTØT=R(1)+RE(2)
00140        I(1)=V(1)/RTØT
00150        V(2)=V(1)-I(1)*R(1)
00160        I(2)=V(2)/R(2)
00170        DØ 6 J=4,N,2
00180        I(J-1)=I(J-3)-I(J-2)
00190        V(J)=V(J-2)-I(J-1)*R(J-1)
00200        IF(J.EQ.N)GØ TØ 10
00210      6 I(J)=V(J)/R(J)
00220     11 PRINT 7
00230        PRINT,(I(J),J=1,N)
00240        PRINT 8
00250        PRINT,(V(J),J=2,N,2)
00260        STØP
00270     10 IF(R(N).GT.0.)GØ TØ 6
00280        I(N)=I(N-1)
00290        V(N)=0.
00300        GØ TØ 11
00310      7 FØRMAT(15HBRANCH CURRENTS)
00320      8 FØRMAT(13HNØDE VØLTAGES)
00330        END
```

Fig. 9.12.7 Program LADD2.

can accomplish this either by placing a link of zero resistance *across* resistor R_n or by *replacing* R_n with the zero resistance link. We will adopt the former course, as shown in Fig. 9.12.6*a*. *As far as the computer program is concerned*, we would enter a value of zero for the 6-ohm resistor shown in Fig. 9.12.6*b*. In any case, the voltage at the last node is zero, and all the current entering this node flows through the zero resistance link.

The program shown in Fig. 9.12.2 is not suited for this case because division by zero would result if $R_n = 0$ (see line 00200). However, we can modify this program by using two logical IF statements. The required modifications have been incorporated in the program in Fig. 9.12.7. The modified program is more general than the original program because it will solve the network for either of the two cases.

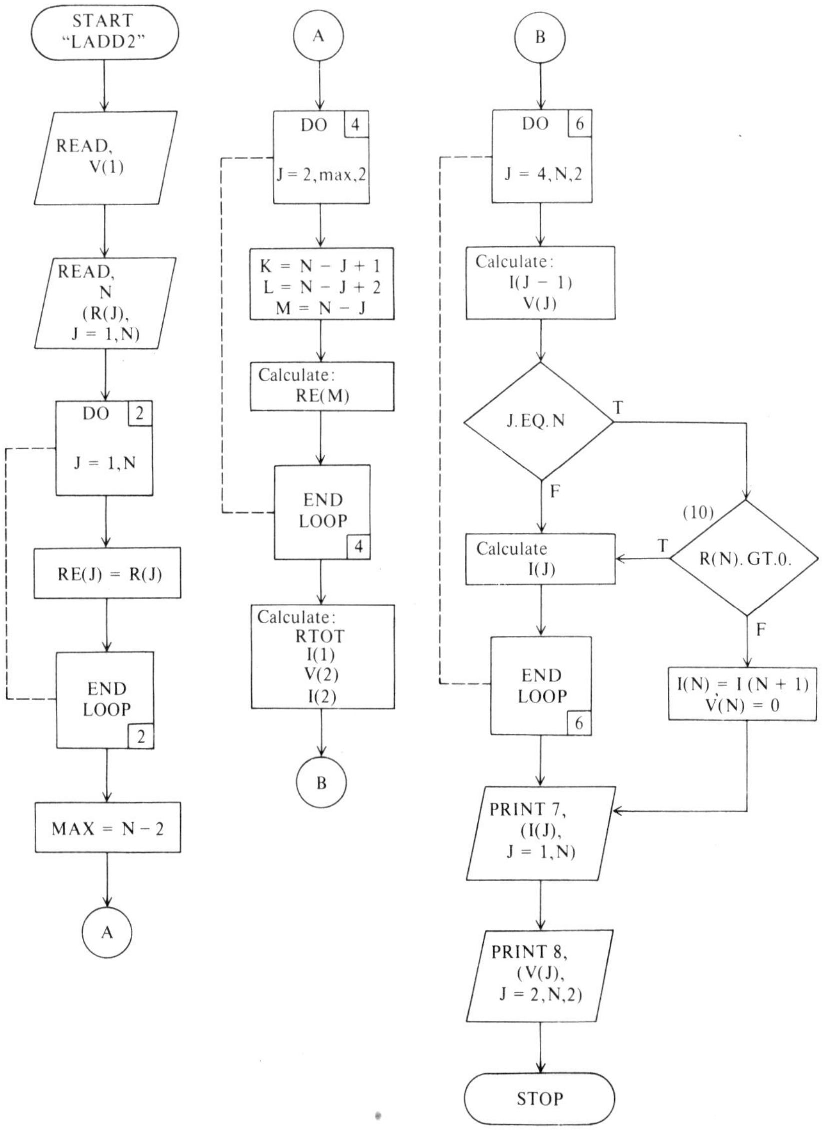

Fig. 9.12.8 Flow chart of program LADD2.

```
INPUT:00030
 ?18.


INPUT:00040
 ?6,15.,4.5,6.,4.5,3.,6.

BRANCH CURRENTS
          1.0000        0.6667        0.3333        0.2222        0.1111
          0.1111
NØDE VØLTAGES
          3.0000        1.0000        0.6667
STØP



INPUT:00030
 ?18.


INPUT:00040
 ?6,15.,4.5,6.,4.5,3.,0.


BRANCH CURRENTS
          1.0082        0.6393        0.3689        0.1475        0.2213
          0.2213
NØDE VØLTAGES
          2.8770        0.6639        0.0000
STØP
```

Fig. 9.12.9 Trial runs for LADD2.

Let us discuss the revisions. A logical IF statement has been added (line 00200) which tests the value of J, the index of the DO loop which begins on line 00170. When J = N, the statement is .TRUE., and the program branches to statement 10 on line 00270. If R(N) = 0., current I(N) is set equal to I(N − 1), and voltage V(N) is set to 0. Then the program branches to statement 11, and all the results are printed. However, if R(N) ≠ 0., the program branches back to statement 6 (line 00210), current I(N) is calculated, and, as the DO loop has been satisfied, the program branches to statement 11, and all results are printed. The flow chart in Fig. 9.12.8 shows these operations.

Two trial runs are shown in Fig. 9.12.9. The first trial run solves the network when R(N) ≠ 0, and yields the same answers as obtained in the run of Fig. 9.12.4. In the second trial run we test the short-circuit case when R(N) = 0. A more general version of this program is discussed in Chap. 12.

9.13 CYCLIC VALVE OPERATION

In this section we will construct a program to illustrate the use of logical constants and logical variables. The program deals with the position (open or closed) of valves in a hydraulic circuit after a given number of *operation cycles* have been completed. For illustration purposes we have assumed the circuit contains 100 valves, all in the open position before the sequence of operation cycles is started. During each operation, the position of certain valves is *reversed*; that is, if one of the selected valves is closed, it will be changed to the open position, and vice versa. The valves selected for reversal during a particular cycle are determined by the *number* of the operation. During the *first operation* cycle *every* valve is reversed; during the *second operation* cycle *every second valve* is reversed (i.e., valves 2, 4, 6, 8, etc.); during the *third operation* cycle *every third valve* is reversed (i.e., valves 3, 6, 9, 12, etc.); and so on. The problem was to determine the position of all the valves in the circuit after *N* cycles. It was also decided that if *N* were a large number, it would be interesting to inspect the positions of all the valves after the 25th, 50th, and 75th cycles. However, after *N* cycles, no matter how large, the problem was to be terminated.

Figure 9.13.1 is a flow chart of the procedure used to obtain a solution. The program itself is shown in Fig. 9.13.2. On line 00030, a logical IF statement limits the size of the number of operations to 100. (It would be meaningless, for example, for the 101st operation cycle to attempt to reverse the nonexistent 101st valve.) If N is greater than 100, N is set equal to 100 and the program advances to DO loop 1 on line 00040; if N is *not* greater than 100, no change is made and the program moves to line 00040.

The DO loop on lines 00040 and 00050 sets each element of the array V, which was declared to be of *logical* type in the type declaration statement on line 00010, equal to a value .TRUE. . If we follow the convention assumed for this problem that .TRUE. is equivalent to open and .FALSE. is equivalent to closed, the initialization procedure of DO loop 1 is tantamount to setting all valves in the open position prior to starting the sequence of operation cycles.

DO loop 2, lines 00060 through 00200, determines the operation cycle I that is to be performed. The nested DO loop 3, lines 00070 through 00120, controls the

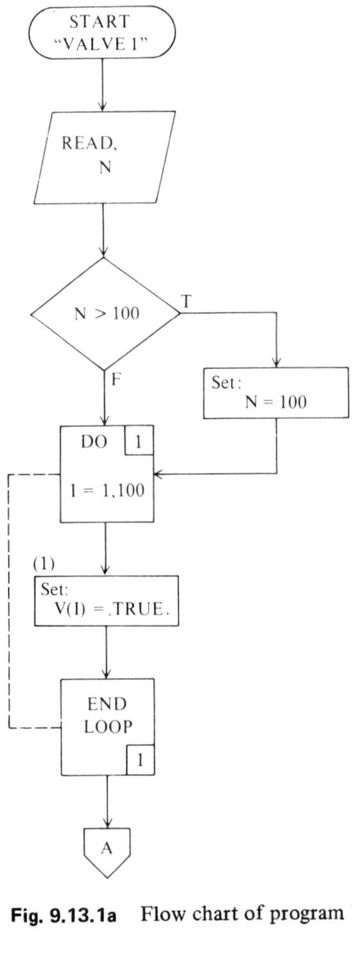

Fig. 9.13.1a Flow chart of program VALVE1.

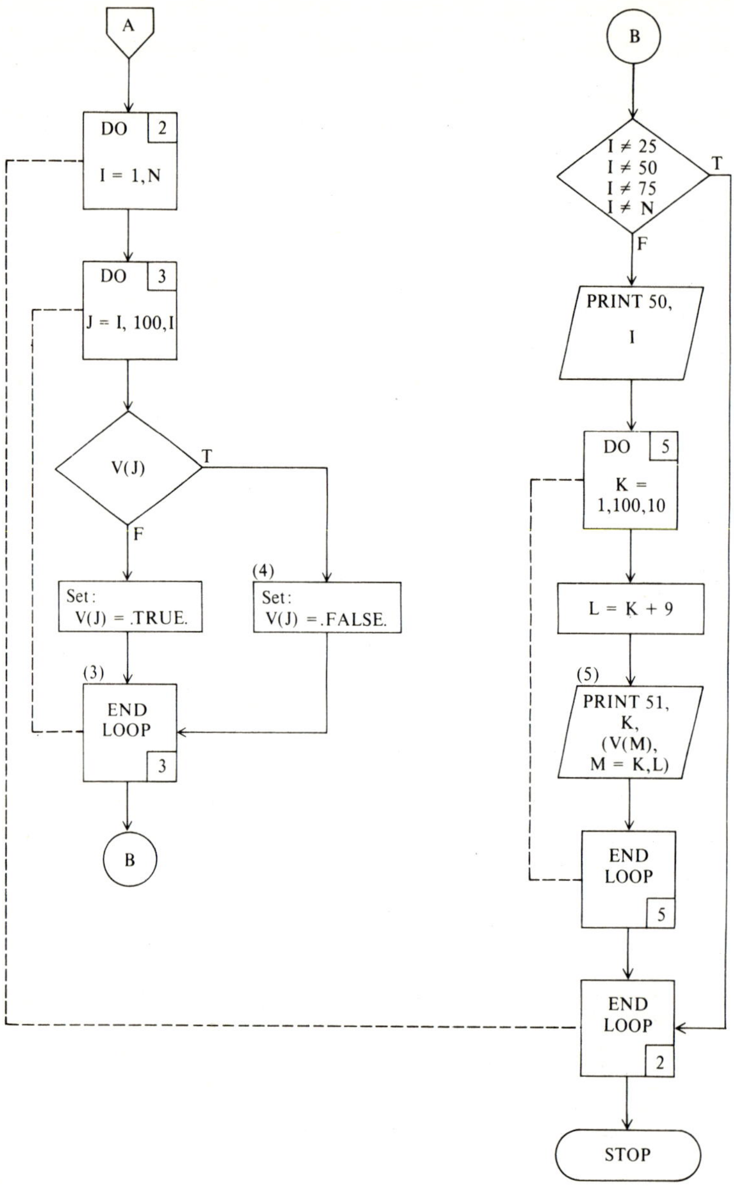

Fig. 9.13.1b Flow chart of program VALVE1.

```
00001C     VALVE1
00002C     PRØGRAM TØ DETERMINE THE PØSITIØNS ØF 100 VALVES IN
00003C     A HYDRAULIC CIRCUIT AFTER EVERY 25TH ØPERATIØN AND AFTER
00004C     "N" ØPERATIØNS.  ("T"=ØPEN; "F"="CLØSED)
00010      LØGICAL V(100)
00020      READ,N
00030      IF(N.GT.100) N=100
00040      DØ 1 I=1,100
00050    1 V(I)=.TRUE.
00060      DØ 2 I=1,N
00070      DØ 3 J=I,100,I
00080      IF(V(J))GØ TØ 4
00090      V(J)=.TRUE.
00100      GØ TØ 3
00110    4 V(J)=.FALSE.
00120    3 CØNTINUE
00130      IF(I.NE.25.AND.I.NE.50.AND.I.NE.75.AND.I.NE.N)GØ TØ 2
00140      PRINT 50,I
00150   50 FØRMAT(/9HØPERATIØNI4/)
00160      DØ 5 K=1,100,10
00170      L=K+9
00180    5 PRINT 51,K,(V(M),M=K,L)
00190   51 FØRMAT(I4,2X10L3/)
00200    2 CØNTINUE
00210      STØP
00220      END
```

Fig. 9.13.2 Program VALVE1.

selection of the *i*th valve (and every *i*th valve thereafter) by setting J equal to the current value of I and incrementing J in steps equal to I. The reader should note that if N had been permitted to assume values greater than 100 (see discussion above regarding line 00030), an invalid set of DO loop parameters would have eventually been produced. That is, if *N* were greater than 100, I would eventually reach a value equal to 101 and the DO statement on line 00070 would be interpreted by the computer to be DO 3 J = *101*,100,*101*. As discussed in Sec. 5.2, the test value must not be smaller than the initial value in a DO statement.

The logical IF statement on line 00080 interrogates the value of V(J). If the value is found to be .TRUE., control is transferred to statement 4 on line 00110, where the value is changed to .FALSE. . If the value of V(J) is found to be .FALSE. (line 00080), control passes to line 00090 where the value is reset to .TRUE. . When every *i*th valve has been reversed, that is, when DO loop 3 has been satisfied, the program advances to the logical IF statement on line 00130. If I has a value *other than* 25, or 50, or 75, or N, the IF statement argument is *true* and the statement GO TO 2 is executed. This transfer statement skips over the output statements to the end of DO loop 2 where I is incremented to produce the next operation cycle. However, if I *is* equal to *any one* of the values tested (i.e., 25, or 50, or 75, or N), the IF statement argument is *false* and control passes to the PRINT statement on line 00140. At this point the number of the operation cycle is printed, followed by a matrix, arranged in rows of 10, showing the position of each of the valves. The results are printed under control of an L type conversion, using a specification (line 00190) of 10L3. The values are displayed, in the normal manner, as T and F, representing, respectively, the values .TRUE. and .FALSE. and, therefore, the valve positions open and closed.

The output of a sample run, in which a value of 38 was entered for N, is shown in Fig. 9.13.3.

```
INPUT:00020
 ?38

ØPERATIØN  25

   1    F  T  T  F  T  T  T  T  F  T

  11    T  T  T  T  T  F  T  T  T  T

  21    T  T  T  T  F  F  F  F  F  F

  31    F  F  F  F  F  T  F  F  F  F

  41    F  F  F  F  F  F  F  F  T  F

  51    F  T  F  T  F  T  F  T  F  T

  61    F  T  F  F  F  T  F  T  F  T

  71    F  T  F  T  F  T  F  F  F  T

  81    F  T  F  F  F  T  T  T  F  F

  91    F  T  T  T  F  F  F  T  T  F

ØPERATIØN  38

   1    F  T  T  F  T  T  T  T  F  T

  11    T  T  T  T  T  F  T  T  T  T

  21    T  T  T  T  F  T  T  T  T  T

  31    T  T  T  T  T  F  T  T  F  F

  41    F  F  F  F  F  F  F  F  T  F

  51    F  F  F  F  F  F  F  F  F  F

  61    F  F  F  T  F  F  F  F  F  F

  71    F  F  F  F  F  F  F  T  F  T

  81    T  T  F  T  F  T  F  T  F  T

  91    F  T  F  T  F  T  F  T  F  F

STØP
```

Fig. 9.13.3 Trial run for VALVE1.

PROBLEMS

9-1. In Boolean algebra the EXCLUSIVE OR means either A or B but not both. One logical statement which illustrates the EXCLUSIVE OR is

$$Y = A \cdot \overline{B} + B \cdot \overline{A}$$

where · means .AND.

+ means .OR.

$\overline{A}$ means .NOT.A

$\overline{B}$ means .NOT.B

The logical statement given above must be read: "If A is true and B is false, or if B is true and A is false, then Y is true." A table which shows the values of Y as A and B assume all possible values is called a *truth table.* The truth table for the logical statement given above is:

A	B	Y
F	F	F
F	T	T
T	F	T
T	T	F

Write a FORTRAN program which evaluates the logical statement given above for all possible values of A and B and prints the truth table shown above.

9-2. Write a program which evaluates the logical statement shown below for all possible values of A and B and prints a truth table such as shown in Prob. 9-1.

$$Y = (A + B)\cdot(\bar{A} + \bar{B})$$

The results should show that the logical statement in this problem is equivalent to that given in Prob. 9-1.

9-3. Modify the program of Sec. 9.12 so that the resistances may be impedances *and* the last element *may* be short-circuited. (Note that we are testing for a short-circuited impedance, which may be complex, and therefore the program must test both parts of the complex value to determine if they are zero.)

Sample data

$$Z_1 = 0 + j2$$
$$Z_2 = 0 - j2$$
$$Z_3 = 1 + j0$$
$$Z_4 = 0 + j1$$
$$V_1 = 4 + j0$$

9-4. The following diagram represents the water supply system in a factory:

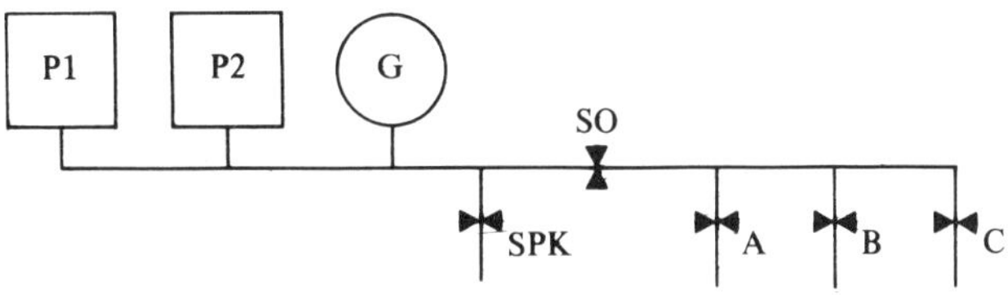

where P1, P2 are pumps, G is a pressure gage, SPK is a valve supplying the factory sprinkler system, SO is a shutoff valve to the process operations, and A, B, C are valves supplying three separate processes. The requirements of the system are:

1. P1 is to run if one or more of valves A, B, C are open.
2. P2 is to run if P1 is running and G shows that the line pressure has dropped below normal.
3. SPK is to be opened if a fire occurs.
4. Both P1 and P2 are to run if SPK is open.
5. SO is to be closed if SPK is open and G shows that the line pressure has dropped.

The states of the various components of the system can be represented by the symbols T and F in a computer printout as shown below:

Component	*State*	*Symbol*
Output		
Pumps P1 and P2	Running	T
	Stopped	F
Valve S0	Open	T
	Closed	F
Input		
Valves A, B, C, SPK	Open	T
	Closed	F
Pressure gage G	Normal	T
	Low	F

Write a program that will cycle through all the possible combinations of conditions of the five input devices and will determine the condition of each of the output devices. The program should print the state of each element of the system for each cycle.

10
The Use of Functions and Subprograms

In several of the programs discussed in the earlier chapters it was necessary to repeat program statements at different places in the program when the same type of calculation was to be made with different arguments. This repetition of similar instructions can be avoided in FORTRAN programming by using functions that are defined and executed within the main program, or by using subprograms, defined in addenda to the main program, that are executed whenever "called" by the main program. The types of functions and subprograms that are available to the programmer and the rules governing their use are described in this chapter.[1] Several problems are discussed in which this feature of the FORTRAN language is employed in programming a solution.

10.1 EQUIVALENCE STATEMENT

The *EQUIVALENCE statement* is a nonexecutable statement of the form

$$\text{EQUIVALENCE } (v_1, v_2, v_3, \ldots), (w_1, w_2, w_3, \ldots)$$

where $v_1, v_2, v_3, \ldots$ and $w_1, w_2, w_3, \ldots$ represent lists of variables, subscripted variables with *constant* subscripts, or array names. Each list must contain at least two names. Any number of lists may be equivalenced.

The EQUIVALENCE statement causes the names in each list to refer to the same storage location in the computer memory. Obviously, the variables in each list must all be in the same mode.

[1] The BLOCK DATA statement is discussed in Sec. 11.4.

The EQUIVALENCE statement can be used to "patch" program segments that were written by different programmers into one main program without renaming the storage locations.[1] It can also be used to conserve storage space in the computer.[2]

The statement

EQUIVALENCE (X,SET,B),(Y,TEST),(ITEM,JOB)

would cause three storage locations to be reserved and referenced by any of the following indicated names:

Location	*Names*
1	X or SET or B
2	Y or TEST
3	ITEM or JOB

Arrays can be equivalenced either by naming a *particular element* of the array in one of the variable lists or by merely using the *name of the array* in one of the lists.

When a particular element of an array is named in an EQUIVALENCE statement, *that element* is equivalenced and elements with lower numbered subscripts occupy preceding storage locations, while higher numbered elements occupy succeeding storage locations.

When the array name is used in an EQUIVALENCE statement, the *first element* is equivalenced and the following elements occupy succeeding storage locations.

For example, the statements

DIMENSION X(3),Y(5)
EQUIVALENCE (SUM,X,Y(2))

would cause five storage locations to be reserved and referred to by any of the following indicated names:

Location	*Names*
1	Y(1)
2	Y(2) or X(1) or SUM
3	Y(3) or X(2)
4	Y(4) or X(3)
5	Y(5)

In a multidimensional array, the *first* subscript is varied most rapidly, then the second, etc. An array dimensioned as

DIMENSION A(3,3)

[1] For example, if two programmers, working independently, were to use the symbolic names X and V for voltage in preparing a program, their work could be combined in a single program *without* revising either segment by inserting the statement EQUIVALENCE (X,V) at the beginning of the combined program.

[2] If a programmer were to use storage locations X, Y, and Z in obtaining an answer to be stored in W (and had no further use for the values stored in X, Y, and Z), and later in the program used locations A, B, and C in a similar manner to obtain and store a result in U, he would ordinarily not be concerned about the additional storage requirements. However, if the program were so large that it became necessary to conserve storage, he could either *reuse* the locations X, Y, and Z themselves in evaluating U, or he could, as an afterthought, simply insert EQUIVALENCE (X,A), (Y,B), (Z,C) early in the program to accomplish the same result.

would be stored in the order:

A(1,1) A(2,1) A(3,1) A(1,2) A(2,2) A(3,2)
A(1,3) A(2,3) A(3,3)

10.2 SUBPROGRAMS

A *subprogram* is a set of computer instructions written in either FORTRAN or machine language that is intended to perform a specific operation as part of a main program. The operation may be evaluating some function, sorting a set of variables according to some scheme, performing some special series of calculations, etc. All these operations, of course, could be written as FORTRAN instructions in the main program at the point where they were required. However, if a particular operation were used several times in a main program, the complete set of instructions for effecting the operation would have to be repeated at each point of use. Subprograms eliminate this need for repetition and make it possible to write the set of instructions once, and then refer to it each time it is needed in the main program.

The types of subprograms that are used in the FORTRAN language are:

1. Compiler-supplied functions
2. Statement functions
3. Function subprograms
4. Subroutine subprograms

1. *Compiler-supplied functions* were discussed in Chap. 1 (see Secs. 1.21 to 1.23).
2. The *statement function* or *arithmetic function* consists of a *single* FORTRAN statement, located in the main program, that *defines* a particular function that is to be evaluated (see Sec. 10.3). When it is *used* in the main program, a statement function is *called implicitly* by using the function name assigned to the function in the defining statement in a FORTRAN expression. [This is the same calling procedure as is used for a compiler-supplied function (see Sec. 1.21).] An arithmetic function returns a *single value* to the main program.
3. The *function subprogram* is a complete FORTRAN program consisting of one or more executable statements that is appended to the main program. It is used where the single-statement limitation of the arithmetic function makes it difficult to describe the steps necessary for a function evaluation. A function subprogram has the following parts:

   a A FUNCTION defining statement (see Sec. 10.5).
   b A DIMENSION statement if any of the arguments are subscripted (see Sec. 7.2).
   c A COMMON statement if values are to be transmitted implicitly between the main program and the subprogram (see Sec. 10.8).
   d As many FORTRAN statements as are required to perform the intended operations. (*No* FUNCTION, SUBROUTINE, or BLOCK DATA statements may be used.)
   e At least one RETURN statement (see Sec. 10.4).
   f As many ENTRY statements as are required (see Sec. 10.9).
   g An END statement.

   The function subprogram is called by the main program in the same manner as the arithmetic function, i.e., by using the assigned name of the function in a

FORTRAN expression. This implicit call returns a *single value* to the main program.

4. The *subroutine subprogram* is a complete FORTRAN program consisting of one or more statements that is appended to the main program. It can be used to return *one or more values* to the main program, or to perform some operation (e.g., sorting) other than the evaluation of variables. A subroutine subprogram consists of the following parts:

   a A SUBROUTINE defining statement (see Sec. 10.6).
   b A DIMENSION statement if any arrays are used (see Sec. 7.2).
   c A COMMON statement if values are to be transmitted implicitly between the main program and the subprogram (see Sec. 10.8).
   d As many FORTRAN statements as are required to perform the intended operations. (*No* FUNCTION, SUBROUTINE, or BLOCK DATA statements may be used.)
   e At least one RETURN statement (see Sec. 10.4).
   f As many ENTRY statements as are needed (see Sec. 10.9).
   g An END statement.

   The subroutine subprogram is called explicitly by the main program with a CALL statement (see Sec. 10.7).

In the following sections we will consider the various types of subprograms in greater detail.

10.3 ARITHMETIC FUNCTION DEFINING STATEMENT

The *arithmetic function* or *statement function* is a nonexecutable FORTRAN statement having the form

$$name\ (v_1, v_2, v_3, \ldots, v_m) = exp$$

where *name* is the name assigned to the function by the programmer, and $v_1, v_2, v_3, \ldots, v_m$ are nonsubscripted *dummy arguments* used in *defining* the arithmetic expression *exp.*[1] The *name* of the function must not be the same as that of any compiler-supplied function. The usual procedure for naming variables is used to indicate the *mode* of the function value.

The arithmetic function statement serves to *define* the function and must appear in the source program *before* the first use of the function. The function is *used* or *called* by writing the name of the function in an expression or replacement statement in the main program. When the function is used, the *actual arguments* to be used in the function are placed in the parentheses after the function name. The *actual arguments* must agree with the *dummy arguments* in *mode*, and must occupy positions in the list that correspond with the positions of the dummy arguments they are replacing.
The *actual arguments* to be substituted may be:

1. Expressions
2. Subscripted variables
3. Compiler-supplied functions

[1] That is, the function may be defined using symbolic names other than those of the variables which will actually be used during the execution of the program.

The following are examples of the use of the arithmetic function:

1. If the following statement defines a function

   TEST(A,B) = A*A–B*B

   (where A and B are dummy arguments), the statement

   RATIO = TEST(X,Y)/TEST(Y,Z)

   (where X and Y *and* Y and Z are the actual arguments) will calculate the value of

   $$\frac{X^2 - Y^2}{Y^2 - Z^2}$$

   and store the value in location RATIO.
2. If the following statement defines a function

   SAMP(P,Q) = SQRT(P+Q)

   the statement

   ANS = C*SAMP(X,SIN(Y))

   will calculate the value of

   $$C\sqrt{X + \sin Y}$$

   and store it in location ANS.

10.4 RETURN STATEMENT

The *RETURN statement* is an executable statement used in connection with subprograms. This statement causes the computer to transfer control to the *calling program* and resume execution at the statement immediately following the *calling statement.*

10.5 FUNCTION DEFINING STATEMENT

The *FUNCTION statement* is a nonexecutable statement of the form

FUNCTION *name* $(v_1, v_2, \ldots, v_m)$

where *name* is the name assigned to the function by the programmer, and $v_1, v_2, \ldots, v_m$ are the names of the *dummy arguments* used in writing the subroutine.

The *name* of the FUNCTION subprogram must not be the same as that of any compiler-supplied function in the system library. The usual procedure for naming variables is employed to indicate, implicitly, the mode of the value that is to be returned by the function. Alternatively, the mode of the function can be explicitly declared as part of the defining statement. That is,

REAL FUNCTION *name* $(v_1, v_2, \ldots, v_m)$

or

INTEGER FUNCTION *name* $(v_1, v_2, \ldots, v_m)$

would return real or integer values, respectively, regardless of the initial characters in the names assigned to the functions.

The *dummy arguments* listed in the FUNCTION defining statement may be:

1. Nonsubscripted variables
2. Names of arrays
3. Names of *other* functions (see Sec. 10.10)

There must be at least one argument in the list.

The FUNCTION statement is always the first statement in a function subprogram that describes the operations to be performed in evaluating a particular function. The *name* assigned to the function must appear at least once in the subprogram on the *left side* of a replacement statement. This procedure is necessary so that the value of the function can be returned to the main program.

A function subprogram is called implicitly by writing the assigned name of the function [or one of its alternative names (see Sec. 10.9)] in an expression or replacement statement in the main program. When the function subprogram is called, the *actual arguments* to be used are placed in the parentheses following the function name. The actual arguments must agree, in mode and position in the list, with the dummy arguments. The actual arguments may be:

1. Arithmetic expressions
2. Subscripted variables
3. Other functions (see Sec. 10.10)

The following is an example of the use of the FUNCTION subprogram:

If the subprogram is defined by the statements

```
FUNCTION DEMO (ARRAY,X,Y,I)
DIMENSION ARRAY(10)
SUM = X*X+Y*Y
DEMO = ARRAY(I)/SUM
RETURN
END
```

and if the main program contains the statements

```
DIMENSION B(5)
ANSWER = C*DEMO(B,9.0,C,3)
```

the function subprogram will evaluate

```
SUM = 9.0*9.0+C*C
DEMO = B(3)/SUM
```

and return this value to the expression being used to evaluate ANSWER.

10.6 SUBROUTINE DEFINING STATEMENT

The *SUBROUTINE statement* is a nonexecutable statement of the form

SUBROUTINE *name* $(v_1, v_2, \ldots, v_m)$

where *name* is the name of the subroutine and $v_1, v_2, \ldots, v_m$ are the names of the dummy arguments used in defining the subroutine.

Unlike the FUNCTION statement, it is not essential that a list of arguments be part of a SUBROUTINE statement because a subroutine subprogram may perform operations other than those which calculate and assign new values to variables.[1] Also, a subroutine subprogram that actually does calculations may employ the COMMON statement (see Sec. 10.8) to transmit the values of the arguments between the main program and the subprogram, and again no list of arguments is needed. In those

[1] For example, in a program in which several tables of the same type are to be printed, a SUBROUTINE subprogram could be used to print the table heading. This would only be advantageous if the heading contained several lines.

instances where a list of dummy arguments *is* included in the SUBROUTINE statement, the arguments may be:

1. Nonsubscripted variables
2. Names of arrays
3. Names of *other* functions (see Sec. 10.10)

The *name* assigned to the SUBROUTINE must be unique. However, the initial character of the *name* does not have to imply a particular data mode because the subprogram name is not used for returning a value to the main program. All values to be returned must be designated by dummy arguments in the list, except where COMMON storage is used for transmittal.

The SUBROUTINE subprogram is called explicitly by means of the CALL statement (see Sec. 10.7). When a subprogram is called, the list of *actual arguments* to be substituted in the subprogram must agree with the list of *dummy arguments* in mode and position and may contain:

1. Expressions
2. Subscripted variables
3. Array names
4. Other functions (see Sec. 10.10)

The following is an example of the use of the SUBROUTINE subprogram:

If a subprogram is defined by the statements

```
   SUBROUTINE CALC (A,B,N,SUM,SUMSQ)
   DIMENSION A(10),B(10)
   SUM = 0.
   SUMSQ = 0.
   DO 5 I=1,N
   SUM = SUM+A(I)*B(I)
5  SUMSQ = SUMSQ + (A(I)*B(I))**2
   RETURN
   END
```

and the calling program contains the statements

```
   DIMENSION TEST1(5),TEST2(5)
   CALL CALC (TEST1,TEST2,3,X,Y)
```

the subroutine will use the first *three* elements of each of the arrays, TEST1 and TEST2, and return the sum of the element products to the main program in storage location X, and the sum of the squares of the element products to location Y.

10.7 CALL STATEMENT

The *CALL statement* is an executable statement of the form:

CALL *name* $(a_1, a_2, a_3, \ldots)$

where *name* is the name of a SUBROUTINE subprogram and $a_1, a_2, a_3, \ldots$ is the list of arguments to be used in the subroutine.

This statement causes the computer to transfer control to the *start* of the *called subprogram* where execution proceeds. The CALL statement can also be used to transfer to some point *within* a called subprogram if such a point has been defined by an ENTRY statement (see Sec. 10.9). When used in this manner, the *name* in the CALL statement is the *label of the ENTRY point.*

Some examples of the CALL statement are

```
CALL TEST (J2,X,Y,COSF(Z))
CALL SUM (X,Y,XY)
```

10.8 COMMON STATEMENT[1]

The *COMMON statement* is a nonexecutable statement of the form

COMMON $v_1, v_2, v_3, \ldots, v_n$

where $v_1, v_2, v_3, \ldots, v_n$ is a list of variable names or array names.

The COMMON statement causes the names in the list to be assigned to storage locations in a special section of the computer memory called the *common block.* The storage locations are assigned in the same sequence as that in which the names are listed in the statement.

PART A

Since the common block storage area is accessible to the *main program* as well as the *subprograms*, the COMMON statement provides a means of implicitly communicating data between segments of the programs.

For example, if a main program contained the statement

```
COMMON A,B,C,M,N
```

and one of the subprograms "called" by the main program contained the statements

```
SUBROUTINE GRASS
COMMON X,Y,Z,I
Z = X**I
RETURN
END
```

then:

1. The value stored in A would be used as the value of X in the subprogram.
2. Y is not used in this subprogram; therefore it is a dummy and acts as a "spacer" in the statement (presumably B and N are used in some other subprogram).
3. The value stored in M would be used as the value of I in the subprogram.
4. The calculated value of Z would be stored in location C when control was returned to the main program.

Usually only one COMMON statement is used in any program segment. However, the reader should note that *allocation of storage in the common block is made consecutively.* If more than one COMMON statement is used in a segment,[2] the storage positions assigned to the variables listed in the *second* COMMON statement will *follow* the positions assigned to the variables listed in the *first* COMMON statement. For example, if the main program contained the statements

```
COMMON A, B, X, Y
. . . . . . . . . .
COMMON U, V, W
```

[1] As in earlier chapters, Part B of this section may be omitted on a first reading.
[2] This procedure is not recommended.

the allocation of storage space would be the same as if the following *single* statement had been used:

COMMON A, B, X, Y, U, V, W

If an array name is listed in a COMMON statement, either the array must have been dimensioned in a *previous* DIMENSION statement (Sec. 7.2) or the array must be dimensioned in the COMMON statement. If the latter method is used, the array size must *not* be specified in a DIMENSION statement.

For example, the storage allocation produced by the set of statements

DIMENSION A(10), B(5)
COMMON A, B, X, Y, Z

could also be produced by the single statement

COMMON A(10), B(5), X, Y, Z

The distinctive functions performed by the EQUIVALENCE and COMMON statements should be noted by the reader. An EQUIVALENCE statement assigns two or more variables *within the same program* to a single storage location. The COMMON statement assigns two or more variables in *different programs* (i.e., main and subprograms) to a single storage location in a *special portion* of the computer memory.

Two or more variables *assigned to common storage* should *not* be equivalenced. However, a variable in a program segment may be equivalenced to *another variable* in the *same segment* even though the second variable is stored in the common storage area.

PART B

The common storage area described in Part A is generally referred to as ***blank** common.* There is also an additional storage feature called *labeled common.* The labeled common differs from blank common in that any number of special segments of the common storage area may be designated and addressed by labels that are chosen by the programmer.

The general form of the *labeled* COMMON statement is

COMMON/$label_1$/v_1, v_2, v_m/$label_2$/v_1, v_2, . . ., v_n

where $label_1$ and $label_2$ (enclosed between slashes) are names[1] assigned to special segments of the common storage block, and the v's are lists of variable names or array names.

This statement can be used to store values in both blank and labeled common. The *blank* common is designated by omitting the label name from between a pair of slashes. For example,

COMMON//A,B,C/AREA/X,Y,Z

would locate variables A, B, and C in blank common, and variables X, Y, and Z in a segment of the common block labeled AREA.

To illustrate the use of the *labeled* COMMON statement, let us consider a

[1] The names assigned as labels of the special storage segments are composed of strings of one to six alphameric characters in which the first character must be alphabetic.

program consisting of a main program and three subprograms called SUB1, SUB2, and SUB3. The common storage block is to be used to implicitly transfer values between program segments. Let us also assume that the set of main program values used by one of the subprograms is different from the sets used by the other subprograms.

For example, suppose the main program contains the variable names

A, B, C, D, E, F, G, P, Q, R

and subprogram SUB1 has been written using the variable names

X1, Y1, Z1

Let us further suppose that when SUB1 is called by the main program, it must operate with the values stored in locations A, B, and C in the main program. That is, in SUB1, variable X1 must be assigned the contents of A; Y1 must be assigned the contents of B; and Z1 must be assigned the value stored in C. To summarize:

SUB1 is written with variables X1, Y1, Z1 and requires the values from A, B, C.

Let us also assume that

SUB2 is written with variables W2, X2, Y2, Z2 and requires the values from D, E, F, G.

SUB3 is written with variables V3, W3, X3, Y3, Z3 and requires the values from A, F, P, Q, R.

Implicit transmission of data between the program segments could be achieved by including the COMMON statements shown in the program outlined below:

```
MAIN PROGRAM
COMMON//A,B,C/SET1/D,E/SET2/F,G/AREA/P,Q,R
. . . . . . . . . . . . . . . . . . . . . . . . . . . . . . . . . . . . . . . . .
. . . . . . . . . . . . . . . . . . . . . . . . . . . . . . . . . . . . . . . . .
END

SUBROUTINE SUB1
COMMON X1,Y1,Z1
. . . . . . . . . . . . . . . . .
. . . . . . . . . . . . . . . . .
END

SUBROUTINE SUB2
COMMON/SET1/W2,X2/SET2/Y2,Z2
. . . . . . . . . . . . . . . . . . . . . . . . . . . . .
. . . . . . . . . . . . . . . . . . . . . . . . . . . . .
END

SUBROUTINE SUB3
COMMON//V3/SET2/W3/AREA/X3,Y3,Z3
. . . . . . . . . . . . . . . . . . . . . . . . . . . . . . .
. . . . . . . . . . . . . . . . . . . . . . . . . . . . . . .
END
```

The COMMON statement in the main program would place A, B, and C in the *blank* common block; D and E in a *labeled* common block identified as SET1; F and G in a *labeled* common block identified as SET2; and P, Q, and R in a *labeled* common block identified as AREA.

In subprogram SUB1 the statement:

COMMON X1,Y1,Z1

refers to the *blank* common block. When this subprogram is called by the main program, the following assignments will be made:[1]

X1 = A
Y1 = B
Z1 = C

In SUB2 the statement

COMMON/SET1/W2,X2/SET2/Y2,Z2

refers to two areas in the *labeled* common block. When this subprogram is called by the main program, the following assignments will be made:

W2 = D
X2 = E
Y2 = F
Z2 = G

In SUB3 the statement

COMMON//V3/SET2/W3/AREA/X3,Y3,Z3

refers to the *blank* common and two areas in the *labeled* common block. When this subprogram is called, the following assignments will be made:

V3 = A
W3 = F
X3 = P
Y3 = Q
Z3 = R

The relationships between all the variable names used in this example are summarized below:

	COMMON storage block									
Program	*Blank COMMON*			*Labeled COMMON*						
	(Unnamed)			SET1		SET2		AREA		
Main	A	B	C	D	E	F	G	P	Q	R
SUB1	X1	Y1	Z1							
SUB2				W2	X2	Y2	Z2			
SUB3	V3					W3		X3	Y3	Z3

[1] The *reverse* assignment will be made on the RETURN from the subprogram to the main program. That is,

A = X1
B = Y1
C = Z1

One of the advantages of using labeled common is the fact that dummy variables do not have to be placed in the subprogram COMMON statements to act as "spacers."

10.9 ENTRY STATEMENT[1]

The *ENTRY statement* is a nonexecutable statement having the form

ENTRY *name* $(v_1, v_2, \ldots, v_m)$

where *name* is the label assigned to the entry point by the programmer, and $v_1, v_2, \ldots, v_m$ are dummy arguments.

The ENTRY statement is used in writing function and subroutine subprograms to make provision for transferring from the main program to a point *within* the subprogram. It should be noted that the form of the ENTRY statement is similar to that of the FUNCTION and SUBROUTINE defining statements.

Transfer to a particular entry point is effected by calling the subprogram, either implicitly or explicitly, with the name that has been assigned to the desired point of entry. Obviously, the names assigned to all entry points must be distinctive and the rules used for naming subprograms must be applied.

When a *call* to an entry point is made in the main program, the *actual arguments* in the list must correspond, in mode and position, with the *dummy arguments* listed in the ENTRY statement. However, it is not necessary that the list of dummy arguments in an ENTRY statement be identical with the list of dummy arguments in the FUNCTION or SUBROUTINE statement that defines the normal entry point to the subprogram.

To illustrate the use of the ENTRY statement, consider a main program which contains, among other statements, the two calling statements

```
CALL SAMPLE (X,J,Z)
CALL ALIAS (Y,W)
```

and has appended to it the subprogram

```
SUBROUTINE SAMPLE (A,I,C)
B = A**I
ENTRY ALIAS (B,C)
C = SQRT (B)
RETURN
END
```

The statement

```
CALL SAMPLE (X,J,Z)
```

would produce action to evaluate the expression

```
SQRT(X**J)
```

and return this value to storage location Z in the main program. Note that the ENTRY statement, being nonexecutable, does not affect the sequence of calculations. The program steps are followed in normal order, skipping over the ENTRY statement, until the RETURN statement is reached.

[1] This feature is not part of ANSI X3.9 1966.

The main program statement

```
CALL ALIAS (Y,W)
```

would cause the subprogram to evaluate SQRT(Y) and return the value to storage location W.

The above example illustrates how the ENTRY statement is used with a subroutine subprogram. Such calls must always be explicit. As mentioned in Sec. 10.2, calls to function subprograms are made implicitly. An example of how this type of call operates with alternate entry points is shown below.

If a main program contains, among other statements, the statements

```
10 Z = TEST(X,Y)
. . . . . . . . . . . . . . . .
15 ANS = CALC(W)
```

and appended to the main program is the function subprogram

```
FUNCTION TEST(A,B)
C = A**2 + B**2
ENTRY CALC(C)
CALC = EXP(C + 25.)
TEST = SIN(CALC)
RETURN
END
```

then execution of statement 10 in the main program will transmit the values stored in X and Y to the function subprogram at the normal entry point where they will be assigned to A and B. The ENTRY statement is nonexecutable and therefore the expression

```
SIN(EXP(X**2 + Y**2 + 25.))
```

will be evaluated and assigned to TEST. When the RETURN statement is encountered, this value will be returned to the main program and stored in location Z.

Execution of statement 15 in the main program will transmit the value of W to the function subprogram at the entry point named CALC, where the value will be assigned to C. The expression

```
EXP(W+25.)
```

will be evaluated and stored in CALC. In the next executable statement in the subprogram, TEST will have the value of SIN(CALC) assigned to it, but when the RETURN statement is executed, it will transmit only the value stored in CALC back to the main program. That is, the *return from the subprogram is through the same entry point used to access it.* In the main program the value stored in CALC will be assigned to the storage location named ANS.

The reader should note that the ENTRY statement discussed here is not part of *standard* FORTRAN, but is available as a special feature in several FORTRAN compilers. The programmer should consult the operator's manual for the computer being used.

10.10 EXTERNAL STATEMENT

The *EXTERNAL statement* is a nonexecutable declaration statement with the form:

EXTERNAL $f_1, f_2, f_3, \ldots, f_m$

where $f_1, f_2, f_3, \ldots, f_m$ are the names of subprograms used as *actual arguments* in calling *other* subprograms. The actual arguments referred to here can be:

1. Compiler-supplied functions (see Secs. 1.21 to 1.23)
2. Function subprograms (see Secs. 10.2 and 10.5)
3. Subroutine subprograms (see Secs. 10.2 and 10.6)

The following examples illustrate the use of this statement.

EXAMPLE 1

Consider a main program containing, in part, the statements

```
EXTERNAL SIN, COS
CALL COMPON (VECTOR, THETA, SIN, Y)
CALL COMPON (VECTOR, THETA, COS, X)
```

and, appended to the main program, the subprogram

```
SUBROUTINE COMPON (R,A,F,C)
C = R * F (A)
RETURN
END
```

Assuming THETA is expressed in radians, the *first* CALL statement in the main program will cause the subprogram COMPON to calculate the value of

```
VECTOR*SIN(THETA)
```

and return the value to Y. The value of SIN(THETA) will be obtained by branching to the compiler-generated subprogram for evaluating the sine function, just as for any call for a compiler-supplied function.

The *second* CALL in the main program will cause the subroutine subprogram to evaluate

```
VECTOR*COS(THETA)
```

and return that value to location X in the main program.

Note that if the subprogram were

```
SUBROUTINE COMP (R,F,C)
C = R*F
RETURN
END
```

and the main program CALL were

```
CALL COMP(VECTOR, SIN(THETA), Y)
```

the label SIN would *not* have to appear in an EXTERNAL statement because the actual argument in the CALL is the *value* of SIN(THETA) rather than the *name* of the function SIN.

The above illustration used compiler-supplied functions as the external function arguments. The programmer may also write his own external functions, as shown in Example 2.

EXAMPLE 2

If a main program contains, among others, the statements

```
EXTERNAL MULT, DIV
CALL CALC(P,Q,MULT,PROD)
CALL CALC(P,Q,DIV,QUOT)
```

and has appended to it the three subprograms

```
SUBROUTINE CALC(X,Y,F,Z)
Z = F(X,Y)
RETURN
END

FUNCTION MULT(A,B)
MULT = A*B
RETURN
END

FUNCTION DIV(A,C)
DIV = A/C
RETURN
END
```

the *first* CALL statement in the main program will transmit the values P and Q to X and Y, transmit the function name MULT to F, and cause the subroutine subprogram CALC to evaluate

```
MULT(P,Q)
```

But since MULT is a programmer-written subprogram, a further transmittal of the values P and Q to A and B in the function subprogram will take place, and MULT will be evaluated as P*Q. This value will be returned to the subroutine subprogram, which, in turn, will return the value to location PROD in the main program.

The *second* CALL in the main program will produce a similar train of events, transmitting P and Q to X and Y in the subroutine CALC, and from there to A and C in the function DIV. As a result of the return sequence, the value of P/Q will ultimately be stored in location QUOT in the main program.

This example was designed to illustrate the use of the EXTERNAL statement. In addition to illustrating the mechanism for using external subprograms as arguments, the example also demonstrates the complete independence of the variable names used in subprograms from the names used in the main program and in other subprograms. This further emphasizes the fact that subprograms are complete programs whose connection with the main program is the linkage produced by the *calling* procedure and the subprogram-*defining* statement.

10.11 FLOW CHARTING SUBROUTINES

The *predefined process block* shown below

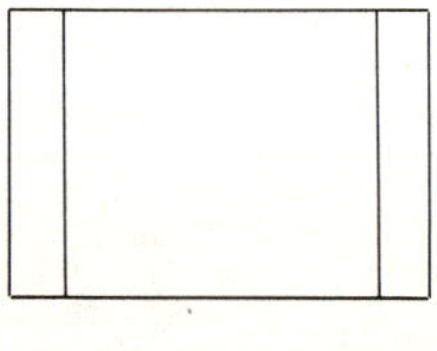

is used with the statements

1. CALL (Sec. 10.7)
2. ENTRY (Sec. 10.9)

to indicate a call to a subroutine appended to the program. When used in a flow chart, this symbol implies a *branch* to the subroutine named in the block and a *return* to this same block when execution of the subprogram has been completed. No path lines are used in the flow chart to indicate this operation.

In addition to the subroutine name, the block also contains the list of arguments being transmitted.

10.12 COULOMB'S LAW[1] (Arithmetic Statement Function)

The electrostatic force exerted by point charge Q_1 on point charge Q_T is given by Coulomb's law, and, in vector notation, is

$$\bar{F}_{1T} = \frac{Q_1 Q_T \bar{a}_R}{4\pi\epsilon_0\epsilon_R R_{1T}^2} \tag{10.12.1}$$

$$\text{where } \bar{a}_R = \frac{\bar{R}_{1T}}{R_{1T}} \tag{10.12.2}$$

The symbols used in the equation are defined as follows.

Symbol	Name	Meaning	Units
$\bar{F}_{1T}$		Vector force on charge Q_T due to charge Q_1	N
Q_k	Q1,QT	Charge	C
ϵ_0	E0	Permittivity of vacuum	F/m
ϵ_R	ER	Relative permittivity of a medium	
$\bar{a}_R$		Unit vector from Q_1 to Q_T	
$\bar{R}_{1T}$		Vector from Q_1 to Q_T	m
R_{1T}	RMAG(R1,R2)	$\lvert\bar{R}_{1T}\rvert$	m

For simplicity, we will assume that the charges lie in the plane $z = 0$. In this case, in the cartesian coordinate system $\bar{R}_{1T} = (x_T - x_1)\,\bar{a}_x + (y_T - y_1)\,\bar{a}_y$. In this formula, (x_T, y_T) and (x_1, y_1) are the x and y coordinates of charge Q_T and Q_1, respectively, and $\bar{a}_x$ and $\bar{a}_y$ are unit vectors pointing toward increasing x and y coordinate values, respectively. We can calculate the unit vector $\bar{a}_R$ by use of Eq. (10.12.2).

$$\bar{a}_R = \frac{\bar{R}_{1T}}{R_{1T}} = \frac{(x_T - x_1)\,\bar{a}_x}{R_{1T}} + \frac{(y_T - y_1)\,\bar{a}_y}{R_{1T}} \tag{10.12.3}$$

$$\text{where } R_{1T} = \sqrt{(x_T - x_1)^2 + (y_T - y_1)^2} \tag{10.12.4}$$

[1] E. D. Torre and C. V. Longo, "The Electromagnetic Field," Allyn and Bacon, Boston, 1969, p. 19.

If we substitute Eqs. (10.12.3) and (10.12.4) into Eq. (10.12.1), we can express the force exerted by charge Q_1 on charge Q_T as two scalar equations, one for the component of force in the x direction, F_{1x}, and the other for the component of force in the y direction, F_{1y}:

$$F_{1x} = \frac{Q_1 Q_T (x_T - x_1)}{4\pi\epsilon_0\epsilon_R R_{1T}^3} \qquad (10.12.5)$$

$$F_{1y} = \frac{Q_1 Q_T (y_T - y_1)}{4\pi\epsilon_0\epsilon_R R_{1T}^3} \qquad (10.12.6)$$

There is no z component of force because of the assumption that the charges lie in the plane $z = 0$. If there are k charges, $Q_1, Q_2, \ldots, Q_k$, and Q_T in the plane $z = 0$, the force on Q_T can also be expressed as two scalar equations. The total force on Q_T in the x direction is the sum of the k components of force in the x direction, and the total force on Q_T in the y direction is the sum of the k components of force in the y direction:

$$F_{\text{tot}(x)} = F_{1x} + F_{2x} + F_{3x} + \cdots + F_{kx} \qquad (10.12.7)$$

$$F_{\text{tot}(y)} = F_{1y} + F_{2y} + F_{3y} + \cdots + F_{ky} \qquad (10.12.8)$$

The components F_{kx} and F_{ky} are calculated using Eqs. (10.12.5) and (10.12.6) with the appropriate x,y coordinates.

Equations (10.12.4) through (10.12.8) are in a form suitable for computation. In the computer program to be discussed below, the following additional names are used:

Symbol	*Name*
x_k, y_k	CKX, CKY
$x_T - x_k$	RX(C)
$y_T - y_k$	RY(C)
R_{kT}	RMAG(R1,R2)
$F_{\text{tot}(x)}$	FTOTX
$F_{\text{tot}(y)}$	FTOTY
F_{kx}	FKX
F_{ky}	FKY
$\dfrac{Q_k Q_T}{4\pi\epsilon_0\epsilon_R R_k T^3}$	FORCE(Q,R)

Figure 10.12.1 shows a computer program which calculates the electrostatic force exerted by three point charges Q_1, Q_2, Q_3 on a fourth point charge Q_T. It is assumed that all the charges are in the plane $z = 0$. The data to be entered into the program are:

1. The relative permittivity of the medium in which the charges reside
2. The magnitude of each charge, with appropriate polarity sign
3. The x,y coordinates of each charge

```
00001C   CØUL1
00002C   THIS PRØGRAM CALCULATES THE ELECTRØSTATIC FØRCE
00003C   EXERTED BY THREE CHARGES ØN CHARGE QT. IT IS
00004C   ASSUMED THAT ALL CHARGES ARE IN THE PLANE Z=0.
00010    FØRCE(Q,R)=Q*QT/(4.*PI*ER*EØ*R**3)
00020    RX(C)=(CTX-C)
00030    RY(C)=(CTY-C)
00040    RMAG(R1,R2)=SQRT(R1**2+R2**2)
00050    READ,QT,CTX,CTY,ER
00060    READ,Q1,Q2,Q3
00070    READ,C1X,C1Y,C2X,C2Y,C3X,C3Y
00080    EØ=8.854E-12
00090    PI=3.1416
00100    F1X=FØRCE(Q1,RMAG(RX(C1X),RY(C1Y)))*RX(C1X)
00110    F1Y=FØRCE(Q1,RMAG(RX(C1X),RY(C1Y)))*RY(C1Y)
00120    F2X=FØRCE(Q2,RMAG(RX(C2X),RY(C2Y)))*RX(C2X)
00130    F2Y=FØRCE(Q2,RMAG(RX(C2X),RY(C2Y)))*RY(C2Y)
00140    F3X=FØRCE(Q3,RMAG(RX(C3X),RY(C3Y)))*RX(C3X)
00150    F3Y=FØRCE(Q3,RMAG(RX(C3X),RY(C3Y)))*RY(C3Y)
00160    FTØTX=F1X+F2X+F3X
00170    FTØTY=F1Y+F2Y+F3Y
00180    PRINT 2,FTØTX,FTØTY
00190  2 FØRMAT("FTØTX",E16.6,"     FTØTY",E16.6//)
00200    STØP
00210    END
```

Fig. 10.12.1 Program COUL1.

The program consists, essentially, of four arithmetic statement functions (see Sec. 10.3) which are used to calculate the components F_{kx} and F_{ky}. These arithmetic statement functions are

1. (Line 00060)

   FORCE(Q,R)=Q*QT/(4.*PI*ER*E0*R**3)

The dummy arguments are Q and R. FORCE(Q,R) will be evaluated for each charge with an appropriate value of R (the magnitude of the vector from each charge to QT). R will be assigned values by the arithmetic statement function RMAG(R1,R2) discussed below. In order to obtain a particular component of force, FORCE(Q,R) must be multiplied by a suitable difference of coordinates [see Eqs. (10.12.5) and (10.12.6)].

2. (Lines 00070 and 00080)

   RX(C) = (CTX − C)
   RY(C) = (CTY − C)

RX(C) denotes the difference of coordinates $(x_T - x_k)$, and RY(C) denotes the difference of coordinates $(y_T - y_k)$. C, the dummy argument, represents a particular x coordinate or a particular y coordinate, and CTX and CTY represent the x and y coordinates, respectively, or charge QT.

3. (Line 00090)

   RMAG(R1,R2) = SQRT(R1**2+R2**2)

The dummy arguments are R1 and R2. The arithmetic statement function RMAG(R1,R2) evaluates R_{kT} in accordance with Eq. (10.12.4).

In the program, each component of force is obtained by multiplying FORCE(Q,R) by RX(C) or by RY(C) depending on whether an x or a y component of

```
INPUT:00050
?12.566E-12,0,2,1

INPUT:00060
?8.854E-12,8.854E-12,8.854E-12

INPUT:00070
?0,0,1,0,2,0

FTØTX     -.177825E-12    FTØTY      0.517257E-12


STØP
```

Fig. 10.12.2 Trial run for COUL1.

force is desired [see Eqs. (10.12.5) and (10.12.6)]. For example, on line 00100, F1X is calculated by multiplying FORCE(Q1,R) by RX(C), where the value of C is x_1 and the value of R is obtained by use of RMAG(R1,R2). The values of R1 and R2 are obtained by means of RX(C) and RY(C). Therefore, in this case, the value of RX(C) is $(x_T - x_1)$, and the value of RY(C) is $(y_T - y_1)$. A trial run is shown in Fig. 10.12.2.

10.13 COULOMB'S LAW (Function and Subroutine Subprograms)

In Sec. 10.12 we discussed a program which calculates the electrostatic forces of a system of charges. That program used several arithmetic statement functions to compute the answers. In this section, we will discuss several alternative programs for achieving the same results. Three of the alternative programs employ either function or subroutine subprograms. The reader will observe that the various methods used in the

```
00001C      CØUL2
00002C      THIS PRØGRAM CALCULATES THE ELECTRØSTATIC FØRCE
00003C      EXERTED BY THREE CHARGES ØN CHARGE QT. IT IS
00004C      ASSUMED THAT ALL CHARGES ARE IN THE PLANE Z=0.
00010       READ,QT,CTX,CTY,ER
00020       READ,Q1,Q2,Q3
00030       READ,C1X,C1Y,C2X,C2Y,C3X,C3Y
00040       F1X=FØRCEX(Q1,C1X,C1Y,CTX,CTY,ER,QT)
00050       F2X=FØRCEX(Q2,C2X,C2Y,CTX,CTY,ER,QT)
00060       F3X=FØRCEX(Q3,C3X,C3Y,CTX,CTY,ER,QT)
00070       F1Y=FØRCEY(Q1,C1X,C1Y,CTX,CTY,ER,QT)
00080       F2Y=FØRCEY(Q2,C2X,C2Y,CTX,CTY,ER,QT)
00090       F3Y=FØRCEY(Q3,C3X,C3Y,CTX,CTY,ER,QT)
00100       FTØTX=F1X+F2X+F3X
00110       FTØTY=F1Y+F2Y+F3Y
00120       PRINT 2,FTØTX,FTØTY
00130     2 FØRMAT("FTØTX",E16.6,"     FTØTY",E16.6//)
00140       STØP
00150       END
00160       FUNCTIØN FØRCEX(Q,CX,CY,CTX,CTY,ER,QT)
00170       PI=3.1416
00180       EØ=8.854E-12
00190       R=SQRT((CTX-CX)**2+(CTY-CY)**2)
00200       FØRCEX=(Q*QT)*(CTX-CX)/(4.*PI*ER*EØ*R**3)
00210       RETURN
00220       END
00230       FUNCTIØN FØRCEY(Q,CX,CY,CTX,CTY,ER,QT)
00240       PI=3.1416
00250       EØ=8.854E-12
00260       R=SQRT((CTX-CX)**2+(CTY-CY)**2)
00270       FØRCEY=(Q*QT)*(CTY-CY)/(4.*PI*ER*EØ*R**3)
00280       RETURN
00290       END
```

Fig. 10.13.1 Program COUL2.

```
INPUT:00010
 ?12.566E-12,0,2,1

INPUT:00020
 ?8.854E-12,8.854E-12,8.854E-12

INPUT:00030
 ?0,0,1,0,2,0

FTØTX     -.177825E-12     FTØTY     0.517257E-12


STØP
```

Fig. 10.13.2 Trial run for COUL2.

programs are not equally advantageous with respect to the particular problem solved here; however, they serve to illustrate some of the features of subprograms.

Figure 10.13.1 shows a program in which *two function subprograms* are used. The function subprograms are called FORCEX and FORCEY, and are used to calculate the x and y force components, respectively. It will be noted that each subprogram is written as a group of statements following the main program. Each subprogram begins with the caption FUNCTION followed by the function name and

```
C    CØUL3
C    THIS PRØGRAM CALCULATES THE ELECTRØSTATIC FØRCE
C    EXERTED BY THREE CHARGES ØN CHARGE QT. IT IS
C    ASSUMED THAT ALL CHARGES ARE IN THE PLANE Z=0.
     READ,QT,CTX,CTY,ER
     READ,Q1,Q2,Q3
     READ,C1X,C1Y,C2X,C2Y,C3X,C3Y
     CALL FØRCE(Q1,C1X,C1Y,CTX,CTY,ER,QT,F1X,F1Y)
     CALL FØRCE(Q2,C2X,C2Y,CTX,CTY,ER,QT,F2X,F2Y)
     CALL FØRCE(Q3,C3X,C3Y,CTX,CTY,ER,QT,F3X,F3Y)
     FTØTX=F1X+F2X+F3X
     FTØTY=F1Y+F2Y+F3Y
     PRINT 2,FTØTX,FTØTY
   2 FØRMAT("FTØTX",E16.6,"    FTØTY",E16.6//)
     STØP
     END
     SUBRØUTINE FØRCE(Q,CX,CY,CTX,CTY,ER,QT,FX,FY)
     PI=3.1416
     E0=8.854E-12
     R=SQRT((CTX-CX)**2+(CTY-CY)**2)
     FX=(Q*QT)*(CTX-CX)/(4.*PI*ER*E0*R**3)
     FY=(Q*QT)*(CTY-CY)/(4.*PI*ER*E0*R**3)
     RETURN
     END
```

Fig. 10.13.3 Program COUL3.

```
INPUT:00010
 ?12.566E-12,0,2,1

INPUT:00020
 ?8.854E-12,8.854E-12,8.854E-12

INPUT:00030
 ?0,0,1,0,2,0

FTØTX     -.177825E-12     FTØTY     0.517257E-12


STØP
```

Fig. 10.13.4 Trial run for COUL3.

by several arguments enclosed in parentheses. For example, the arguments of FORCEX (line 00160) are Q, CX, CY, CTX, CTY, ER, and QT. The calculations performed by subprogram FORCEX are defined by the statements on lines 00190 and 00200, and the subprogram concludes with RETURN and END statements. FORCEX is called three times by the main program (lines 00040, 00050, and 00060), thereby supplying values for F1X, F2X, and F3X. Each time FORCEX is called, values are

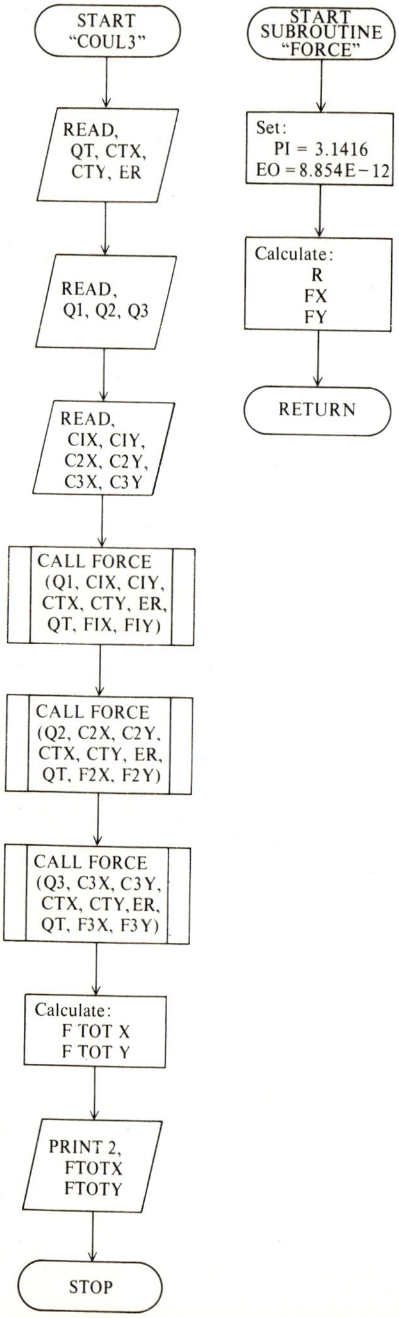

Fig. 10.13.5 Flow chart of program COUL3.

```
00001C    CØUL4
00002C    THIS PRØGRAM CALCULATES THE ELECTRØSTATIC FØRCE
00003C    EXERTED BY THREE CHARGES ØN CHARGE QT. IT IS
00004C    ASSUMED THAT ALL CHARGES ARE IN THE PLANE Z=0.
00010     DIMENSIØN Q(10),CX(10),CY(10)
00020     CØMMØN Q,CX,CY,CTX,CTY,ER,QT,FX,FY,I
00030     READ,N,(Q(I),I=1,N)
00040     READ,(CX(I),I=1,N)
00050     READ,(CY(I),I=1,N)
00060     READ,QT,CTX,CTY,ER
00070     FTØTX=0.
00080     FTØTY=0.
00090     DØ 5 I=1,N
00100     CALL FØRCE
00110     FTØTX=FTØTX+FX
00120   5 FTØTY=FTØTY+FY
00130     PRINT 2,FTØTX,FTØTY
00140   2 FØRMAT("FTØTX",E16.6,"    FTØTY",E16.6//)
00150     STØP
00160     END
00170     SUBRØUTINE FØRCE
00180     DIMENSIØN Q(10),CX(10),CY(10)
00190     CØMMØN Q,CX,CY,CTX,CTY,ER,QT,FX,FY,I
00200     PI=3.1416
00210     EØ=8.854E-12
00220     R=SQRT((CTX-CX(I))**2+(CTY-CY(I))**2)
00230     FX=(Q(I)*QT)*(CTX-CX(I))/(4.*PI*ER*EØ*R**3)
00240     FY=(Q(I)*QT)*(CTY-CY(I))/(4.*PI*ER*EØ*R**3)
00250     RETURN
00260     END
```

Fig. 10.13.6 Program COUL4.

substituted for the dummy arguments Q, CX, and CY. For example, on line 00040 a value is supplied for F1X by calling FORCEX, with Q1, C1X, and C1Y substituting for Q, CX, and CY, respectively. Similar comments apply to FORCEY. A trial run is shown in Fig. 10.13.2.

Let us consider another method of writing the program. In this version, shown in Fig. 10.13.3, a *subroutine subprogram* is used. Unlike a function subprogram, a subroutine subprogram can return several values to the main program. In this case, the subroutine subprogram, called FORCE, returns x and y force components, called FX and FY, respectively, to the main program. Like the function subprogram, data is transmitted between the main program and the subroutine by arguments enclosed in parentheses. However, the method of calling the subroutine subprogram differs from the method of calling the function subprogram. For example, on line 00040 , there is the statement

CALL FORCE(Q1, C1X, C1Y, CTX, CTY, ER, QT, F1X, F1Y)

The reader will observe that Q1, C1X, and C1Y replace the dummy arguments Q, CX, and CY, which appear in the subroutine-defining statement on line 00130. The CALL statement causes the program to branch to the subroutine. After execution is completed, values of FX and FY are returned and stored in F1X and F1Y, respectively. Therefore, in this program one subroutine subprogram replaces two function subprograms. A trial run is shown in Fig. 10.13.4, and a flow chart for the program is shown in Fig. 10.13.5.

An alternate method of using the subroutine subprogram is shown in Fig. 10.13.6. In this program, the variables passed between the main program and the subprogram are listed in COMMON statements on lines 00020 and 00190. (The COMMON statement is explained in Sec. 10.8.) This program is more general than the

```
INPUT:00030
 ?3,8.854E-12,8.854E-12,8.854E-12

INPUT:00040
 ?0,1,2

INPUT:00050
 ?0,0,0

INPUT:00060
 ?12.566E-12,0,2,1

FTØTX      -.177825E-12     FTØTY      0.517257E-12


STØP
```

Fig. 10.13.7 Trial run for COUL4.

previous programs because arrays are used to store the values of charges $Q_1, Q_2, Q_3, \ldots, Q_n$ and their respective x and y coordinates. As written, the program is limited to 10 charges (in addition to QT). It is necessary that the DIMENSION statement appear in the subprogram as well as in the main program.

The subroutine subprogram FORCE is called N times by the DO loop. The use of the COMMON statement makes it possible to omit the arguments in the CALL statement (see line 00100). However, it is necessary to include I, the index of the DO loop in the COMMON statements. For example, when I = 1, Q(1), CX(1), and CY(1) are passed to the subprogram and the corresponding values of FX and FY are returned to the main program. A trial run is shown in Fig. 10.13.7.

10.14 HYPERGEOMETRIC PROBABILITY (Function Subprogram)

In Sec. 5.9 we discussed the hypergeometric probability distribution and a program for calculating it. We are now in a position to write a somewhat more compact program

```
C        HYPER2
C        THIS PRØGRAM CALCULATES HYPERGEØMETRIC PRØBABILITY.
C        N IS THE TØTAL PØPULATIØN. M IS THE NUMBER ØF ØNE
C        TYPE. NS IS THE SAMPLE SIZE. K IS THE NUMBER ØF
C        M IN NS.
         PRINT 2
       2 FØRMAT("ENTER VALUES FØR N,M,NS,AND K")
         READ,N,M,NS,K
         D=FACT(M)/(FACT(K)*FACT(M-K))
         NMSK=N-M-NS+K
         H=FACT(N-M)/(FACT(NS-K)*FACT(NMSK))
         U=FACT(N)/(FACT(NS)*FACT(N-NS))
         PK=D*H/U
         PRINT 3,PK
       3 FØRMAT("PK=",F6.2//)
         STØP
         END
         FUNCTIØN FACT(J)
         IF(J-1) 10,10,11
      10 FACT=1.
         RETURN
      11 X=J
         FACT=1.
       8 FACT=FACT*X
         X=X-1.
         IF(X-1.)7,7,8
       7 RETURN
         END
```

Fig. 10.14.1 Program HYPER2.

```
ENTER VALUES FØR N,M,NS,AND K
INPUT:00030
 ?30,20,5,3

PK=   0.36


STØP
```

Fig. 10.14.2 Trial run for HYPER2.

for this calculation by using a function subprogram for evaluating the factorial. The revised program is shown in Fig. 10.14.1. The same names are used for the variables in this program as were used in the program discussed in Sec. 5.9. This program shows that the function subprogram is very useful when a particular set of calculations, in this case the calculation of the factorial, is made repeatedly. The trial run in Fig. 10.14.2 was made using the same data as before.

10.15 ELLIPTIC INTEGRALS OF THE FIRST KIND

In certain engineering problems it is necessary to evaluate integrals of the form

$$K(k) = \int_0^{\pi/2} \frac{d\phi}{\sqrt{1 - k^2 \sin^2 \phi}} \tag{10.15.1}$$

This integral is known as the complete elliptic integral of the first kind, and is usually denoted, as indicated above, by $K(k)$. The parameter k is determined by the geometry of the problem under consideration.

One method of evaluating $K(k)$ is by the use of a polynomial approximation.[1] In this section we will discuss a simple program which evaluates the elliptic integral of the first kind by means of this polynomial approximation. Two examples of the use of this program are given in the exercises at the end of the chapter.

The program is shown in Fig. 10.15.1, and is seen to consist of a main program and a FUNCTION subprogram. The symbol k has been declared to be of REAL type

```
C    ELLIP1
C    THIS PRØGRAM EVALUATES THE CØMPLETE ELLIPTIC
C    INTEGRAL ØF THE FIRST KIND.
     REAL K
     READ,K
     ANS=EL1(K)
     PRINT 5,K,ANS
   5 FØRMAT("K(",F7.4,")=",E15.4//)
     STØP
     END
     FUNCTIØN EL1(G)
     U=1.-G**2
     IF(U.EQ.0.) GØ TØ 10
     EL1=1.3862944+U*(.1119723+.0725296*U)-ALØG(U)
&    *(.5+U*(.1213478+.0288729*U))
     RETURN
  10 EL1=.1E+76
     RETURN
     END
```

Fig. 10.15.1 Program ELLIP1.

[1] M. Abramowitz and I. Stegun (eds.), "Handbook of Mathematical Functions," Applied Mathematics Series, 55, U.S. Government Printing Office, Washington, D.C., 1964, p. 591.

```
INPUT:00020
 ?.5

K( 0.5000)=       0.1686E+01


STØP


INPUT:00020
 ?1.

K( 1.0000)=       0.1000E+76


STØP
```

Fig. 10.15.2 Trial runs for ELLIP1.

by the declaration statement on line 00010, in order to retain the customary notation. The evaluation of $K(k)$ is carried out by the subprogram EL1(G), where G is a dummy variable. The polynomial approximation to the elliptic integral of the first kind is on lines 00110 and 00120. Notice that the polynomial is written in terms of a parameter U, where $U = 1 - k^2$.

Note that if *k equals 1,* U is 0, the logical IF statement on line 00100 is .TRUE., and the program branches to statement 10 on line 00140. The actual value of EL1 is infinity, but the program assigns a value of 10^{75}. Return to the main program in this case is effected by the statement on line 00150. If *k is less than 1,* the function is evaluated by the statement on lines 00110 and 00120 and the RETURN statement on line 00130 is executed. A trial run is shown in Fig. 10.15.2.

10.16 RELATIVE HUMIDITY

Relative humidity can be determined quite easily from the dry-bulb and wet-bulb temperatures by using a psychrometric chart. One of the major difficulties with using the chart is the fact that it must necessarily be constructed for a particular barometric pressure, usually standard pressure at sea level. Correction of the chart readings to the actual elevation and atmospheric pressure conditions is tedious. However, if the saturated vapor conditions at the wet- and dry-bulb temperatures can be found, the Carrier equation makes it possible to find the actual vapor pressure for the existing atmospheric pressure, and from this value the relative humidity can be calculated. The engineer customarily *reads* the saturated vapor pressures from steam tables. In this section we will program the computer to *evaluate* the saturation pressures at the wet- and dry-bulb temperatures using the following function:[1]

$$p_s = p_c \exp\left[\tau(10^{-5})(t_c - t) \sum_{i=1}^{8} F_i(0.65 - 0.01t)^{i-1}\right] \tag{10.16.1}$$

where p_s = saturation pressure, lb/in^2 absolute
t = saturation temperature, °C
p_c = critical pressure = 3246.0525 lb/in^2 absolute
t_c = critical temperature = 374.136°C
τ = $1000/T$ 1/°K
T = saturation temperature (absolute), °K

[1] J. H. Keenan, F. G. Keyes, et al., "Steam Tables," Wiley, New York, 1969, p. 141.

and the function coefficients F_i are:

$$\begin{aligned}
F_1 &= -741.9242 \\
F_2 &= -29.72100 \\
F_3 &= -11.55286 \\
F_4 &= -0.8685635 \\
F_5 &= 0.1094098 \\
F_6 &= 0.439993 \\
F_7 &= 0.2520658 \\
F_8 &= 0.05218684
\end{aligned}$$

At this point, it is worthwhile to note that the technique of *evaluating* the pressures *within the program* is much more efficient than devising some scheme for *storing* the steam tables and then having the computer "look up" the appropriate value when the temperature is entered as an argument. When functions of this type [Eq. (10.16.1)] had to be evaluated by hand computation, tabulated values were indispensable. With the computer serving as a high-speed calculating machine, the usual approach to the solution of many problems can frequently be altered to produce more efficient methods of attack. The engineer who is new to computer programming should bear this in mind. Too often an inefficient program procedure is adopted because the programmer adheres to traditional calculation methods.

The saturation pressure at the wet-bulb temperature, determined with Eq. (10.16.1), can be used in the Carrier equation[1] to find the vapor pressure

$$p_v = p_{wb} - \frac{(p_a - p_{wb})(t_d - t_w)}{(2831. - 1.43t_w)} \tag{10.16.2}$$

where p_v = actual vapor pressure, lb/in^2 absolute
p_{wb} = saturation pressure at the wet-bulb temperature, lb/in^2 absolute
p_a = atmospheric pressure, lb/in^2 absolute
t_d = dry-bulb temperature, °F
t_w = wet-bulb temperature, °F

With the actual vapor pressure calculated from Eq. (10.16.2), and the saturation pressure at the dry-bulb temperature calculated from Eq. (10.16.1), we can evaluate the relative humidity from its defining equation

$$R = \frac{p_v}{p_{db}} 100 \tag{10.16.3}$$

where R = relative humidity, %
p_v = actual vapor pressure, lb/in^2 absolute
p_{db} = saturation pressure at the dry-bulb temperature, lb/in^2 absolute

The algebraic symbols used in the above equations and their corresponding names in the computer program are shown in Table 10.16.1. A flow chart showing the calculation procedure is given in Fig. 10.16.1 and the computer program is listed in Fig. 10.16.2.

The required data are entered by the READ statement on line 00010. The CALL statement on line 00020 transmits the dry-bulb temperature to subroutine subprogram

[1] W. H. Severns and J. R. Fellows, "Air Conditioning and Refrigeration," Wiley, New York, 1958, p. 21.

Table 10.16.1

Algebraic symbol	*Program name*	*Meaning*	*Units*
p_s	PS	Vapor pressure at saturation	lb/in^2 absolute
p_c	PC	Critical pressure = 3246.0525	lb/in^2 absolute
t	C	Saturation temperature	°C
...	T	Saturation temperature	°F
t_c	TC	Critical temperature = 374.136	°C
τ	...	1000/absolute temperature	1/°K
F_i	F(I)	Function coefficients	
p_v	P	Actual vapor pressure	lb/in^2 absolute
p_{wb}	PW	Saturation pressure at the wet-bulb temperature	lb/in^2 absolute
p_a	AP	Atmospheric pressure	lb/in^2 absolute
t_d	DBT	Dry-bulb temperature	°F
t_w	WBT	Wet-bulb temperature	°F
R	RH	Relative humidity	%
p_{db}	PD	Saturation pressure at the dry-bulb temperature	lb/in^2 absolute

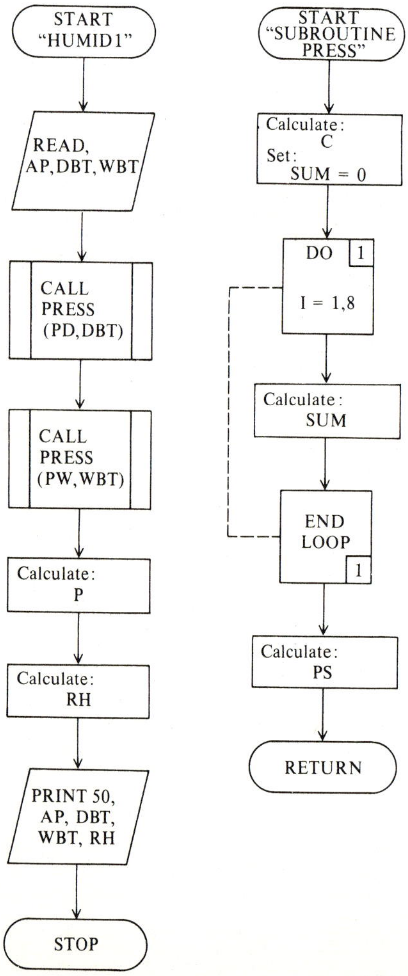

Fig. 10.16.1 Flow chart of program HUMID1.

```
C      HUMID1
C      PRØGRAM TØ CALCULATE THE RELATIVE HUMIDITY GIVEN THE
C      DRY BULB AND WET BULB TEMPERATURES IN DEGREES FAHRENHEIT
C      AND THE BARØMETRIC PRESSURE IN PSIA.
       READ,AP,DBT,WBT
       CALL PRESS(PD,DBT)
       CALL PRESS(PW,WBT)
       P=PW-(AP-PW)*(DBT-WBT)/(2831.-1.43*WBT)
       RH=P*100./PD
       PRINT 50,AP,DBT,WBT,RH
    50 FØRMAT(/22HATMØSPHERIC PRESSURE =F8.3,5H PSIA/
&       22HDRY BULB TEMPERATURE =F6.1,3X18HDEGREES FAHRENHEIT/
&       22HWET BULB TEMPERATURE =F6.1,3X18HDEGREES FAHRENHEIT//
&       3X19HRELATIVE HUMIDITY =F6.1,9H  PERCENT/)
       STØP
       END
       SUBRØUTINE PRESS(PS,T)
       DIMENSIØN F(8)
       DATA F/-741.9242,-29.721,-11.55286,-0.8685635,0.1094098,
&       0.439993,0.2520658,0.05218684/,TC,PC/374.136,3246.0525/
       C=5.*(T-32.)/9.
       SUM=0.0
       DØ 1 I=1,8
     1 SUM=SUM+F(I)*(0.65-0.01*C)**(I-1)
       PS=PC*EXP((TC-C)/100./(C+273.15)*SUM)
       RETURN
       END
```

Fig. 10.16.2 Program HUMID1.

```
INPUT:00010
 ?14.3,80.,60.


ATMØSPHERIC PRESSURE =   14.300 PSIA
DRY BULB TEMPERATURE =   80.0    DEGREES FAHRENHEIT
WET BULB TEMPERATURE =   60.0    DEGREES FAHRENHEIT

   RELATIVE HUMIDITY =   30.6   PERCENT

STØP
```

```
INPUT:00010
 ?14.7,80.,60.


ATMØSPHERIC PRESSURE =   14.700 PSIA
DRY BULB TEMPERATURE =   80.0    DEGREES FAHRENHEIT
WET BULB TEMPERATURE =   60.0    DEGREES FAHRENHEIT

   RELATIVE HUMIDITY =   30.1   PERCENT

STØP
```

```
INPUT:00010
 ?15.1,80.,60.


ATMØSPHERIC PRESSURE =   15.100 PSIA
DRY BULB TEMPERATURE =   80.0    DEGREES FAHRENHEIT
WET BULB TEMPERATURE =   60.0    DEGREES FAHRENHEIT

   RELATIVE HUMIDITY =   29.5   PERCENT

STØP
```

Fig. 10.16.3 Trial runs for HUMID1.

PRESS on line 00130, where the value is stored in location T. The DATA statement on line 00150 initializes the eight function coefficients, F(1), F(2), . . ., F(8), and also the variables TC and PC. Note that the array F was dimensioned on line 00140 and, therefore, merely naming the array in the DATA statement implied that values were to be assigned to *each element* of the array. The list of eight constants (enclosed between a pair of slashes) following the array name represents the values assigned to F(1), F(2), etc.

Note that Eq. (10.16.1) requires that the temperature be expressed in degrees Celsius. The dry-bulb temperature transmitted to the subroutine and stored in location T was in degrees Fahrenheit; therefore, before the function can be evaluated, it is necessary to make the conversion shown on line 00170.

The DO loop on lines 00190 and 00200 carries out the summation procedure indicated in Eq. (10.16.1), and on line 00210 the saturation pressure is calculated. This value is returned to line 00020 in the main program and stored there in location PD.

On line 00030 a second CALL statement transmits the wet-bulb temperature to subroutine PRESS, and eventually the saturation pressure at that temperature is returned and stored in location PW.

Execution of the replacement statement on line 00040 calculates the actual vapor pressure of the air according to Eq. (10.16.2), and stores the value in location P. On line 00050 the relative humidity is calculated. The PRINT statement on line 00060 "echoes" the input data and prints the value of the relative humidity. Several trial runs of the program showing the effect of variations in the atmospheric pressure are included in Fig. 10.16.3.

10.17 REGRESSION LINES FOR TRAFFIC-FLOW DATA

There are many situations in which an engineer wishes to determine the parameters a and b of a linear relation between variables x and y: $y = a + bx$. Let us assume an experiment has been performed and data have been collected and plotted. Because of experimental error, it is often difficult to *fit* a straight line to the data. One method of determining such a line, called the *regression line*, is by the *method of least squares.*[1]

Suppose we have n data points (x_i, y_i) and wish to represent the data by a straight line. By use of the method of least squares, the coefficients a and b are found to be

$$a = \frac{\Sigma y_i \, \Sigma x_i^2 - \Sigma x_i \, \Sigma x_i y_i}{\Delta} \tag{10.17.1}$$

$$b = \frac{n \Sigma x_i y_i - \Sigma x_i \, \Sigma y_i}{\Delta} \tag{10.17.2}$$

$$\text{where } \Delta = n \Sigma x_i^2 - (\Sigma x_i)^2 \tag{10.17.3}$$

A program for carrying out these calculations is shown in Fig. 10.17.2, and its flow chart is shown in Fig. 10.17.1. The program consists of two parts: a main program and SUBROUTINE LEAST. The main program and the subprogram exchange data by COMMON statements (lines 00010 and 00130). The reader should note that no DIMENSION statements are needed for the variables X(I) and Y(I) because they

[1] W. Volk, "Applied Statistics for Engineers," 2d ed., McGraw-Hill, New York, 1969, p. 261.

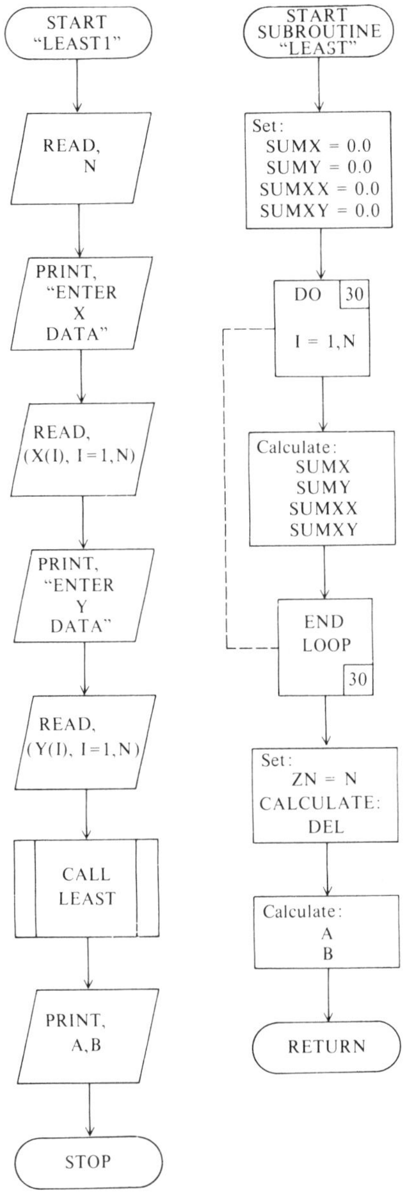

Fig. 10.17.1 Flow chart of program LEAST1.

are dimensioned in the COMMON statements. The main program is quite short. It consists, essentially, of three READ statements, a CALL statement, and PRINT statements. All calculations are carried out in the subprogram. The four summations required by Eqs. (10.17.1) to (10.17.3) are carried out by means of the DO loop which begins on line 00180. These summations are stored in four variables which are given the descriptive names SUMX, SUMY, SUMXY, and SUMXX; then DEL is calculated; and, finally, A and B are calculated. A trial run is shown in Fig. 10.17.3.

Let us briefly consider the application of the least-squares method to data

```
C     LEAST1
      CØMMØN X(20),Y(20),N,A,B
      READ,N
      PRINT,"ENTER X DATA"
      READ,(X(I),I=1,N)
      PRINT,"ENTER Y DATA"
      READ,(Y(I),I=1,N)
      CALL LEAST
      PRINT,A,B
      STØP
      END
      SUBRØUTINE LEAST
      CØMMØN X(20),Y(20),N,A,B
      SUMX=0.
      SUMY=0.
      SUMXX=0.
      SUMXY=0.
      DØ 30 I=1,N
      SUMX=SUMX+X(I)
      SUMY=SUMY+Y(I)
      SUMXX=SUMXX+X(I)**2
   30 SUMXY=SUMXY+X(I)*Y(I)
      ZN=N
      DEL=ZN*SUMXX-SUMX**2
      A=(SUMY*SUMXX-SUMX*SUMXY)/DEL
      B=(ZN*SUMXY-SUMX*SUMY)/DEL
      RETURN
      END
```

Fig. 10.17.2 Program LEAST1.

collected in the study of traffic flow.[1] Two variables used to describe traffic flow are u, the speed of vehicles expressed in miles per hour, and d, the density of vehicles expressed in vehicles per mile. Various functional relations between u and d have been proposed by traffic engineers. We will consider only the following exponential relation (proposed by Greenberg)[2]

$$d = d_m \, e^{-u/u_m} \tag{10.17.4}$$

where d_m and u_m are parameters to be determined.

If we make a logarithmic transformation of the equation above, we obtain

$$\ln d = \ln d_m - \frac{1}{u_m} u \tag{10.17.5}$$

```
INPUT:00020
 ?6

ENTER X DATA
INPUT:00040
 ?3.05,4.29,5.41,6.7,8.71,10.55

ENTER Y DATA
INPUT:00060
 ?4.09,5.8,6.8,8.11,10.45,12.69

        0.7500          1.1222
STØP
```

Fig. 10.17.3 Trial run for LEAST1.

[1] D. R. Drew, "Traffic Flow Theory and Control," McGraw-Hill, New York, 1968, p. 310.

[2] M. Wohl and B. V. Martin, "Traffic System Analysis," McGraw-Hill, New York, 1967, p. 332.

```
C     LEAST2
      CØMMØN X(20),Y(20),N,A,B
      READ,N
      PRINT,"ENTER X DATA"
      READ,(X(I),I=1,N)
      PRINT,"ENTER Y DATA"
      READ,(Y(I),I=1,N)
      DØ 1 I=1,N
    1 Y(I)=ALØG(Y(I))
      CALL LEAST
      UM=-1./B
      DM=EXP(A)
      PRINT 7,A,B
    7 FØRMAT("A=",F9.4,6X,"B=",F9.4/)
      PRINT 8,UM,DM
    8 FØRMAT("UM=",F9.4,6X,"DM=",F9.4//)
      STØP
      END
      SUBRØUTINE LEAST
      CØMMØN X(20),Y(20),N,A,B
      SUMX=0.
      SUMY=0.
      SUMXX=0.
      SUMXY=0.
      DØ 30 I=1,N
      SUMX=SUMX+X(I)
      SUMY=SUMY+Y(I)
      SUMXX=SUMXX+X(I)**2
   30 SUMXY=SUMXY+X(I)*Y(I)
      ZN=N
      DEL=ZN*SUMXX-SUMX**2
      A=(SUMY*SUMXX-SUMX*SUMXY)/DEL
      B=(ZN*SUMXY-SUMX*SUMY)/DEL
      RETURN
      END
```

Fig. 10.17.4 Program LEAST2.

The values of the constants d_m and u_m can be determined by making a least-squares regression fit to the available data. Let us compare Eq. (10.17.5) to the "standard" form: $y = a + bx$. It is seen that

$$y = \ln d \tag{10.17.6}$$

$$a = \ln d_m \tag{10.17.7}$$

$$b = -\frac{1}{u_m} \tag{10.17.8}$$

$$x = u \tag{10.17.9}$$

The values of a and b can be determined by means of a program similar to the program discussed above (Fig. 10.17.2). However, *before* subprogram LEAST is called, the natural logarithm of each value of density must be calculated [Eq. (10.17.6)]. After LEAST has been called, and the values of a and b have been returned to the main program, the parameters d_m and u_m can be calculated as follows:

$$d_m = e^a \tag{10.17.10}$$

$$u_m = -\frac{1}{b} \tag{10.17.11}$$

The revised program is shown in Fig. 10.17.4. In this program, we continue to use the names X and Y for the experimental data entered into the program. As this program is intended for use in the study of traffic flow, we will enter values of speed for X and values of density for Y. After the data have been entered, a logarithmic

```
INPUT:00020
 ?6

ENTER X DATA
INPUT:00040
 ?32,23,17,15,14,10

ENTER Y DATA
INPUT:00060
 ?34,60,88,94,96,129

A=     5.4462          B=     -.0596

UM=   16.7886          DM=  231.8720


STØP
```

Fig. 10.17.5 Trial run for LEAST2.

transformation of the density data is carried out by a DO loop (lines 00070, 00080). After the transformation has been performed, subprogram LEAST is called, and returns the values of A and B. Then parameters u_m and d_m (denoted UM and DM in the program) are calculated. Finally, the values of A, B, UM, and DM are printed. A trial run is shown in Fig. 10.17.5. In this trial run, some of the data given by Greenberg have been used. From the results of the trial run we can write the following relation between u and d:

$$d = 231.9\, e^{-u/16.8}$$

10.18 HIGH-FREQUENCY TRANSMISSION LINE[1]

In this section we will discuss several transmission-line calculations which are facilitated by use of the computer. The program contains a subroutine subprogram involving complex quantities. The discussion will be restricted to lossless transmission lines terminated by a complex load impedance Z_r. No transmission lines are truly lossless, but high-frequency transmission lines usually approach this ideal.[2] The characteristic impedance Z_0 of a lossless transmission line is a resistance, and $Z_0 = +\sqrt{L/C}$ (see Sec. 4.13).

The program to be discussed calculates the wavelength λ, the phase constant β, the standing wave ratio S, the reflection coefficient ρ, and the reciprocal Y_0 of the characteristic impedance of a lossless transmission line. It also calculates the input admittance Y_s and the input impedance Z_s as a function of distance z *from the end of the line.* The equations for these parameters are

$$\lambda = v/f \qquad \text{m} \tag{10.18.1}$$

where v = velocity, m/s

f = frequency, Hz

$$\beta = \frac{2\pi}{\lambda}, \ \text{rad/m} \tag{10.18.2}$$

$$\rho = \frac{Z_r - Z_0}{Z_r + Z_0} \tag{10.18.3}$$

[1] D. T. Paris and F. K. Hurd, "Basic Electromagnetic Theory," McGraw-Hill, New York, 1969, p. 407.

[2] A line is considered to be lossless when the quantity αl (which appears in the wave propagation equations) is very small. This condition usually occurs at frequencies of 50 MHz or higher.

$$S = \frac{1+|\rho|}{1-|\rho|} \tag{10.18.4}$$

$$Z_s = Z_0 \frac{1+\rho e^{-j2\beta z}}{1-\rho e^{-j2\beta z}}, \text{ ohms} \tag{10.18.5}$$

$$Y_s = \frac{1}{Z_s}, \text{ mhos} \tag{10.18.6}$$

The symbols are defined in Table 10.18.1.

The program for calculating the various transmission-line parameters mentioned above is shown in Fig. 10.18.1. Data are entered into the computer for frequency, wave velocity, characteristic impedance of the line, and complex load impedance. On lines 00060, 00070, and 00080, the wavelength, phase constant, and reciprocal of the characteristic impedance YO, respectively, are calculated. The significance of YO is discussed below. The reflection coefficient RHO is calculated by subroutine REFLEC, and the value of RHO is returned to the main program (in rectangular form) when the statement CALL REFLEC (on line 00100) is executed. The reader should note that there is a COMMON statement in the main program on line 00020 and in the subroutine on line 00380. The COMMON variables ZR and RHO are complex, and must be declared complex in *both* the main program and in the subroutine subprogram.[1]

After a value of RHO is returned to the main program, the reciprocal of the standing wave ratio, RECIP, is calculated (line 00110). It is preferable to calculate the standing wave ratio SWR by taking the reciprocal of RECIP. The reason is that the magnitude of ρ will always be in the range

$$0 \leq |\rho| \leq 1$$

Table 10.18.1

Symbol	*Program name*	*Meaning*	*Units*
Z_0	ZO	Characteristic impedance	ohms
...	YO	Reciprocal of ZO	mhos
Z_r	ZR	Load impedance	ohms
v	VEL	Wave velocity	m/s
f	F	Frequency	Hz
λ	XLAMDA	Wavelength	m
β	BETA	Phase constant	rad/s
S	SWR	Standing wave ratio	
...	RECIP	Reciprocal of SWR	
ρ	RHO	Reflection coefficient	
z	Z	Distance from the load	m
...	TAU	Product of phase constant and distance from load	rad
Z_s	ZINZ	Input impedance as a function of distance from load	ohms
...	ZREAL	Real part of ZINZ	ohms
...	ZIMAG	Imaginary part of ZINZ	ohms
Y_s	YINZ	Reciprocal of ZINZ	mhos
...	G	Real part of YINZ	mhos
...	B	Imaginary part of YINZ	mhos
$e^{-j2\beta z}$	X	A factor in Eq. (10.18.5), in rectangular form	rad

[1] If it is desired to express ρ in polar form as well as in rectangular form, the program discussed in Sec. 4.12 can be added to subroutine REFLEC.

```
00001C      LINE1
00002C      THIS PRØGRAM CALCULATES THE FØLLØWING PARAMETERS
00003C      FØR A LØSSLESS LINE:WAVELENGTH ; PHASE CØNSTANT;
00004C      STANDING WAVE RATIØ ; REFLECTIØN CØEFFICIENT ;
00005C      AND INPUT ADMITTANCE AND IMPEDANCE AS A FUNCTIØN
00006C      ØF DISTANCE FRØM THE LØAD.
00010       CØMPLEX ZR,RHØ,ZINZ,YINZ,X
00020       CØMMØN ZØ,ZR,RHØ
00030       READ,F,ZØ,VEL
00040       READ,ZR
00050       PI=3.1416
00060       XLAMDA=VEL/F
00070       BETA=2.*PI/XLAMDA
00080       YØ=1./ZØ
00090       PRINT,"YØ=",YØ
00100       CALL REFLEC
00110       RECIP=(1.-CABS(RHØ))/(1.+CABS(RHØ))
00120       IF(RECIP) 20,20,21
00130    20 SWR=1.E75
00140       GØ TØ 25
00150    21 SWR=1./RECIP
00160    25 PRINT,"STANDING WAVE RATIØ=",SWR
00170       PRINT 30
00180    30 FØRMAT(7X1HG10X1HB10X5HZREAL7X5HZIMAG6X4HDEG.4X5HMETER/)
00190       DØ 10 I=1,100
00200       U=I-1
00210       Z=U/100.
00220       TAU=BETA*Z
00230       X=CMPLX(CØS(2.*TAU),SIN(-2.*TAU))
00240       ZINZ=ZØ*(1.+RHØ*X)/(1.-RHØ*X)
00250       YINZ=1./ZINZ
00260       TAUDEG=57.296*TAU
00270    50 FØRMAT(2(E12.3,E12.3),2F8.3)
00280       PRINT 50,YINZ,ZINZ,TAUDEG,Z
00290       IF(TAU-3.1) 10,11,11
00300    10 CØNTINUE
00310    11 PRINT 40,RHØ
00320    40 FØRMAT(4HRHØ=F8.3,F8.3)
00330       PRINT,"WAVELENGTH=",XLAMDA
00340       PRINT,"PHASE CØNSTANT=",BETA
00350       STØP
00360       END
00370       SUBRØUTINE REFLEC
00380       CØMPLEX ZR,RHØ
00390       CØMMØN ZØ,ZR,RHØ
00400       RHØ=(ZR-ZØ)/(ZR+ZØ)
00410       RETURN
00420       END
```

Fig. 10.18.1 Program LINE1.

When $|\rho| = 1$, the standing wave ratio is infinite, but its reciprocal is zero. In the program, when RECIP = 0., SWR is arbitrarily assigned a value of 1.E75 (see lines 00120 and 00130).

The DO loop which begins on line 00190 calculates the input impedance of the line ZINZ [Eq. (10.18.5)], and the input admittance YINZ at varying distances from the load. The distance Z and the phase angle TAU are generated by use of the DO loop (see lines 00200 through 00220). The DO loop is halted when the phase angle TAU is approximately 3.1 rad, or when Z = 1 m.[1] It is often convenient to express the distance from the end of the line in electrical degrees as well as in meters.[2] For this reason, in the printout (see Fig. 10.18.2), the input impedance and admittance are

[1] The range of the distance and the steps of its variation were chosen to suit the trial example. There may be circumstances under which it is desirable to modify these quantities.

[2] A third way of expressing distance from the load is in terms of wavelengths.

```
INPUT:00030
 ?3.E08, 300., 3.E08

INPUT:00040
 ?75.,75.

YØ=          0.0033
STANDING WAVE RATIØ=           4.2656
      G              B              ZREAL           ZIMAG          DEG.        METER

   0.667E-02     -.667E-02     0.750E+02     0.750E+02      0.000      0.000
   0.522E-02     -.632E-02     0.777E+02     0.941E+02      3.600      0.010
   0.415E-02     -.582E-02     0.812E+02     0.113E+03      7.200      0.020
   0.336E-02     -.529E-02     0.855E+02     0.135E+03     10.800      0.030
   0.278E-02     -.478E-02     0.909E+02     0.156E+03     14.400      0.040
   0.234E-02     -.431E-02     0.975E+02     0.179E+03     18.000      0.050
   0.201E-02     -.387E-02     0.105E+03     0.203E+03     21.600      0.060
   0.175E-02     -.347E-02     0.115E+03     0.230E+03     25.200      0.070
   0.155E-02     -.311E-02     0.128E+03     0.258E+03     28.800      0.080
   0.138E-02     -.278E-02     0.144E+03     0.288E+03     32.400      0.090
   0.125E-02     -.247E-02     0.163E+03     0.322E+03     36.000      0.100
   0.114E-02     -.219E-02     0.188E+03     0.358E+03     39.600      0.110
   0.106E-02     -.192E-02     0.220E+03     0.399E+03     43.200      0.120
   0.993E-03     -.167E-02     0.263E+03     0.442E+03     46.800      0.130
   0.936E-03     -.143E-02     0.319E+03     0.489E+03     50.400      0.140
   0.891E-03     -.120E-02     0.396E+03     0.536E+03     54.000      0.150
   0.855E-03     -.989E-03     0.500E+03     0.579E+03     57.600      0.160
   0.827E-03     -.779E-03     0.641E+03     0.604E+03     61.200      0.170
   0.806E-03     -.573E-03     0.824E+03     0.586E+03     64.800      0.180
   0.792E-03     -.371E-03     0.103E+04     0.486E+03     68.400      0.190
   0.784E-03     -.172E-03     0.121E+04     0.267E+03     72.000      0.200
   0.782E-03     0.260E-04     0.128E+04     -.425E+02     75.600      0.210
   0.785E-03     0.224E-03     0.117E+04     -.336E+03     79.200      0.220
   0.795E-03     0.424E-03     0.979E+03     -.522E+03     82.801      0.230
   0.811E-03     0.627E-03     0.772E+03     -.597E+03     86.401      0.240
   0.833E-03     0.833E-03     0.600E+03     -.600E+03     90.001      0.250
   0.863E-03     0.104E-02     0.469E+03     -.569E+03     93.601      0.260
   0.902E-03     0.127E-02     0.373E+03     -.524E+03     97.201      0.270
   0.950E-03     0.150E-02     0.303E+03     -.476E+03    100.801      0.280
   0.101E-02     0.174E-02     0.250E+03     -.430E+03    104.401      0.290
   0.108E-02     0.199E-02     0.211E+03     -.388E+03    108.001      0.300
   0.117E-02     0.226E-02     0.181E+03     -.348E+03    111.601      0.310
   0.128E-02     0.255E-02     0.157E+03     -.313E+03    115.201      0.320
   0.142E-02     0.286E-02     0.139E+03     -.280E+03    118.801      0.330
   0.159E-02     0.320E-02     0.124E+03     -.250E+03    122.401      0.340
   0.181E-02     0.358E-02     0.112E+03     -.223E+03    126.001      0.350
   0.209E-02     0.398E-02     0.103E+03     -.197E+03    129.601      0.360
   0.245E-02     0.443E-02     0.956E+02     -.173E+03    133.201      0.370
   0.292E-02     0.491E-02     0.893E+02     -.150E+03    136.801      0.380
   0.355E-02     0.543E-02     0.843E+02     -.129E+03    140.401      0.390
   0.440E-02     0.596E-02     0.802E+02     -.108E+03    144.001      0.400
   0.556E-02     0.643E-02     0.769E+02     -.891E+02    147.601      0.410
   0.712E-02     0.671E-02     0.744E+02     -.701E+02    151.201      0.420
   0.916E-02     0.651E-02     0.725E+02     -.516E+02    154.801      0.430
   0.115E-01     0.540E-02     0.713E+02     -.334E+02    158.401      0.440
   0.135E-01     0.297E-02     0.705E+02     -.155E+02    162.001      0.450
   0.142E-01     -.472E-03     0.703E+02     0.234E+01    165.601      0.460
   0.131E-01     -.374E-02     0.707E+02     0.202E+02    169.201      0.470
   0.108E-01     -.580E-02     0.715E+02     0.382E+02    172.801      0.480
   0.858E-02     -.663E-02     0.730E+02     0.564E+02    176.401      0.490
   0.667E-02     -.667E-02     0.750E+02     0.750E+02    180.001      0.500
RHØ=    -.538     0.308
WAVELENGTH=          1.0000
PHASE CØNSTANT=           6.2832
STØP
```

Fig. 10.18.2 Trial run for LINE1.

tabulated as functions of distance in meters and of electrical degrees.

It was previously mentioned that the reciprocal Y_0 of the characteristic impedance is of importance in some applications. One such application is in connection with maximum power transfer to a load. If a transmission line is not terminated in its characteristic impedance, reflections occur, and maximum power transfer is not achieved. In order to minimize reflections on a transmission line, a reactance (called a *stub*) is placed in parallel with the line at any point[1] where G, the real part of the input admittance, is equal to YO, the reciprocal of the characteristic impedance.[2] The magnitude of the reactance which is placed in parallel with the line at the given point should be equal to the magnitude of B, the imaginary part of the input admittance at that point, and opposite in sign. For example, in the trial run shown in Fig. 10.18.2, a transmission line of 300-ohm characteristic impedance is terminated by a load of value $75 + j75$ ohms. Therefore, Y_0 is approximately 0.333×10^{-2} mho. The reader will observe that at a distance of 0.03 m from the load, $G = 0.336 \times 10^{-2}$ mho and $B = -0.529 \times 10^{-2}$ mho. Therefore, a reactance of value $+0.529 \times 10^{-2}$ mho should be placed at a distance of 0.03 m to minimize reflections. The reader can observe from the data of the trial run of Fig. 10.18.2 that there is a second possible stubbing point between 0.38 and 0.39 m. If the latter point were to be chosen, a reactance of -0.529×10^{-2} mho would be used.

PROBLEMS

10-1. Rewrite the program of Fig. 4.12.1 using an arithmetic statement function ANG(B,A) to evaluate 57.296*ATAN(ABS(B/A)) wherever the latter expression is required.

10-2. During a computer output operation, a very long table of results is to be printed. On the first page of the output, the programmer wants to print the title of the table with the column headings two lines below the title and the results below the column headings. On the second page (and all succeeding pages) only the column headings are to be printed.

Write a subroutine subprogram, using an ENTRY statement, that will accomplish the desired result. Show how this program should be called by the main program.

10-3. In many branches of engineering one may wish to calculate the variation of some transfer function as radian frequency varies. The results are ordinarily plotted to a logarithmic scale to encompass a broad range of frequencies.

Write a subroutine subprogram for changing the index parameters of a DO loop as was done in Fig. 7.7.2. The input to the program should be the number of cycles desired. Start with

MIN = 1
MAX= 9
INC = 1

and increase these values 10 times each cycle.

10-4. A normalized vector $\bar{a}$ is a vector of unit magnitude which is obtained from another vector $\bar{A}$ by dividing each component of $\bar{A}$ by the magnitude of $\bar{A}$. For example, if

$$\bar{A} = [1, -1, 3, 2]$$

$$|\bar{A}| = \sqrt{15}$$

$$\bar{a} = \left[\frac{1}{\sqrt{15}}, \frac{-1}{\sqrt{15}}, \frac{3}{\sqrt{15}}, \frac{2}{\sqrt{15}}\right]$$

write a subprogram UNIT(A,AU) which receives the components of $\bar{A}$ from a main program and returns the components of $\bar{a}$ to the main program.

Sample data

Set	*N*	$\bar{A}$
1	4	[1, −1, 3, 2]
2	6	[2, 3, 6, −5, 4, −7]

[1] The position at which the stub is placed is called a stubbing point.

[2] D. T. Paris and F. K. Hurd, "Basic Electromagnetic Theory," McGraw-Hill, New York, p. 420.

10-5. Suppose that, on the average, each year 13 bearings of a certain type fail in a machine and that a fresh supply of bearings is delivered every four weeks. If the failure of bearings follows a Poisson distribution (see Prob. 5-5), how large an inventory of bearings n must be maintained to ensure to a probability of 0.995 that the machine is not inoperative because of a shortage of replacement bearings? In this problem the average rate of failure per week $\nu = \frac{13}{52}$, and with $t = 4$ weeks, $\lambda = \nu t = 1$. Write a FORTRAN program that calculates the value of n from the following formula:

$$\sum_{k=1}^{n} \frac{\lambda^k e^{-\lambda}}{k!} = 0.995$$

10-6. Given two vectors, each of dimension n,

$$\bar{A} = [A_1, A_2, A_3, \ldots, A_n]$$
$$\bar{B} = [B_1, B_2, B_3, \ldots, B_n]$$

the scalar product of these vectors, $\bar{A} \cdot \bar{B}$ is defined as

$$\bar{A} \cdot \bar{B} = C = A_1 B_1 + A_2 B_2 + A_3 B_3 + \ldots + A_n B_n$$

If the scalar product C of $\bar{A}$ and $\bar{B}$ is zero, $\bar{A}$ and $\bar{B}$ are said to be orthogonal vectors.

(*a*) Write a subprogram which receives the components of $\bar{A}$ and $\bar{B}$ from a main program and returns the value of C to the main program.

(*b*) Write a main program which calls the subprogram. If $|C| \leqslant 1 \times 10^{-7}$, the message A AND B ARE ORTHOGONAL should be printed; if $|C| > 1 \times 10^{-7}$, the value of C should be printed.

Sample data

Set	*N*	$\bar{A}$	$\bar{B}$
1	5	[1, 3, 1, 0, −6]	[4, 1, 4, 2, −3]
2	3	[6, −2, 4]	[1, 7, 2]
3	4	[5, −3, 5, −3]	[4, 4, 4, 9]

10-7. The permutation symbol ϵ_{ijk} is defined by the following rules:

$\epsilon_{ijk} = 0$ if two or more subscripts are equal
$\epsilon_{ijk} = +1$ if ijk are in cyclic order clockwise
$\epsilon_{ijk} = -1$ otherwise

For example,

$$\epsilon_{112} = \epsilon_{122} = 0$$
$$\epsilon_{123} = \epsilon_{231} = \epsilon_{312} = +1$$
$$\epsilon_{321} = \epsilon_{132} = \epsilon_{213} = -1$$

Write a FORTRAN subprogram which receives the values of i, j, and k from a main program and returns the value of ϵ_{ijk} to the main program. The main program should print a table of the values the permutation symbol assumes as i, j, and k are varied.

10-8. The cross product of two three-dimensional vectors

$$\bar{A} = [A_1, A_2, A_3]$$
$$\bar{B} = [B_1, B_2, B_3]$$

is defined by

$$\bar{A} \times \bar{B} = \sum_{i=1}^{3} \sum_{j=1}^{3} \sum_{k=1}^{3} \epsilon_{ijk} A_j B_k \bar{a}_i$$

where the three components of the cross product are

$$(A_2 B_3 - A_3 B_2)$$
$$(A_3 B_1 - A_1 B_3)$$
$$(A_1 B_2 - A_2 B_1)$$

and $\bar{a}_i$ is a unit vector.

Write a program which calculates the three components of two three-dimensional vectors $\overline{A}$ and $\overline{B}$. Use the subprogram written for Prob. 10-7 to return the value of the permutation symbol.

Sample data

Set	$\overline{A}$	$\overline{B}$
1	[1, 4, −2]	[2, −3, −1]
2	[27, 16, 95]	[43, −9, −4]

10-9. The Belleville washer spring is used for carrying high loads with small deflections:

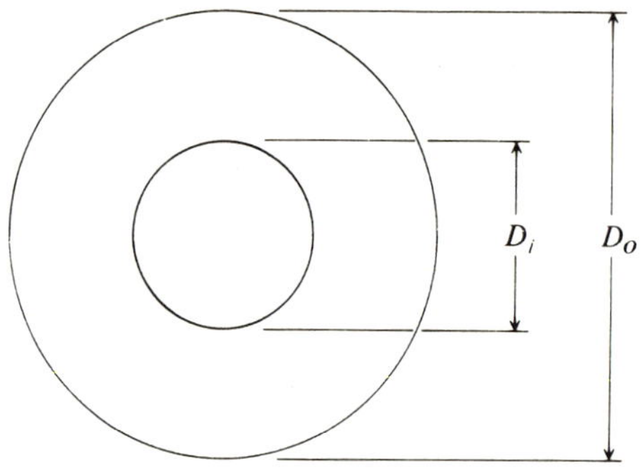

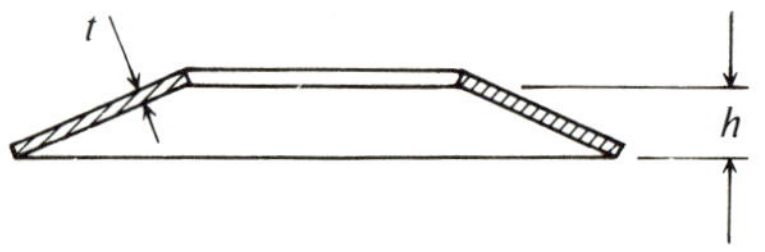

The design equations for the Belleville spring are:

$$P = \frac{4E\delta}{(1-\nu^2)K_1 D_o{}^2}\left[(h-\delta)(h-\frac{\delta}{2})t + t^3\right] \tag{1}$$

$$\sigma = \frac{4E\delta}{(1-\nu^2)K_1 D_o{}^2}\left[K_2\left(h-\frac{\delta}{2}\right) + K_3 t\right] \tag{2}$$

with $$K_1 = \frac{6}{\pi \ln \alpha}\;\frac{(\alpha-1)^2}{\alpha^2} \tag{3}$$

$$K_2 = \frac{6}{\pi \ln(\alpha)}\left[\frac{\alpha-1}{\ln(\alpha)} - 1\right] \tag{4}$$

$$K_3 = \frac{6}{\pi \ln(\alpha)}\;\frac{\alpha-1}{2} \tag{5}$$

$$\alpha = \frac{D_o}{D_i} \tag{6}$$

where P = applied load, lb
E = modulus of elasticity, lb/in^2
δ = deflection, in
ν = Poisson's ratio
h = inside height of spring, in
t = thickness of spring, in
D_o = outside diameter, in
D_i = inside diameter, in
σ = maximum tensile stress, lb/in^2

Write a FORTRAN program that will accept the spring dimensions and the physical properties of the material (including the allowable tensile stress) and will print a table listing the required loading and the stress produced for various deflections. Start with a deflection of 0.001 in and increase in steps of 0.001 in until the stress produced exceeds the allowable stress or until the spring has been pressed into a flat position.

Suggestion: Use a subprogram to calculate the load and stress. Evaluate the design parameters in the main program and transmit them to the subprogram with a COMMON statement.

Sample data

Set	D_o	D_i	h	t	E	ν	σ
1	2.	1.	0.125	0.05	3×10^7	0.3	2×10^5
2	3.	1.25	0.100	0.06	3×10^7	0.3	2×10^5

10-10. The capacitance between flat conductors lying on a substrate of high dielectric constant is of interest in microelectronic circuits:

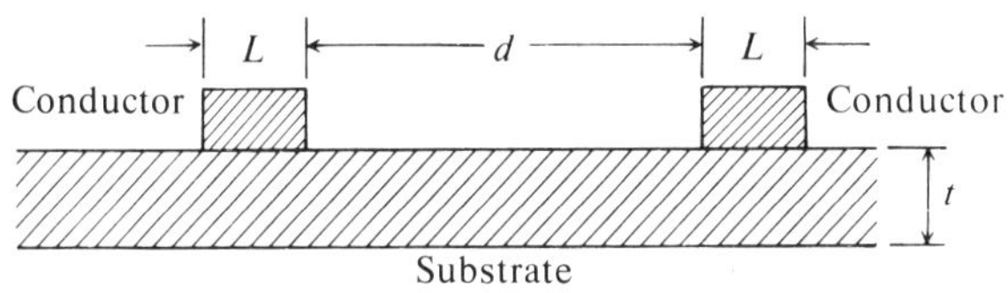

The capacitance C is

$$C = \frac{0.1125\epsilon_r K(k_1)}{K(k_2)} \quad \text{pF/in}$$

where $k_1 = \sqrt{1 - \left(\frac{a}{b}\right)^2}$

$$k_2 = \left(\frac{a}{b}\right)^2$$

$$a = \frac{\pi}{2t} \tanh \frac{\pi d}{4t}$$

$$b = \frac{\pi}{2t} \tanh \left(\frac{\pi L}{2t} + \frac{\pi d}{4t}\right)$$

ϵ_r = relative permittivity

$K(k_1), K(k_2)$ = complete elliptic integrals of the first kind with arguments k_1 and k_2, respectively

Write a FORTRAN program which calculates C. The program should use the FUNCTION subprogram which was discussed in Sec. 10.15.

Sample data

$\epsilon_r = 1773.$
$t = 0.001$ in
$L = 0.001$ in
$d = 0.002$ in

10-11. The scalar electric potential at any point (r,ϕ,z) due to a filamentary ring of charge lying in the plane $z = 0$ and centered at the origin is

$$V = \frac{\rho R}{\pi \epsilon_0 \epsilon_r \sqrt{(r+R)^2 + z^2}} K(k) \tag{1}$$

where

$$k = \frac{2\sqrt{rR}}{\sqrt{(r+R)^2 + z^2}} \tag{2}$$

V = scalar electric potential, V
ϵ_0 = permittivity of vacuum = 8.854×10^{-12} F/m
ϵ_r = relative permittivity
r, z = coordinates of field point in circular coordinate system, m
R = radius of ring charge, m
ρ = charge density, C/m
$K(k)$ = complete elliptic integral of the first kind

Notice that V is independent of the angular coordinate ϕ, and therefore ϕ does not appear in Eq. (1).

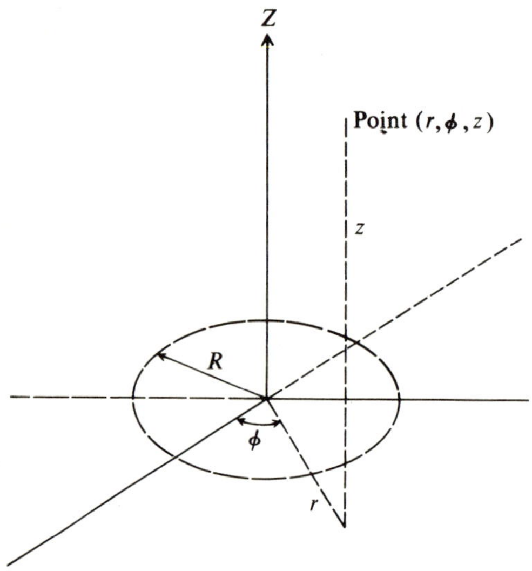

Write a FORTRAN program which calculates and prints the value of the electrical potential V. The program should use the FUNCTION subprogram discussed in Sec. 10.15.

Sample data

R = 0.2 m
r = 0.2 m
z = 0.3 m
ϵ_r = 1.0
ρ = 27.8×10^{-12} F/m

10-12. $E(XY)$, the expectation of the product of two *discrete* random variables X and Y, is

$$E(XY) = \sum_i \sum_j (x_i y_j) p(x_i, y_j)$$

where x_i, y_j = values of random variables
$p(x_i, y_j)$ = joint probability density function

Write a subprogram EXY which receives the maximum values of indices I and J, the values of X(I) and Y(J), and the joint probability density PXY(I,J) from a main program and returns the value of the expectation to the main program. Note that, in general, the range of subscript I will differ from the range of subscript J.

Suggestions: Use a COMMON statement for transmitting the required values of X(I), Y(J). and PXY(I,J) to the subprogram.

Sample data

I	J	X(I)	Y(J)	PXY(I,J)
1	1	51	10	0.1
1	2	51	20	0.1
1	3	51	30	0
2	1	52	10	0.1
2	2	52	20	0.3
2	3	52	30	0.2
3	1	53	10	0.1
3	2	53	20	0.1
3	3	53	30	0

11
Data-handling Techniques

11.1 ALPHAMERIC DATA[1]

Alphameric data or *Hollerith* data consist of literal information represented as strings of alphabetic, numeric, and special symbol characters. These data are read by the computer (input) or written by the computer (output) according to two FORMAT specifications: the *Hollerith specification* (see Sec. 2.9) and the *A specification* (see Sec. 11.2). The former specification permits the storage and subsequent recovery of *alphameric constants*, which are literal data that do not change during program execution. These formats are most convenient for printing messages, table headings, etc., along with the calculated results.

The A specification can also be used for handling alphameric constants, but it has the further advantage of allowing alphameric data to be stored as *alphameric variables* and to be referenced by variable names, just as is done with numeric and logical variables. This feature permits the processing of alphameric data during program execution. *No* type declaration statement is needed to define these variables. The alphameric data is stored either as integer or real numbers and is processed within the computer as ordinary numeric data. The conversion *from* alphameric data on input and *to* alphameric data on output is controlled by utilization of the A specification.

[1] "Alphameric" is an acronym formed from "alphabetic" and "numeric." It signifies that data may contain both alphabetic and numeric information.

11.2 A SPECIFICATION

The *A specification* is an *alphameric* specification for *alphabetic* and *numeric* data and has the form

Aw

where w is an integer constant indicating the number of characters to be processed.

The A specification is used for handling the same type of data as the H specification but has the advantage of making it possible to *process* the data while it is in storage. That is, words can be sorted alphabetically, catalog numbers compared, etc.

Alphameric data read into the computer under the A specification is assigned to storage locations that are named in the same manner as other FORTRAN variables. When used for input, the computer stores the w characters scanned by the reader in the storage location referenced by the assigned variable name. Assuming that a variable, i.e., a *storage location*, can store a *maximum* of *six* alphameric characters, the following rules must be followed:

1. If w is less than or equal to 6, the entire field of w characters will be placed in the designated storage location and *left-justified*; the $(6 - w)$ remaining spaces will be filled with *blanks*.
2. If w is greater than 6, the 6 *rightmost* characters in the field of w characters will be placed in storage; the $(w - 6)$ *leftmost* characters will be dropped.

Example of A specification input

If a card with bINCREMENTb0FbXbbbbb punched in the first 20 columns is read under control of

```
   READ 20, A,B,C
20 FORMAT (A6,A4,A10)
```

then:

bINCRE will be stored in location A. (Rule 1 applies. The six characters read completely fill the space available in storage.)

MENTbb will be stored in location B. (Rule 1 applies. The four characters read by the computer are left-justified in the storage space available and $(6 - 4)$ blanks are added.)

Xbbbbb will be stored in location C. (Rule 2 applies. Of the 10 characters read by the computer, the 6 *rightmost* are retained and stored in the storage space available.)

When used for output the computer prints the alphameric characters found in the storage location identified by the variable named in the list. Again assuming that a maximum of six alphameric characters can be stored in one location, the following rules are followed:

1. If w is less than 6, the w *leftmost* characters are printed.
2. If w is greater than 6, all six characters found in the storage location are printed and the group is *preceded by* $(w - 6)$ blanks.

Example of A specification output

If storage location A contains bINCRE, storage location B contains MENTbb, and storage location C contains Xbbbbb, the statements

```
    PRINT 25, B,A,C
 25 FORMAT (A3,A4,A10)
```

will print the following line:

```
MENbINCbbbbXbbbbb
 3   4     10
```

11.3 DATA STATEMENT

The *FORTRAN DATA statement* is a nonexecutable declaration statement having the general form

DATA *list 1*/C11, C12, . . . , C1m/, *list 2*/C21, C22, . . . , C2n/

where *list 1* and *list 2* are ordered lists containing *m* and *n* variable names, array names, or subscripted variables with *constant* subscripts, and C11, C12, . . . , C1*m* and C21, C22, . . . , C2*n* are correspondingly ordered lists of *constant data values.* A list may contain any number of variables, and any number of lists may be included in a DATA statement. Each half of a pair of lists, i.e., the list of variable names and the list of constants, must contain the same number of items.

This statement serves to *set* or *initialize* the values of the listed variables at the values specified by the corresponding string of constrants. For example, the statements

```
DIMENSION C(10)
COMPLEX W
DATA  A,B,I/2.0, 3E+7, 25/, C(4), W/32.1, (2.0, 1.0)/
```

produce the same effect as the series of initial assignment statements reading

```
   A = 2.0
   B = 3E+7
   I = 25
C(4) = 32.1
   W = (2.0, 1.0)
```

The reader should not conclude from this example that the data must be grouped in any special way. The groupings used above were arbitrary.[1]

If subsequent program statements required any change in the values stored in these locations, the *new* values would replace these *initialized* values. The DATA statement only *sets* the values of the variables at the time when program execution is first begun.

In initializing the elements of an array (or portions of an array) *an implied loop* (see Sec. 7.3) may be used in the DATA statement. An example of this procedure is

```
DIMENSION A(10)
DATA (A(I), I = 3,7)/2., 8.7, 6.1, 0., 10./
```

[1] The programmer may have originally set the values for A, B, and I, and at some later time discovered the necessity for initializing C(4) and W.

which initializes certain elements of the array A as follows:

A(3) = 2.
A(4) = 8.7
A(5) = 6.1
A(6) = 0.
A(7) = 10.

Where several variables in a list must be set at the same value, a *repetition counter* can be used to indicate the number of times a particular constant is to be repeated in the list. For example, in the statements

```
DIMENSION B(100)
DATA (B(J), J = 1, 100)/75*2., 25*0./
```

the 75 and the 25 are repetition counters. This is a concise notation that is equivalent to stringing out 75 twos followed by 25 zeros in the list of constants. The effect produced here is the initialization of the first 75 elements of B at a value of 2., and the last 25 elements at a value of 0.

It should be noted that the constant data values *are not* necessarily limited to numerical data. For example, the statements

```
LOGICAL X, Y, Z
DATA X, Y, Z/.TRUE., 2*.FALSE./
```

are equivalent to the assignments

```
X = .TRUE.
Y = .FALSE.
Z = .FALSE.
```

where *logical constants* (see Sec. 9.1) have been used to initialize the logical variables X, Y, and Z.

Similarly, *alphameric* constants can be used to initialize either real or integer variables. The alphameric or literal data are represented by use of a Hollerith specification as follows:

```
DATA A, B/4HTEXT, 5HNOTES/
```

This statement would store the literal message TEXT in storage location A, and NOTES in location B.

The DATA statement *cannot* be used with variables that are stored in a common block (see Sec. 10.8). A special subprogram called BLOCK DATA, discussed below, is used to initialize variables in this category.

11.4 BLOCK DATA STATEMENT

The *BLOCK DATA statement* is the defining statement for a special subprogram used to *initialize* the value of variables stored in a *labeled* section of common storage. This initialization occurs at the time the program is being compiled and may be overridden by replacement statements during program execution.

The subprogram following the BLOCK DATA statement may contain only the following nonexecutable statements:

1. COMMON
2. DIMENSION
3. EQUIVALENCE
4. Type declaration
5. DATA

The subprogram must be terminated with an END statement, and, as stated earlier, the COMMON statement *must* refer to a *labeled* block in common storage. All the variables in the labeled common block must be named in the subprogram COMMON statement, even if some of them are not being initialized.

The DATA statement in the subprogram lists the particular variables in each labeled block that are to be assigned values. Variable names from one block, and their assigned values, may be separated by a comma from the list of variables stored in another common block.

To illustrate the use of this FORTRAN feature, consider a main program having the following variables assigned to common storage:

```
COMMON//A,B,C/SECT1/Q,R,S,T,U/SECT2/X,Y,Z/SECT3/V(25)
```

Note that variables A, B, and C are stored in *blank* common and therefore *cannot* be initialized; SECT1, SECT2, and SECT3 are the names assigned to the labeled sections of common, and variables in these sections *can* be initialized.

Let us assume that it is necessary to initialize only the variables R, S, U, and V(6). The following BLOCK DATA subprogram, appended to the main program, will accomplish the initialization:

```
BLOCK DATA
COMMON/SECT1/Q,R,S,T,U/SECT3/V
DIMENSION V(25)
DATA U,R,S/100.,10.,0.5/,V(6)/30./
END
```

Note that the portion of common storage labeled SECT1 contains the variables Q, R, S, T, U. Since R, S, U are three of the variables to be initialized, the COMMON statement in the BLOCK DATA subprogram *names* SECT1 and lists *all* the variables in SECT1. The DATA statement in the subprogram lists *only* those variables in SECT1 that are to be initialized, followed by the values to be assigned. Observe that the order of listing does not have to correspond with the order of appearance of the variables in the COMMON statement.

None of the variables in SECT2 of labeled common storage are to be initialized, and therefore SECT2 is not mentioned in the subprogram.

SECT3 in the COMMON statement in the main program specifies the 25 elements of array V. The COMMON statement in the subprogram lists all these elements by naming the array which is dimensioned in a separate statement. In the DATA statement, *only* the element that is to be initialized, V(6), is listed.

11.5 VAN DER WAAL'S EQUATION (Data Statement)

In Sec. 6.1, a program for calculating the volume of carbon dioxide was discussed. In that program (Fig. 6.1.1) two gas constants A and B were assigned values of 3.59 and

```
00001C      VAN2
00010       DATA A,B,R/3.59,.0427,.08205/
00020       READ,C,P
00030       T=273.+C
00040       F3=R*T
00050       DØ 5 I=1,10000
00060       X=I
00070       V=X/1000.
00080       F1=P+A/V**2
00090       F2=V-B
00100       IF(F1*F2-F3) 5,6,6
00110     5 CØNTINUE
00120       PRINT,"V IS GREATER THAN 10 LITERS"
00130       STØP
00140     6 PRINT,
00150       PRINT,"V=",V,"    LITERS"
00155       PRINT,
00160       STØP
00170       END
```

Fig. 11.5.1 Program VAN2.

```
INPUT:00020
?32.,75.


V=        0.1120    LITERS

STØP
```

Fig. 11.5.2 Trial run for VAN2.

0.0427, respectively, by assignment statements on lines 00030 and 00040. The same result is achieved by a DATA statement in a revised version of the program shown in Fig. 11.5.1. A sample run of this revised program is shown in Fig. 11.5.2.

11.6 PLOTTING A GRAPH: PART 1

One useful application of A FORMAT is in programs which plot graphs. In this section, we will discuss a simple example of such a program. Other examples will be presented later.

The flow chart and its program are shown in Figs. 11.6.1 and 11.6.2, respectively. Plotting of points is achieved by means of the constants SPACE, POINT, XAXIS, and the subscripted variable GRAPH. Alphameric data are read into the computer by means of the A specification (lines 00030 and 00040). If the reader examines the trial run shown in Fig. 11.6.3, it will be observed that a blank space, a point (.), and an exclamation mark (!) were assigned to SPACE, POINT, and XAXIS, respectively.

The variable GRAPH is initialized by the DO loop on lines 00070 and 00080. The plotting procedure is implemented by the DO loop which begins on line 00090. The test value of the index I of this loop determines the number of points which are to be plotted; in this illustration, 40 points were used. The function $y = f(x)$ which is to be evaluated is on line 00100. Let us analyze the operation of this DO loop.

At the first iteration, I = 1 and X = 0. First, Y is calculated. Then, the *integer* variable K is assigned a numerical value *proportional* to Y. K is scaled in such a way that it can never be negative or zero because it is used as a subscript of GRAPH. Furthermore, K is scaled so that the points of the graph are properly spread out and the graph is centered on the paper. In the present example, when X = 0., Y = 0., and K = 30. On line 00120, GRAPH (30) is assigned the value stored in XAXIS, and on

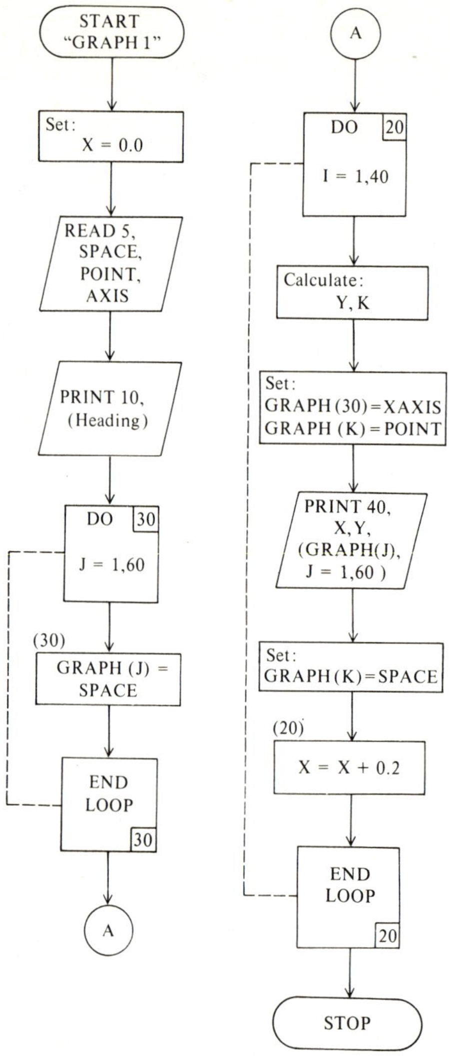

Fig. 11.6.1 Flow chart of program GRAPH1.

line 00130 GRAPH (30) is assigned the value stored in POINT. Obviously, only the latter value will be printed. The assignment statement GRAPH (30) = XAXIS, is executed in each iteration, and causes the symbol for the x axis to be printed. This will not occur if Y happens to be zero, as mentioned above. In this case the symbol assigned to POINT will appear on the x axis. (Notice the first point of the graph.)

The print statement on line 00140 causes the current values of X, Y, and GRAPH to be printed in accordance with the FORMAT statement on the next line. The reader should note that the implied DO loop on line 00140 causes GRAPH (1) through GRAPH (60) to be printed. In the present example, during the first iteration, and until the print statement is executed, GRAPH (30) has the value stored in POINT; GRAPH (1) through GRAPH (29) and GRAPH (31) through GRAPH (60) have the value stored in SPACE. In this case, therefore, only . is printed, and it is printed in the thirtieth space. After the print statement has been executed, GRAPH (K) is again

```
C     GRAPH1
C     THIS PROGRAM PLØTS A GRAPH. THE USER MUST SUPPLY THE
C     SYMBØLS FØR PLØTTING THE PØINTS AND FØR PLØTTING THE
C     X AXIS.
      DIMENSIØN GRAPH(60)
      X=0.
      READ 5,SPACE,PØINT,XAXIS
    5 FØRMAT(3A1)
      PRINT 10
   10 FØRMAT(3X,"X",4X,"Y",28X,"GRAPH")
      DØ 30 J=1,60
   30 GRAPH(J)=SPACE
      DØ 20 I=1,40
      Y=SIN(X)*EXP(-0.1*X)
      K=31.*Y+30.
      GRAPH(30)=XAXIS
      GRAPH(K)=PØINT
      PRINT 40,X,Y,(GRAPH(J),J=1,60)
   40 FØRMAT(F4.1,1X,F6.3,60A1)
      GRAPH(K)=SPACE
   20 X=X+.2
      PRINT,
      STØP
      END
```

Fig. 11.6.2 Program GRAPH1.

```
INPUT:00030
 ? .!

   X    Y                            GRAPH
 0.0  0.000                             .
 0.2  0.195                             !     .
 0.4  0.374                             !          .
 0.6  0.532                             !               .
 0.8  0.662                             !                   .
 1.0  0.761                             !                      .
 1.2  0.827                             !                        .
 1.4  0.857                             !                         .
 1.6  0.852                             !                         .
 1.8  0.813                             !                        .
 2.0  0.744                             !                      .
 2.2  0.649                             !                   .
 2.4  0.531                             !               .
 2.6  0.397                             !           .
 2.8  0.253                             !      .
 3.0  0.105                             !  .
 3.2  -.042                           . !
 3.4  -.182                       .     !
 3.6  -.309                   .         !
 3.8  -.418                .            !
 4.0  -.507             .               !
 4.2  -.573           .                 !
 4.4  -.613         .                   !
 4.6  -.627         .                   !
 4.8  -.616         .                   !
 5.0  -.582          .                  !
 5.2  -.525            .                !
 5.4  -.450               .             !
 5.6  -.361                 .           !
 5.8  -.260                    .        !
 6.0  -.153                        .    !
 6.2  -.045                           . !
 6.4  0.061                             !.
 6.6  0.161                             !   .
 6.8  0.250                             !      .
 7.0  0.326                             !         .
 7.2  0.386                             !          .
 7.4  0.429                             !            .
 7.6  0.453                             !             .
 7.8  0.458                             !             .

STØP
```

Fig. 11.6.3 Trial run for GRAPH1.

initialized at a value of SPACE, and X is incremented. In the second iteration, Y = 0.195 and K = 36. Therefore, . is printed in the thirty-sixth space, and ! is printed in the thirtieth space. In the trial run that is shown in Fig. 11.6.3, the function $y = \sin(x)\, e^{-0.1x}$ is plotted.

The reader has probably observed that it is easier to plot graphs with the independent variable axis extending vertically downward because of the nature of the digital output mechanism.

11.7 PLOTTING A GRAPH: PART 2

In this section we will briefly discuss a second program, shown in Fig. 11.7.1, for plotting a graph. This program differs from the program of the previous section in the following ways:

1. A y axis as well as an x axis is printed. The subscripted variable YAXIS is used to print the y axis. Scales are provided for both axes. The desired y axis scale was estimated by using data from the program of Sec. 6.7 to determine the maximum value of the function to be plotted. This is further discussed below.
2. DATA statements are used to initialize SPACE, POINT, XAXIS, GRAPH, and YAXIS. This technique was discussed in Sec. 11.3.
3. The plotting is accomplished by calling a subroutine named PLOT. The CALL statement for PLOT is within a DO loop in the main program. The maximum value of the index of this DO loop determines the number of points that will be plotted.
4. The function to be plotted is in the subprogram entitled FUNCTION Y(X).

Let us examine SUBROUTINE PLOT. The printing symbols are initialized by the DATA statement on line 00140. The variables GRAPH (J) and YAXIS (J) are

```
00001C    GRAPH2
00002C    THIS PRØGRAM PLØTS A GRAPH WITH AN AXIS SYSTEM.
00010     X=0.
00020     DØ 20 I=1,20
00030     CALL PLØT(X,I)
00040  20 X=X+.2
00050     STØP
00060     END
00070     FUNCTIØN Y(X)
00080     Y=1.-(CØS(.975*X)+.216*SIN(.975*X))/EXP(1.21*X)
00090     RETURN
00100     END
00110     SUBRØUTINE PLØT(X,I)
00120     DIMENSIØN GRAPH(70),YAXIS(70)
00130     DATA SPACE,PØINT,XAXIS/1H ,1H.,1H!/
00140     DATA (GRAPH(J),J=1,67)/67*1H /
00150     DATA (YAXIS(J),J=1,70)/70*1H+/
00160     IF(I.GT.1) GØ TØ 10
00170     PRINT 7,(YAXIS(J),J=1,70)
00180   7 FØRMAT(70A1/)
00190     PRINT 8
00200   8 FØRMAT(5X,"0",8X,".2",8X,".4",8X,".6",8X,".8",8X,"1")
00210  10 K=50.*Y(X)+6.
00220     GRAPH(6)=XAXIS
00230     GRAPH(K)=PØINT
00240     PRINT 40,X,(GRAPH(J),J=6,67)
00250  40 FØRMAT(F5.2,67A1)
00260     GRAPH(K)=SPACE
00270     RETURN
00280     END
```

Fig. 11.7.1 Program GRAPH2.

```
++++++++++++++++++++++++++++++++++++++++++++++++++++++++++++++++++++++
    0        .2        .4        .6        .8        1
0.00.
0.20!         .
0.40!                  .
0.60!                          .
0.80!                                 .
1.00!                                      .
1.20!                                            .
1.40!                                               .
1.60!                                                 .
1.80!                                                  .
2.00!                                                   .
2.20!                                                    .
2.40!                                                    .
2.60!                                                    .
2.80!                                                    .
3.00!                                                    .
3.20!                                                    .
3.40!                                                   .
3.60!                                                   .
3.80!                                                   .
STØP
```

Fig. 11.7.2 Trial run for GRAPH2.

initialized (each time the subroutine is called) by the DATA statement on line 00150 which employs both implied loops and repetition counters discussed in Sec. 11.3. When I, the index of the DO loop within the main program, has a value of 1, the PRINT statement on line 00170 causes the x axis to be printed. However, when I exceeds 1, the logical IF statement on line 00160 causes the program to bypass the PRINT statement. As this particular PRINT statement prints the y axis, it is needed only once. This is so because the y axis is printed in the horizontal direction, and is printed as the first line.

As in the program discussed in the previous section, the variable K is proportional to Y. The remainder of the subprogram is essentially the same as the corresponding segment of the previous program discussed in Sec. 11.6.

In Sec. 6.7 we discussed the transient response of the underdamped shunt compensated network. In the trial run shown in Fig. 6.7.3 the first maximum of the transient response of that network was found to be about 1.03 and occurred at about 2.5 seconds. Using the program under discussion in this section, the response of the same network has been plotted in Fig. 11.7.2 for a time interval of 3.8 seconds. The top of the curve is flattened because of the truncation error that occurs when the numerical value of the function is assigned to the integer variable K.

11.8 SORTING ALPHAMERIC DATA

The manipulation of alphameric data in a computer can be most easily demonstrated by a program that will sort a list of names into alphabetic order. The computer system that was used for this program was designed so that numeric values are assigned by the system to *strings* of alphameric characters. These numeric values are such that AB has a smaller value than AC, which in turn has a smaller value than AD, and so on. Therefore, an alphabetizing process is merely a procedure for sorting a list of values into ascending order.

Because alphameric data are stored internally as numeric values, the programmer has the choice of converting the alphameric string to either a *real* value or an *integer* value. A real value is capable of representing a string of up to *six* alphameric characters; an integer value can be used to represent a string containing up to *three* alphameric characters. In the program used in this section to illustrate the alphabetizing procedure, we have elected to use *integer* values in storage. As this would limit our names to only three letters, an unrealistic number, we have stored the names in a two-dimensional array having N rows and three columns, where N is the number of names in the list. With three columns available for each row, and with each element in the row capable of storing three letters, each name can contain up to nine alphabetic characters. To illustrate this method of storage, let us suppose a two-dimensional array named L is used to store the list of names and the *seventh* name in the list is WILLIAM. This name would be stored in three segments as follows:

L(7,1) would contain the value for WIL
L(7,2) would contain the value for LIA
L(7,3) would contain the value for Mbb

Note that all three of the fields must be filled. In this case the third column element was filled with blanks.

The alphabetizing procedure requires that the *rows* be arranged in ascending order by comparison of the *first column element* in each row. If the computer finds two or more identical first column elements, then a comparison is made of the second column elements. If the first and second column elements in two or more rows are the same, a comparison of the third column elements is made. When a distinction between rows has been made, the rows are arranged in ascending order by a sorting procedure similar to that discussed in Sec. 4.9.

For example, if the *seventh* name in the list is WILLIAM and the *eighth* name is WILFRED, the program would compare L(7,1) and L(8,1) and find them to contain identical numerical values because both represent the string WIL. The program would then compare L(7,2) with L(8,2) and find that the value stored in L(8,2) is the smaller numeric value. [L(7,2) contains LIA and L(8,2) contains FRE.] Because a distinction had been made, it would not be necessary to compare the third column elements, and the program, in this case, would proceed to the sorting process where *all the elements* in the seventh and eighth rows would be exchanged. This would make WILFRED the seventh name on the list and WILLIAM the eighth. If the comparison procedure had shown the names to be in the proper alphabetic order, *no* exchange would have been made.

The procedure described above is outlined in the flow chart of Fig. 11.8.1. The program ALPHA1 is listed in Fig. 11.8.2, and a list of the symbols used in the program is shown in Table 11.8.1. In the program, an integer array L(100,3) is defined in the COMMON statement on line 00010, and therefore the program is capable of sorting a list of up to 100 names containing nine letters each. Storing this array in unlabeled COMMON makes it accessible to both the main program, where the element comparisons are made, and the subprogram SORT, where the interchange of rows is accomplished whenever necessary.

The number of names in the list and the list of names to be sorted are entered by the READ statements on lines 00020 and 00030, respectively. Line 00030 uses implied nested loops to read the names according to FORMAT statement 50. This FORMAT statement, on line 00040, provides for *five* names on an input line, each name occupying *three* array elements, and each element having the specification A3.

The nested DO loops starting on lines 00060 and 00070 and extending to line 00120 perform the comparisons of the segments of the *i*th and $(i + 1)$th words in the

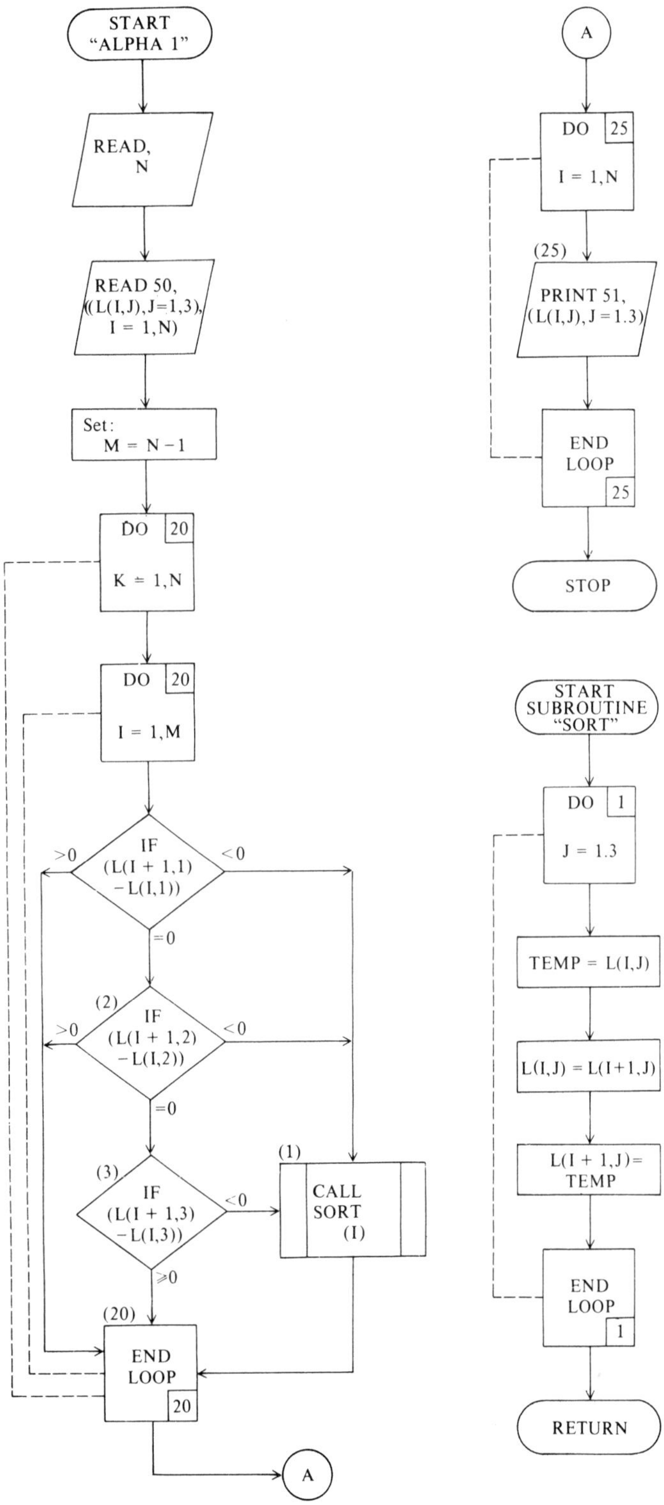

Fig. 11.8.1 Flow chart of program ALPHA1.

```
00001C       ALPHA1
00002C       PRØGRAM TØ ARRANGE A LIST ØF ALPHAMERIC
00003C       VALUES IN INCREASING ØRDER
00010        CØMMØN L(100,3)
00020        READ,N
00030        READ 50,((L(I,J),J=1,3),I=1,N)
00040     50 FØRMAT(15A3)
00050        M=N-1
00060        DØ 20 K=1,N
00070        DØ 20 I=1,M
00080        IF(L(I+1,1)-L(I,1))1,2,20
00090      2 IF(L(I+1,2)-L(I,2))1,3,20
00100      3 IF(L(I+1,3)-L(I,3))1,20,20
00110      1 CALL SØRT (I)
00120     20 CØNTINUE
00130        DØ 25 I=1,N
00140     25 PRINT 51,(L(I,J),J=1,3),(L(I,J),J=1,3)
00150     51 FØRMAT(3X3A3,I15,2I12)
00160        STØP
00170        END
00180        SUBRØUTINE SØRT(I)
00190        CØMMØN L(100,3)
00200        DØ 1 J=1,3
00210        TEMP=L(I,J)
00220        L(I,J)=L(I+1,J)
00230      1 L(I+1,J)=TEMP
00240        RETURN
00250        END
```

Fig. 11.8.2 Program ALPHA1.

list. The IF statement on line 00080 compares the *first* segments of each pair of words; if these are *identical*, the program branches to the IF statement on line 00090 where the *second* segments are compared; if these are also identical, the IF statement on line 00100 is used to compare the *third* segments of the pair. Note that in all of these arithmetic IF statements when the segments tested *are* in the proper sequence, the program jumps to line 00120 and the comparison cycle is continued. If the segments *are not* in the proper sequence, the program branches to line 00110 where the subroutine SORT is called and the positions of the pair of names are reversed on the list. When the IF statement on line 00100 finds the *third* elements in a pair to be identical, the two names must be identical and no action is taken by the computer except to advance to the next pair of names that are to be compared.

The inner loop (line 00070) compares the consecutive *pairs* of words in the list, that is, the *first* and *second* names, then the *second* and *third*, then the *third* and *fourth*, and so on. Note that this loop is executed only (N–1) times to prevent a false comparison between the Nth or last name on the list and the *row* in the array that would follow it. If the array were dimensioned (as was done in our trial run, below) to provide for a larger number of rows than were actually used to contain the list of

Table 11.8.1

Name	*Meaning*
N	Number of names to alphabetize
L(I,J)	Array in which segments of the names are stored
I	Row containing *all* the segments of the *i*th name
J	Column containing *one* of the segments of the *i*th name
M	Next-to-the-last name in the list
K	Index for the *k*th time the entire list is scanned
TEMP	Temporary storage location used during the sorting process

names, the contents of the *row* following the last name on the list would be compared with the numeric equivalent of the last name. As this (N+1)th row could contain *any* storable piece of information, it is impossible to predict the outcome of the sorting process.

In some compilers, when an array is dimensioned, all the elements of the array are automatically initialized at zero or blank. Under these circumstances the (N+1)th row would have a lower numerical value than any name in the list and would ultimately occupy the L(1,1), L(1,2), and L(1,3) positions, that is, the *first* position on the list. The second, third, fourth, etc., positions in the list would contain the correctly alphabetized list. The (N+1)th position would contain the *last* name on the list read into the computer for alphabetizing. At first glance this appears to be a minor problem; however, during the printing operation the *last* name would not be printed because it would be contained in the (N+1)th row.

Regardless of whether or not the compiler initializes a dimensioned array at zero, if all 100 rows were used to contain the list of names, the *row following the last name* would have a subscript larger than the dimensioned size of the array. An attempt to call forth the contents of this location may produce an error message during the execution of the program.

The outer loop of the nested DO loops (line 00060) causes the entire process of comparisons and exchanges to start all over again at the top of the list. Under control of this loop, the sorting process is accomplished N times to ensure that the lowest valued set of elements is stored in locations L(1,1), L(1,2), and L(1,3).

When the array contains the list in alphabetic order, the program prints out the complete list, one name on each line, under control of the DO loop starting on line 00130. To examine the mechanics of alphameric storage, the output operation was programmed to print the numeric value of each of the three segments of the name on the same line with the name. This was accomplished by commanding the computer to print the contents of each row *twice* on the same line. FORMAT statement 51, on line 00150, controlled the alphameric output by specification 3A3 and the numeric output of the same information by specification I15,2I12. The I15 specification was used to provide an additional three spaces between the column of names and the three columns of numeric equivalents.

```
INPUT:00020
 ?12

INPUT:00030
 ?ANTHØNY   ABIGAIL   SEBASTIANSAM        MARIE
 ?JØSEPH    FRED      JØSE       MARY     FREDA
 ?JØSEPHINEJØSIAH

    ABIGAIL             4276809        4669769        4988960
    ANTHØNY             4279892        4738894        5840928
    FRED                4608581        4464672        2105376
    FREDA               4608581        4473120        2105376
    JØSE                4869971        4530208        2105376
    JØSEPH              4869971        4542536        2105376
    JØSEPHINE           4869971        4542536        4804165
    JØSIAH              4869971        4800840        2105376
    MARIE               5062994        4801824        2105376
    MARY                5062994        5840928        2105376
    SAM                 5456205        2105376        2105376
    SEBASTIAN           5457218        4281172        4800846
STØP
```

Fig. 11.8.3 Trial run for ALPHA1.

```
INPUT:00020
 ?6


INPUT:00030
 ?MARK3      70-DE-27 AISI-H16 F7562      92643A

 ?PART 621


    70-DE-27            3616813        4474157        3290912
    92643A              3748406        3420993        2105376
    AISI-H16            4278611        4795720        3225120
    F7562               4601653        3551776        2105376
    MARK3               5062994        4928288        2105376
    PART 621            5259602        5513270        3289376
STØP
```

Fig. 11.8.4 Trial run for ALPHA1.

The results of the first trial run, made with a list of 12 names, are shown in Fig. 11.8.3. The reader should examine the three right-hand columns of the output to obtain an understanding of the way alphameric strings are represented internally in the computer. Note particularly that JOSEPH and JOSEPHINE contain identical *first* and *second* segments, and the determination of their order in the list was made by the values 2105376 (for bbb) and 4804165 (for INE). Similarly, JOSE and JOSEPH were ordered by the values of their respective *second* segments, Ebb and EPH.

The general sorting process described above is *not* limited to alphabetic strings. To demonstrate the generality of the program, a second trial run is included in Fig. 11.8.4. In this example a list of six strings of alphameric characters has been sorted. Because of this generality, the procedure is applicable to lists of part numbers, material code names, drawing numbers, etc.

PROBLEMS

11-1. Probs. 4-8 and 5-12 were concerned with the sonic velocity in one of several selected gases at a given temperature. In both these problems the program required that the user supply the code number of the particular gas in which he was interested, as well as the gas temperature or temperature range. Revise the program written as a solution for Prob. 5-12 so that the user can read-in the literal name of the gas, rather than the code number. (It is suggested that the *names* used for the gases in the program be: AIR, CO2, HE, N2, and O2.)

Sample data

Set 1:

Gas name = CO_2

Temperature range from 40 to 135°F in steps of 5°F

Set 2:

Gas name = AIR

Temperature range from −20 to 130°F in steps of 10°F

11-2. In Prob. 1-64 a formula was given that expresses the required shaft diameter for a specified torsional moment if the allowable shear stress of the material is known.

Write a FORTRAN program that will make it possible to read-in the name of the shaft material and the size of the imposed load and have the computer print the required shaft diameter.

Use a DATA statement to store the names of the materials and their respective allowable stresses in two arrays.

Some typical physical properties are:

Material	*Allowable stress, lb/in*^{2}
Gray C.I.	3,000
1020 steel	6,000
4130 steel	9,000
Al alloy	4,000
Ti alloy	16,000
Nylon	1,200

Sample data

Set	*Material*	*Torsional load*
1	4130 steel	200,000 lb-in
2	Nylon	800 lb-in
3	Ti alloy	200,000 lb-in

11-3. Write a FORTRAN program that will read-in a list of up to 50 material names with up to 10 physical properties for each material, and will sort and print the list in ascending order of value for any one of the properties.

Sample data

Given the following list of materials and properties, have the computer list them in increasing order of specific weights.

Material name	*Specific weight, lb/in*^{3}	*E*	*G*	*Allowable stress, lb/in*^{2}
		lb/in^2 x 10^6		
Gray C.I.	0.26	13.0	5.5	3,000
1020 steel	0.283	29.8	11.5	6,000
4130 steel	0.283	29.8	11.5	9,000
Bronze	0.32	16.0	6.0	5,000
Stainless steel	0.273	29.8	11.5	6,500
Monel	0.319	26.0	9.5	3,700
Inconel	0.307	31.0	11.0	4,100
Mg alloy	0.066	6.5	2.4	4,000
Al alloy	0.100	10.6	4.0	4,000
Ti alloy	0.163	16.0	6.0	16,000
Phenol	0.048	0.71	0.28	900
Polystyrene	0.038	0.50	0.18	800
Nylon	0.041	0.40	0.14	1,200

11-4. In Prob. 1-55 an expression is given for converting a temperature expressed in degrees Fahrenheit to an equivalent temperature in degrees Celsius.

Write a FORTRAN program that will read-in a temperature and the symbol for the temperature scale in which it is expressed according to the format specification F8.2,A3, and will make the necessary conversion and print the result. (That is, if a temperature expressed in degrees Celsius is read by the computer, it will be converted to degrees Fahrenheit, and vice versa.)

Sample data

Set	*Formatted input*
1	bbb75.20bCb
2	bb160.95bFb
3	bb-40.00bFb

11-5. Prob. 9-4 determined the state of eight devices in a fluid circuit for all possible combinations of input conditions. Revise the program written for that problem so a literal message of the state of each device is printed, rather than one of the code symbols T or F.

12 Organizing a Problem for Programming

In the preceding chapters we have discussed the rules of FORTRAN and illustrated their usage. It is evident that familiarity with these rules and the exercise of some ingenuity will enable the reader to write a variety of programs for solving engineering problems. However, there are many problems which the programmer can solve only if he also has a knowledge of certain branches of applied mathematics. For example, programming a problem involving *ABCD* parameters requires competence in matrix multiplication; programming a problem in differential equations requires a knowledge of certain numerical methods.

Let us, therefore, now place the art of programming in a somewhat more realistic setting. Assume that we have an engineering problem to solve, but are unfamiliar with a particular mathematical method needed for its solution. First, we must learn the mathematical method. Then we must decide how to program it. Finally, we have to determine whether the program should be modified to suit the particular engineering problem.

In this chapter we have chosen three mathematical topics to discuss and to apply to engineering problems. These are:

1. Matrix multiplication
2. Regression analysis
3. A topic from the theory of equations.

A study of these topics and the illustrations of their usage should reinforce the reader's knowledge of FORTRAN and how to apply it.

12.1 MATRIX MULTIPLICATION[1]

In this section, we will discuss a program for matrix multiplication. Before doing so, let us recall a few of the rules which are involved in matrix multiplication.

1. Two matrices $\bar{A}$ and $\bar{B}$ are said to be conformable for multiplication in the order $\bar{A}\bar{B}$ if the number of columns of $\bar{A}$ is equal to the number of rows of $\bar{B}$.
2. Let $\bar{C}$ be the matrix which is equal to the product $\bar{A}\bar{B}$, where the order of $\bar{A}$ is $(p \times q)$ and the order of $\bar{B}$ is $(q \times r)$. Any element c_{ij} of matrix $\bar{C}$ is given by the rule

$$c_{ij} = \sum_{k=1}^{q} a_{ik} b_{kj} \tag{12.1.1}$$

To illustrate the procedure, we will calculate the matrix product $\bar{A}\bar{B}$, where:

$$\bar{A} = \begin{bmatrix} 1 & 1 \\ 2 & 4 \\ 3 & 1 \end{bmatrix} \qquad \bar{B} = \begin{bmatrix} 4 & 2 & 3 \\ 2 & 1 & 2 \end{bmatrix}$$

$$\bar{C} = \begin{bmatrix} (1)(4) + (1)(2) & (1)(2) + (1)(1) & (1)(3) + (1)(2) \\ (2)(4) + (4)(2) & (2)(2) + (4)(1) & (2)(3) + (4)(2) \\ (3)(4) + (1)(2) & (3)(2) + (1)(1) & (3)(3) + (2)(1) \end{bmatrix}$$

$$\bar{C} = \begin{bmatrix} 6 & 3 & 5 \\ 16 & 8 & 14 \\ 14 & 7 & 11 \end{bmatrix}$$

A program for multiplying two matrices is shown in Fig. 12.1.1. The names used to indicate the rows and columns of matrix $\bar{A}$ are, respectively, NRA and NCA; the names used to indicate the rows and columns of matrix $\bar{B}$ are, respectively, NRB and NCB. In the remainder of this discussion, we will use the matrices of the example given above. In that example $\bar{A}$ is a matrix of order 3×2; hence NRA = 3 and NCA = 2. $\bar{B}$ is a matrix of order 2×3; hence NRB = 2 and NCB = 3.

The calculation of the product matrix is carried out by the use of three DO loops. The first DO loop, with index I, begins on line 00130; the second DO loop, with index J, is nested within the first DO loop; and the third DO loop, with index K, is nested within the second DO loop. Let us see how the first element, C(1,1), of the product matrix is calculated. At the beginning, I = 1 and J = 1. I and J will remain at these values until the innermost loop, with index K, is satisfied. The maximum value of K is NCA, which, as noted above, is 2. Therefore, the value of C(1,1) is obtained in two passes through the innermost DO loop, with I = 1 and J = 1.

Pass 1 K = 1; initial value of C(1,1) is zero.

C(1,1) = C(1,1) + A(1,1) * B(1,1)
 = 0. + (1)(4)
 = 4.

[1] C. R. Wylie, Jr., "Advanced Engineering Mathematics," 3d ed., McGraw-Hill, New York, 1966, p. 420.

```
C     MATRX1
C     THIS PRØGRAM PERFØRMS MATRIX MULTIPLICATIØN.
      DIMENSIØN A(10,10),B(10,10),C(10,10)
      PRINT,"ENTER THE NØ. ØF RØWS AND CØLUMNS ØF A"
      READ,NRA,NCA
      PRINT,"ENTER A"
      READ,((A(I,J),J=1,NCA),I=1,NRA)
      PRINT,"ENTER THE NØ. ØF RØWS AND CØLUMNS ØF B"
      READ,NRB,NCB
      PRINT,"ENTER B"
      READ,((B(I,J),J=1,NCB),I=1,NRB)
      IF(NCA.EQ.NRB) GØ TØ 15
      PRINT,"MATRICES ARE NØT CØNFØRMABLE"
      STØP
   15 DØ 6 I=1,NRA
      DØ 6 J=1,NCB
      C(I,J)=0.
      DØ 6 K=1,NCA
    6 C(I,J)=C(I,J)+A(I,K)*B(K,J)
      PRINT,"A MATRIX"
      DØ 10 I=1,NRA
   10 PRINT 7,(A(I,J),J=1,NCA)
      PRINT,"B MATRIX"
      DØ 11 I=1,NRB
   11 PRINT 7,(B(I,J),J=1,NCB)
      PRINT,"PRØDUCT MATRIX"
      DØ 12 I=1,NRA
   12 PRINT 7,(C(I,J),J=1,NCB)
    7 FØRMAT(6E12.3)
      STØP
      END
```

Fig. 12.1.1 Program MATRX1.

Pass 2 K = 2

C(1,1) = C(1,1) + A(1,2) * B(2,1)
= 4. + (1)(2)
= 6.

The next element to be calculated is C(1,2). I remains at 1, and J assumes the value 2. I and J remain at these values until the innermost loop is satisfied.

Pass 1 K = 1; initial value of C(1,2) is zero.
C(1,2) = C(1,2) + A(1,1) * B(1,2)
= 0. + (1)(2)
= 2.

Pass 2 K = 2

C(1,2) = C(1,2) + A(1,2) * B(2,2)
= 2. + (1)(1)
= 3.

With I = 1 and J = 3, the element C(1,3) is calculated in a similar fashion. This completes the calculation of the elements of the first row of the product matrix. Next, the index I is set at 2, and remains at that value during the calculation of the elements of the second row of the product matrix. The index J, however, varies from 1 to 3, as in the case of the elements of the first row, and for each value of J, the innermost DO

```
ENTER THE NØ. ØF RØWS AND CØLUMNS ØF A
INPUT:00030
 ?2,3

ENTER A
INPUT:00050
 ?1,2,1,4,0,2

ENTER THE NØ. ØF RØWS AND CØLUMNS ØF B
INPUT:00070
 ?3,2

ENTER B
INPUT:00090
 ?3,-4,1,5,-2,2

A MATRIX
   0.100E+01     0.200E+01     0.100E+01
   0.400E+01     0.000E+00     0.200E+01
B MATRIX
   0.300E+01     -.400E+01
   0.100E+01     0.500E+01
   -.200E+01     0.200E+01
PRØDUCT MATRIX
   0.300E+01     0.800E+01
   0.800E+01     -.120E+02
STØP
```

Fig. 12.1.2 Trial run for MATRX1.

loop is satisfied. Finally, we note that the index I remains at a value of 3 during the calculation of the elements of the third row. In summary:

1. The elements of matrix $\bar{C}$ are calculated row by row, and indices I and J denote the row and column, respectively, of the particular element of $\bar{C}$ which is being calculated.
2. For each value of J, the innermost DO loop is satisfied, and the value of element C(I,J) is thereby calculated.

A trial run using the same data is shown in Fig. 12.1.2.

12.2 UNDAMPED FREE VIBRATION[1]

In this section we will discuss a method for determining the first mode of vibration for a system of springs and masses. The system to be analyzed is undamped and unforced and has N degrees of freedom. We shall employ an iteration method for solving the problem and will limit discussion to a system consisting of masses and springs in series; therefore N is the number of masses in the system. The result of the calculations will be the first natural frequency of the system, that is, the lowest frequency at which resonance will occur.

The *influence coefficient* a_{ij} for the ith mass in the system is equal to the displacement of the ith mass caused by a unit force acting on the jth mass. These coefficients can be determined from the spring constants for the springs used in the

[1] R. Burton, "Vibration and Impact," Addison-Wesley, Reading, Mass., 1958, p. 209.

system. The following simple example shows how influence coefficients can be calculated. Consider the following system:

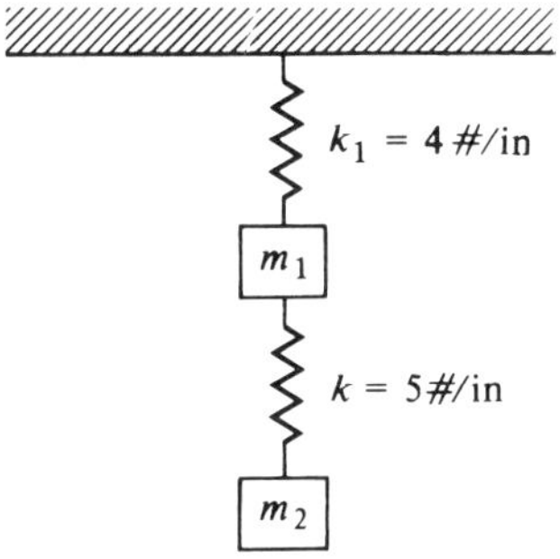

$$a_{11} = \frac{1}{k_1} = \frac{1}{4}$$

$$a_{21} = \frac{1}{k_1} = \frac{1}{4}$$

$$a_{12} = \frac{1}{k_1} = \frac{1}{4}$$

$$a_{22} = \frac{1}{k_1} + \frac{1}{k_2} = \frac{1}{4} + \frac{1}{5} = \frac{9}{20}$$

For a system vibrating with a circular frequency of ω rad/s, the inertia force is given by $m_i\omega^2 x_i$. The sum of each of these inertia forces times the corresponding influence coefficient will give the displacement of each mass in the system. For example, to find the displacement of x_1 we would multiply a_{11} by the inertia force of the first mass and add to this the product of a_{12} by the inertia force of the second mass, and so on. Writing similar displacement equations for each mass in the system would result in the following system of equations:

$$\begin{aligned}
x_1 &= a_{11}(m_1\omega^2 x_1) + a_{12}(m_2\omega^2 x_2) + a_{13}(m_3\omega^2 x_3) + \cdots\cdots\cdots \\
x_2 &= a_{21}(m_1\omega^2 x_1) + a_{22}(m_2\omega^2 x_2) + a_{23}(m_3\omega^2 x_3) + \cdots\cdots\cdots \\
&\cdots\cdots\cdots\cdots\cdots\cdots\cdots\cdots\cdots\cdots \\
x_n &= a_{n1}(m_1\omega^2 x_1) + a_{n2}(m_2\omega^2 x_2) + a_{n3}(m_3\omega^2 x_3) + \cdots + a_{nn}(m_n\omega^2 x_n)
\end{aligned}$$

The factors in the parentheses are the inertia forces. In order to write this system of equations in a more concise form, as a matrix equation, we must first define the component matrices.

The influence coefficient matrix is a square matrix of order n. It is a symmetric matrix; that is, $a_{ij} = a_{ji}$.

$$\bar{A} = \begin{bmatrix} a_{11} & a_{12} & \cdots & a_{1n} \\ a_{21} & a_{22} & \cdots & a_{2n} \\ \cdots & \cdots & \cdots & \cdots \\ a_{n1} & a_{n2} & \cdots & a_{nn} \end{bmatrix}$$

The mass matrix is a *diagonal* matrix of order n:

$$\bar{M} = \begin{bmatrix} m_1 & 0 & 0 & 0 \cdots & 0 \\ 0 & m_2 & 0 & 0 \cdots & 0 \\ 0 & 0 & m_3 & 0 \cdots & 0 \\ \cdots & \cdots & \cdots & \cdots\cdots & \cdots \\ 0 & 0 & 0 & 0 \cdots & m_n \end{bmatrix}$$

The displacement matrix is a column matrix containing n elements:

$$\bar{X} = \begin{bmatrix} x_1 \\ x_2 \\ x_3 \\ \cdot\cdot \\ \cdot\cdot \\ \cdot\cdot \\ x_n \end{bmatrix}$$

Using the above matrices, we can now write the free vibration displacement equation in matrix form:

$$\bar{X} = \omega^2 \bar{A}\bar{M}\bar{X}$$

where ω is the first natural frequency of the system. If we now define the dynamic matrix $\bar{U}$ as

$$\bar{U} = \bar{A}\bar{M}$$

the displacement equation can be rewritten as

$$\frac{1}{\omega^2}\bar{X} = C\bar{U}\bar{X}$$

where $C = 1$.

To solve the above matrix equation by iteration, we will employ the following steps:

1. Initialize the constant C at unity.
2. Set all the elements of the displacement matrix $\bar{X}$ at unity. We will call this the trial matrix.
3. Calculate the value of a new matrix $\bar{T}$ as

   $$\bar{T} = \bar{U}\bar{X}$$

   where $\bar{U}$ is the dynamic matrix and $\bar{X}$ is the trial matrix.
4. Reduce the matrix $\bar{T}$ to the same form as $\bar{X}$ by factoring out of $\bar{T}$ a constant equal to the value of the first term t_1

   $$\bar{T}' = \frac{1}{t_1}\bar{T}$$

   And set

   $$C = (C)(t_1)$$
5. Compare the trial matrix $\bar{X}$ with the factored matrix $\bar{T}'$.

6. If they are different, set $\overline{X}$ equal to $\overline{T}'$ and return to step 3.
7. If they are the same, then $1/\omega^2 = C$, where C is the product of all the factors from each of the iteration steps.

The above list of steps describes the calculation procedure for solving this problem. The reader should note that in step 7 we require that $\overline{X}$ be exactly the same as $\overline{T}'$. In a computer-executed calculation it is unlikely that these two values will ever be identical. Therefore, it is necessary to choose an arbitrary small value ϵ to use as a test for convergence. When the absolute difference between the corresponding elements in $\overline{X}$ and $\overline{T}'$ is less than ϵ, we will accept the condition as exact equality.

With the general plan of calculation outlined, the next step in our programming procedure is to outline the computer program steps that are necessary. The required steps are:

1. Dimension the required arrays.
2. Read-in the number of degrees of freedom and the convergence test value ϵ.
3. Initialize the influence coefficient and mass matrices at zero.
4. Read-in values for the masses and the spring constants, and initialize the mass matrix $\overline{M}$.
5. Calculate the influence coefficient matrix $\overline{A}$.
6. Calculate the dynamic matrix $\overline{U}$.
7. Initialize the trial displacement matrix $\overline{X}$ at unity.
8. Initialize the displacement matrix factor C.
9. Calculate the value of $\overline{T}$.
10. Reduce the new displacement matrix by dividing each element in $\overline{T}$ by the value stored in t_1. (This reduced matrix was called $\overline{T}'$ above.)
11. Set C equal to $C \times t_1$.
12. Compare matrices $\overline{X}$ and $\overline{T}'$.
13. If the corresponding elements of $\overline{X}$ and $\overline{T}'$ have an absolute difference *greater than or equal to* ϵ, set $\overline{X}$ equal to $\overline{T}'$ and return to step 9.
14. If the corresponding elements of $\overline{X}$ and $\overline{T}'$ have an absolute difference *less than* ϵ, calculate the value of ω.

With the required computer calculations in outline form, the next step in solving the problem is the preparation of a flow chart. The chart drawn for the computer steps listed above is shown in Fig. 12.2.1. The FORTRAN program for producing the action described in the flow chart is listed in Fig. 12.2.2, and a tabulation of the variable names used in the program is shown in Table 12.2.1. The flow chart makes it evident that it is not necessary to create a new array for $\overline{T}'$. The elements of $\overline{T}'$ can be stored in the same locations as those used by $\overline{T}$.

Although it is not needed in the calculations, an iteration counter, KOUNT, was added to the program to prevent excessive looping if some "accident" should occur. This is an important control device and should always be built-in if a program is designed to use an iterative procedure. It is especially important during the early stages of program development when a programming error could trap the computer in an infinite loop. However, even after the program has been debugged, in some iterative methods an anomaly in the input data may prevent the convergence of the method, and a limit on the number of loops that may be executed is a worthwhile safeguard.

Before we discuss the program itself, let us review the method of calculating the

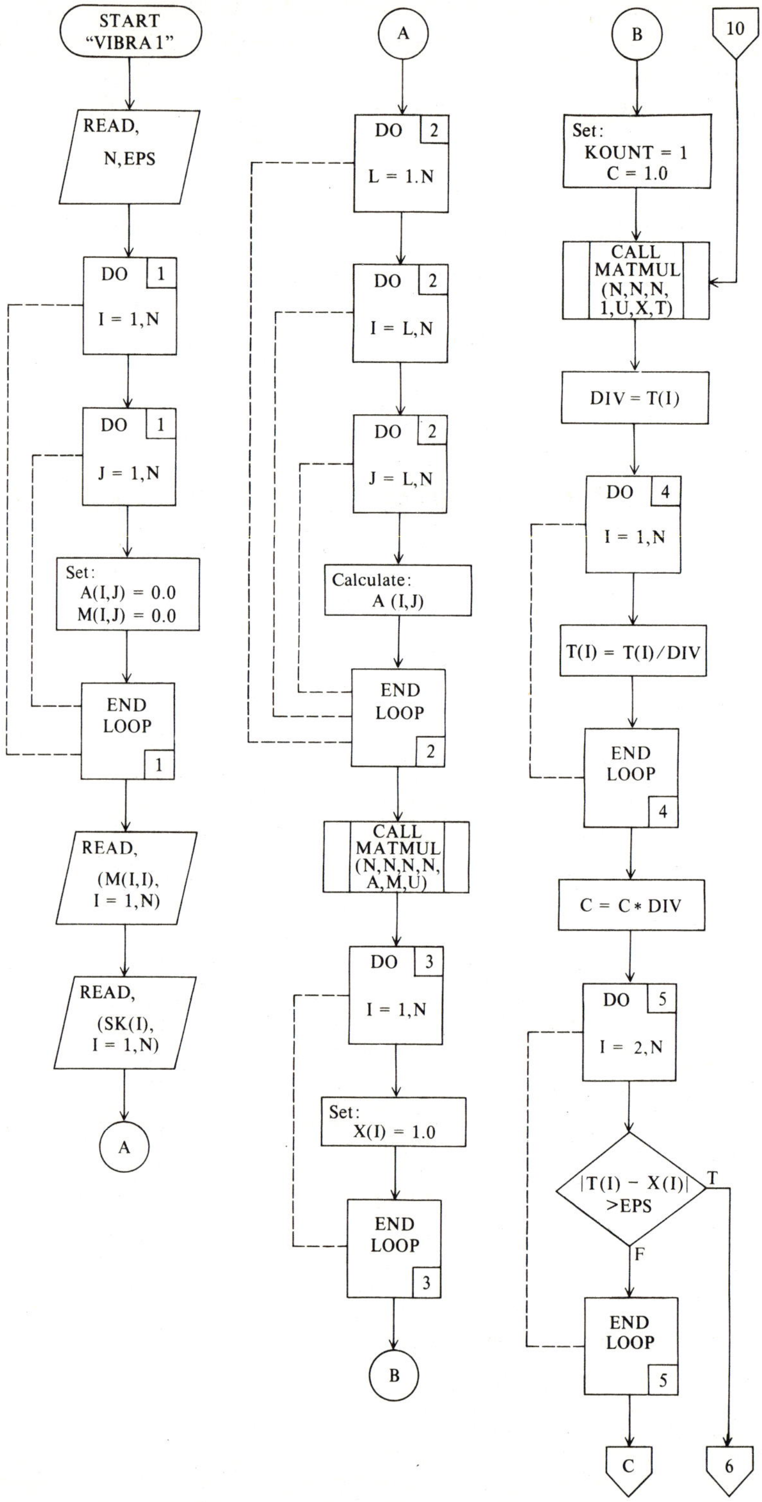
START
"VIBRA 1"
READ,
N,EPS
DO 1
I = 1,N
DO 1
J = 1,N
Set:
A(I,J) = 0.0
M(I,J) = 0.0
END LOOP 1
READ,
(M(I,I),
I = 1,N)
READ,
(SK(I),
I = 1,N)
A
A
DO 2
L = 1.N
DO 2
I = L,N
DO 2
J = L,N
Calculate:
A (I,J)
END LOOP 2
CALL MATMUL (N,N,N,N, A,M,U)
DO 3
I = 1,N
Set:
X(I) = 1.0
END LOOP 3
B
B
10
Set:
KOUNT = 1
C = 1.0
CALL MATMUL (N,N,N, 1,U,X,T)
DIV = T(I)
DO 4
I = 1,N
T(I) = T(I)/DIV
END LOOP 4
C = C * DIV
DO 5
I = 2,N
|T(I) − X(I)| >EPS
T
F
END LOOP 5
C
6

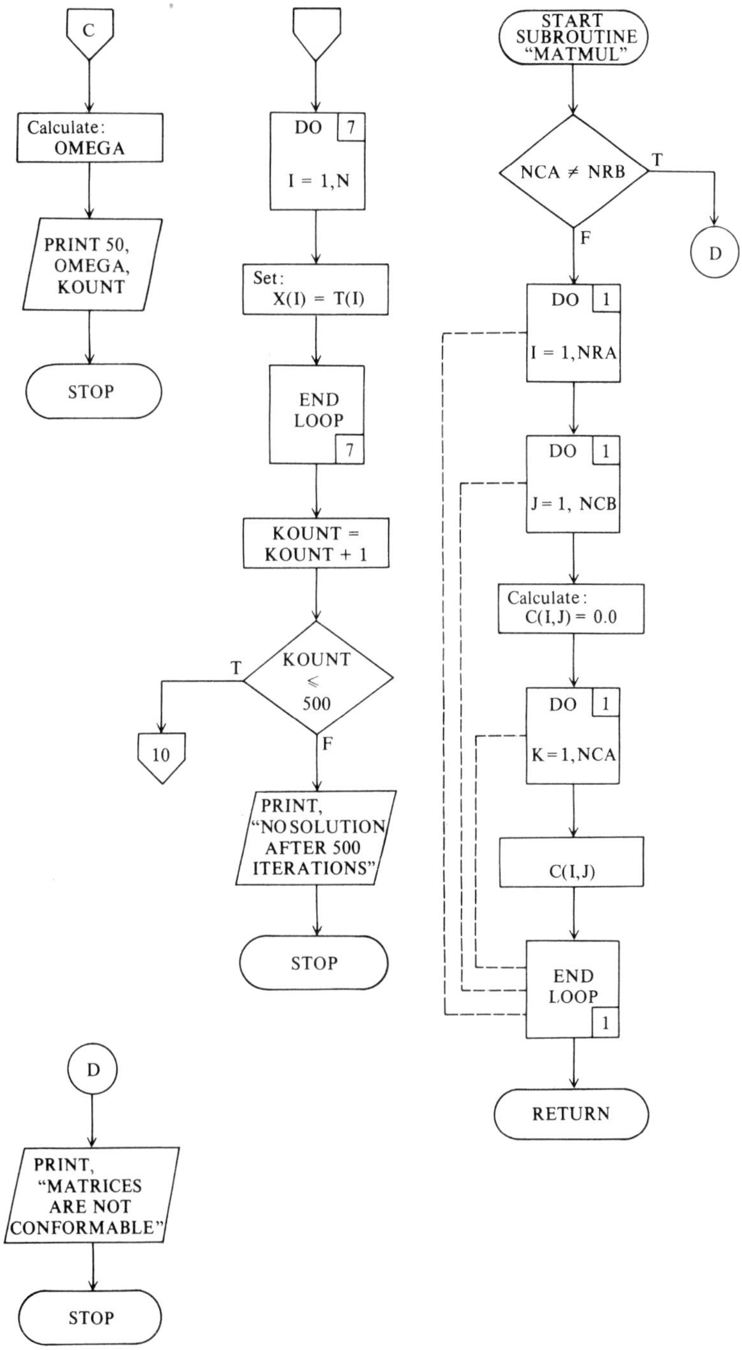

Fig. 12.2.1 Flow chart of program VIBRA1.

Table 12.2.1

Name	*Meaning*	*Units*
N	Number of degrees of freedom	
EPS	Convergence test value	in
A(I,J)	Influence coefficient matrix	in/lb
M(I,J)	Mass matrix	lb-s^2/in
SK(I)	Spring constant array	lb/in
U(I,J)	Dynamic matrix	s^2
X(I)	Trial displacement matrix	in
KOUNT	Iteration counter	
C	Product of successive displacement matrix factors	
T(I)	New displacement matrix	in
DIV	Displacement matrix factor	
OMEGA	Natural frequency	rad/s
I,J,L	DO loop indices	

```
   VIBRA1
   PRØGRAM TØ CALCULATE THE FIRST NATURAL FREQUENCY
   ØF AN UNDAMPED FREELY VIBRATING SYSTEM

   REAL M(10,10)
   DIMENSIØN A(10,10),U(10,10),X(10),T(10),SK(10)
    *
    * READ NUMBER ØF DEGREES ØF FREEDØM ØF
    * SYSTEM AND CØNVERGENCE TEST VALUE
    *
   READ,N,EPS
    *
    * INITIALIZE INFLUENCE CØEFFICIENT
    * AND MASS MATRICES AT ZERØ
    *
   DØ 1 I=1,N
   DØ 1 J=1,N
   A(I,J)=0.0
 1 M(I,J)=0.0
    *
    * READ VALUES FØR MASSES AND SPRING CØNSTANTS
    *
   READ,(M(I,I),I=1,N)
   READ,(SK(I),I=1,N)
    *
    * CALCULATE INFLUENCE CØEFFICIENT MATRIX
    *
   DØ 2 L=1,N
   DØ 2 I=L,N
   DØ 2 J=L,N
 2 A(I,J)=A(I,J)+1./SK(L)
    *
    * CALCULATE "DYNAMIC" MATRIX
    *
   CALL MATMUL(N,N,N,N,A,M,U)
    *
    * INITIALIZE "TRIAL" DISPLACEMENT MATRIX
    *
   DØ 3 I=1,N
 3 X(I)=1.0
    *
    * INITIALIZE ITERATIØN CØUNTER AND
    * DISPLACEMENT MATRIX FACTØR
```

```
00400C      *
00410       KØUNT=1
00420       C=1.0
00430C      *
00440C      * CALCULATE "NEW" DISPLACEMENT MATRIX
00450C      *
00460    10 CALL MATMUL(N,N,N,1,U,X,T)
00470C      *
00480C      * REDUCE "NEW" DISPLACEMENT MATRIX
00490C      *
00500       DIV=T(1)
00510       DØ 4 I=1,N
00520     4 T(I)=T(I)/DIV
00530C      *
00540C      * CALCULATE "NEW" DISPLACEMENT MATRIX FACTØR
00550C      *
00560       C=C*DIV
00570C      *
00580C      * TEST FØR CØNVERGENCE
00590C      *
00600       DØ 5 I=2,N
00610       IF(ABS(T(I)-X(I)).GT.EPS) GØ TØ 6
00620     5 CØNTINUE
00630C      *
00640C      * CALCULATE NATURAL FREQUENCY ØF SYSTEM
00650C      * AND PRINT RESULTS
00660C      *
00670       ØMEGA=SQRT(1./C)
00680       PRINT 50,ØMEGA,KØUNT
00690       STØP
00700    50 FØRMAT(/25HFIRST NATURAL FREQUENCY =F9.5//
00710&       I5,11H ITERATIØNS/)
00720C      *
00730C      * RESET "TRIAL" DISPLACEMENT MATRIX
00740C      *
00750     6 DØ 7 I=1,N
00760     7 X(I)=T(I)
00770C      *
00780C      * INCREMENT AND TEST ITERATIØN CØUNTER
00790C      *
00800       KØUNT=KØUNT+1
00810       IF(KØUNT.LE.500)GØ TØ 10
00820       PRINT,"NØ SØLUTIØN AFTER 500 ITERATIØNS"
00830       STØP
00840       END
00850       SUBRØUTINE MATMUL(NRA,NCA,NRB,NCB,A,B,C)
00860       DIMENSIØN A(10,10),B(10,10),C(10,10)
00870       IF(NCA.NE.NRB) GØ TØ 2
00880       DØ 1 I=1,NRA
00890       DØ 1 J=1,NCB
00900       C(I,J)=0.0
00910       DØ 1 K=1,NCA
00920     1 C(I,J)=C(I,J)+A(I,K)*B(K,J)
00930       RETURN
00940     2 PRINT,"MATRICES ARE NØT CØNFØRMABLE"
00950       STØP
00960       END
```

Fig. 12.2.2 Program VIBRA1.

influence coefficient matrix. Consider the spring system shown in Fig. 12.2.3. The influence coefficient matrix for this system is shown below:

$$\bar{A} = \begin{bmatrix} 1/8 & 1/8 & 1/8 & 1/8 \\ 1/8 & 9/8 & 9/8 & 9/8 \\ 1/8 & 9/8 & 31/24 & 31/24 \\ 1/8 & 9/8 & 31/24 & 55/24 \end{bmatrix}$$

The reader should note the symmetry of this matrix and then examine the nested DO loops (lines 00240 to 00270) used to calculate it. Let us examine the mechanics of the

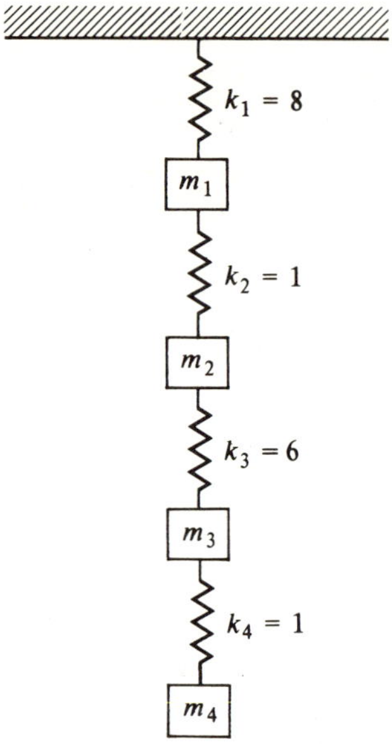

Fig. 12.2.3 Spring-mass system in series.

calculation procedure used. When L is equal to 1, the row and column indices, I and J, respectively, start at a value of 1 and range up to the order of the matrix. The entire matrix is set equal to $1/k_1$, in this case 1/8. During the second cycle, L is equal to 2, and the indices I and J start with a value of 2 and range up to N. Therefore, all the elements in the matrix except those in the first row and the first column are increased by $1/k_2$, in this case, 1/1. During the third cycle I and J start at 3, and $1/k_3$ is added to all the elements in the lower right-hand corner of the matrix, from A(3,3) to A(N,N). This procedure is continued until the entire matrix has been evaluated.

The main program (Fig. 12.2.2) has been annotated by using comment statements between the executable FORTRAN statements in the body of the program. This procedure makes it easier to follow the logic of the calculations. An additional step along these same lines is to completely document the program within itself by defining all the variable names used in the program, describing their function in the calculations, and specifying the units of all values to be used as input. This requires the addition of a considerable number of comment statements to a program and has not been illustrated here because of space limitations.

As the reader can see, the program follows the calculation steps outlined above and flow charted in Fig. 12.2.1. However, there is one exception that may warrant our attention. In the calculation outline, the reduction of the new displacement matrix was to be accomplished by dividing each element of $\overline{T}'$ (or $\overline{T}$; as mentioned above, only one array is needed to store these values) by the value of the first element, T(1). On line 00500, and in the DO loop immediately following, we have produced this result *indirectly* by storing the value of T(1) in DIV and using the latter as the divider in the loop. This step was necessary because without it the first cycle of the DO loop would have reduced the contents of T(1) to 1 and all the subsequent cycles would have divided T(2), T(3), etc., by 1 and produced no change in value.[1] Furthermore, the

[1] A similar problem occurs in the program of Fig. 12.3.2 (see line 00270).

```
INPUT:00070
?4,0.01

INPUT:00190
?10.,20.,30.,40.

INPUT:00200
?70.,75.,70.,75.


FIRST NATURAL FREQUENCY =   0.12391

    4 ITERATIØNS

STØP
```

Fig. 12.2.4 Trial run for VIBRA1.

proposed plan to accumulate the reduction factors in C would have been defeated because in each instance C*T(1) would have yielded a product equal to 1. There are other ways of programming this reduction; this method was chosen (at the expense of requiring an additional storage location) because it is easily understood.

The subroutine subprogram called to perform the required matrix multiplications has been adapted from the program discussed in Sec. 12.1. In Fig. 12.2.2, the subprogram is named MATMUL and is appended to the main program (line 00850).

On line 00610, the test to determine if the *new* displacement matrix is equal to the *trial* displacement matrix is made in a DO loop. As long as the difference of the absolute values is within the acceptable limit (EPS), the loop continues with the comparisons. However, as soon as one pair of corresponding elements fails the test, the program resets the ***trial*** displacement matrix and advances to the next iteration cycle.

The results of a trial run of the program are shown in Fig. 12.2.4. For the data used in the trial the first natural frequency was found to be 0.124 rad/s.

12.3 SYSTEMS OF LINEAR NONHOMOGENEOUS EQUATIONS

A system of n linear nonhomogeneous equations can be expressed in the form

$$\begin{array}{l} a_{11}x_1 + a_{12}x_2 + \cdots + a_{1n}x_n = B_1 \\ a_{21}x_1 + a_{22}x_2 + \cdots + a_{2n}x_n = B_2 \\ \ldots \quad \ldots \quad \ldots \quad \ldots \quad \ldots \\ a_{n1}x_1 + a_{n2}x_2 + \cdots + a_{nn}x_n = B_n \end{array}$$

In the equations above, each x is an unknown quantity, each a is a constant coefficient, and each B is a given quantity. There are several methods for solving this system of equations. The method that will be used here is called *Gauss reduction.*[1]

The object of the method is to reduce a system of n equations to the form

$$\begin{array}{l} x_1 + 0 \qquad\qquad\qquad = S_1 \\ \qquad x_2 + 0 \qquad\qquad = S_2 \\ \ldots \quad \ldots \quad \ldots \quad \ldots \quad \ldots \\ \qquad\qquad 0 + x_n = S_n \end{array}$$

[1] I. S. Sokolnikoff and R. M. Redheffer, "Mathematics of Physics and Modern Engineering," 2d ed., McGraw-Hill, New York, 1966, p. 660.

In this form each equation consists of only one unknown and one constant. Thus the solution is available by inspection. The procedure is easily demonstrated by the following example. We will divide the procedure into three cycles, each of which consists of similar operations.

EXAMPLE 1 Solve the system of equations:

$$4x_1 + 4x_2 - 8x_3 = -20$$

$$2x_1 + x_2 + x_3 = 8$$

$$3x_1 - x_2 + x_3 = 5$$

The first step of cycle 1 is to divide the first equation by a_{11}, the coefficient of x_1. (This coefficient is called the *pivot.*) As a result, a_{11} will be equal to 1. Then, suitable multiples of the first equation are subtracted from the second and third equations so as to eliminate terms $2x_1$ and $3x_1$.

Cycle 1

$$x_1 + x_2 - 2x_3 = -5$$

$$2x_1 + x_2 + x_3 = 8$$

$$3x_1 - x_2 + x_3 = 5$$

$$x_1 + x_2 - 2x_3 = -5$$

$$0 - x_2 + 5x_3 = 18$$

$$0 - 4x_2 + 7x_3 = 20$$

In cycle 2, the second equation is divided by a_{22} (the current pivot), and then suitable multiples of the second equation are subtracted from the first and third equations so as to eliminate the term x_2 from the first equation and the term $-4x_2$ from the third equation.

Cycle 2

$$x_1 + x_2 - 2x_3 = -5$$

$$0 + x_2 - 5x_3 = -18$$

$$0 - 4x_2 + 7x_3 = 20$$

$$x_1 + 0 + 3x_3 = 13$$

$$0 + x_2 - 5x_3 = -18$$

$$0 + 0 - 13x_3 = -52$$

Finally, divide the third equation by a_{33} (the current pivot), and subtract suitable multiples of the third equation from the first and second equations so as to eliminate $3x_3$ and $-5x_3$.

Cycle 3

$$x_1 + 0 + 3x_3 = 13$$
$$0 + x_2 - 5x_3 = -18$$
$$0 + 0 + x_3 = 4$$

$$x_1 + 0 + 0 = 1$$
$$0 + x_2 + 0 = 2$$
$$0 + 0 + x_3 = 4$$

At the completion of the third cycle, the solution of the system of equations is seen to be

$$x_1 = 1$$
$$x_2 = 2$$
$$x_3 = 4$$

Occasionally a difficulty arises in the application of the Gauss reduction method. We will illustrate this with a second example and also show how to overcome the difficulty. Assume that we wish to solve the following system of equations:

$$2x_1 + 4x_2 + 6x_3 + 2x_4 = 12$$
$$3x_1 + 5x_2 + 7x_3 + 9x_4 = 0$$
$$x_1 + 2x_2 + 3x_3 + 4x_4 = 0$$
$$4x_1 + 2x_2 + 2x_3 + x_4 = 0$$

After two cycles of the reduction procedure, the equations appear as follows:

$$x_1 + 0 - x_3 + 13x_4 = -30$$
$$0 + x_2 + 2x_3 - 6x_4 = 18$$
$$0 + 0 + 0 + 3x_4 = -6$$
$$0 + 0 + 2x_3 - 39x_4 = 84$$

At this stage of the procedure, we should divide the third equation by a_{33} and then subtract suitable multiples of the *third equation* from the first, second, and fourth equations so as to eliminate the terms $-x_3$, $2x_3$, and $2x_3$. This is not possible because in the third equation the coefficient of x_3 (the pivot) is zero. However, if we interchange the third and fourth equations, we can continue the reduction.[1] Fortunately, this procedure is valid; at the beginning of any cycle it is always permissible to interchange the current pivot equation with an equation below it. Thus, the difficulty has been overcome. After the interchange, the equations appear as follows:

$$x_1 + 0 - x_3 + 13x_4 = -30$$
$$0 + x_2 + 2x_3 - 6x_4 = 18$$
$$0 + 0 + 2x_3 - 39x_4 = 84$$
$$0 + 0 + 0 + 3x_4 = -6$$

[1] The interchange must always occur between the equation with zero pivot element and an equation *below* it.

After the reduction has been completed, the solution to the equation is found to be

$$x_1 = -1 \qquad x_2 = 0 \qquad x_3 = 3 \qquad x_4 = -2$$

The preceding example shows that it will be necessary to include a procedure for interchanging equations in the computer program which we will write. However, there is an additional reason for interchanging equations. It has been shown that better computational accuracy is achieved if, at the beginning of each cycle of the reduction procedure, that equation is transferred to the pivotal row which yields a pivot element with the largest absolute value.[1] For example, in the illustration given above, at the beginning of the first cycle, we would interchange the first and fourth equations. Thus, at the beginning of the first cycle, the pivot element would be 4 rather than 2. Consistent use of this procedure will, of course, also remove a zero element from the pivot position.

We are now ready to discuss a computer program which solves a system of equations by gaussian reduction. The program is shown in Fig. 12.3.2 and a flow chart is shown in Fig. 12.3.1. The program consists of a short main program which transmits data to SUBROUTINE SOLVE by a COMMON statement (lines 00010 and 00130). The main program consists, essentially, of a READ, a CALL, and a PRINT statement. It should be noted that for a system of N simultaneous equations the data to be *entered* into the program occupy N rows and N + 1 columns (line 00040). The solution to the system of equations, after gaussian reduction has been completed, is to be found in column N + 1. Therefore, only the numbers in column N + 1 need to be printed (see line 00080).

Let us now examine SOLVE, which contains an outer DO loop (line 00150) and several nested loops. Note that ISWAP is set equal to the value of IROW on line 00170. The first nested DO loop (line 00180) performs the interchange of rows discussed above. At the beginning of the *first* cycle the magnitude of coefficient A(1,1) is compared with the magnitude of coefficient A(2,1). If the magnitude of A(1,1) is equal to or greater than the magnitude of A(2,1), the logical IF statement on line 00190 is .TRUE., the program branches to statement 50 on line 00210, and the search for the largest pivot continues. If the magnitude of A(1,1) is *less* than the magnitude of A(2,1), ISWAP is set equal to IN (which has a value of 2 during this first cycle) and the procedure continues until the largest available pivot has been selected. Then, the DO loop, DO 92 (line 00230), is executed. After this loop interchanges the first row with that row which contains the largest pivot, statement 99 is executed. However, *if A(1,1) happens to be the largest available pivot*, the logical IF statement (line 00220) causes the program to jump over DO 92 and execute statement 99 directly.

Statement 99 assigns the value of the current pivot to PIVOT. The purpose of the IF statement on line 00280 is to halt the program if PIVOT is equal to or less than some arbitrary small number (in this program 1×10^{-6}). This will give the programmer the opportunity to check for the possibility of an error in the system of equations. If PIVOT exceeds 1×10^{-6}, the program branches to the DO loop on line 00310. In this DO loop the pivot equation is divided by the value of PIVOT. Next, DO loop 18 is executed. In this DO loop multiples of the pivot equation are subtracted from all other equations so as to carry out the reduction procedure. The logical IF statement on line 00340 prevents the subtraction of the pivot equation from itself. N cycles of the

[1] D. K. Fadeev and V. N. Fadeeva, "Computational Methods of Linear Algebra," Freeman, San Francisco, 1963, p. 138.

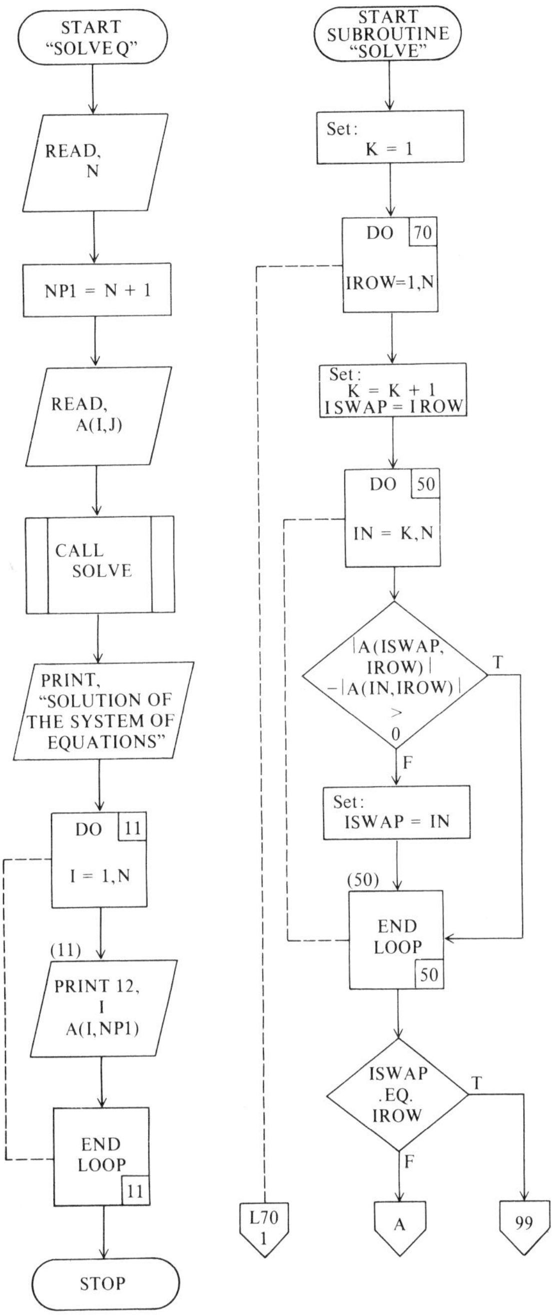

Fig. 12.3.1a Flow chart of program SOLVEQ.

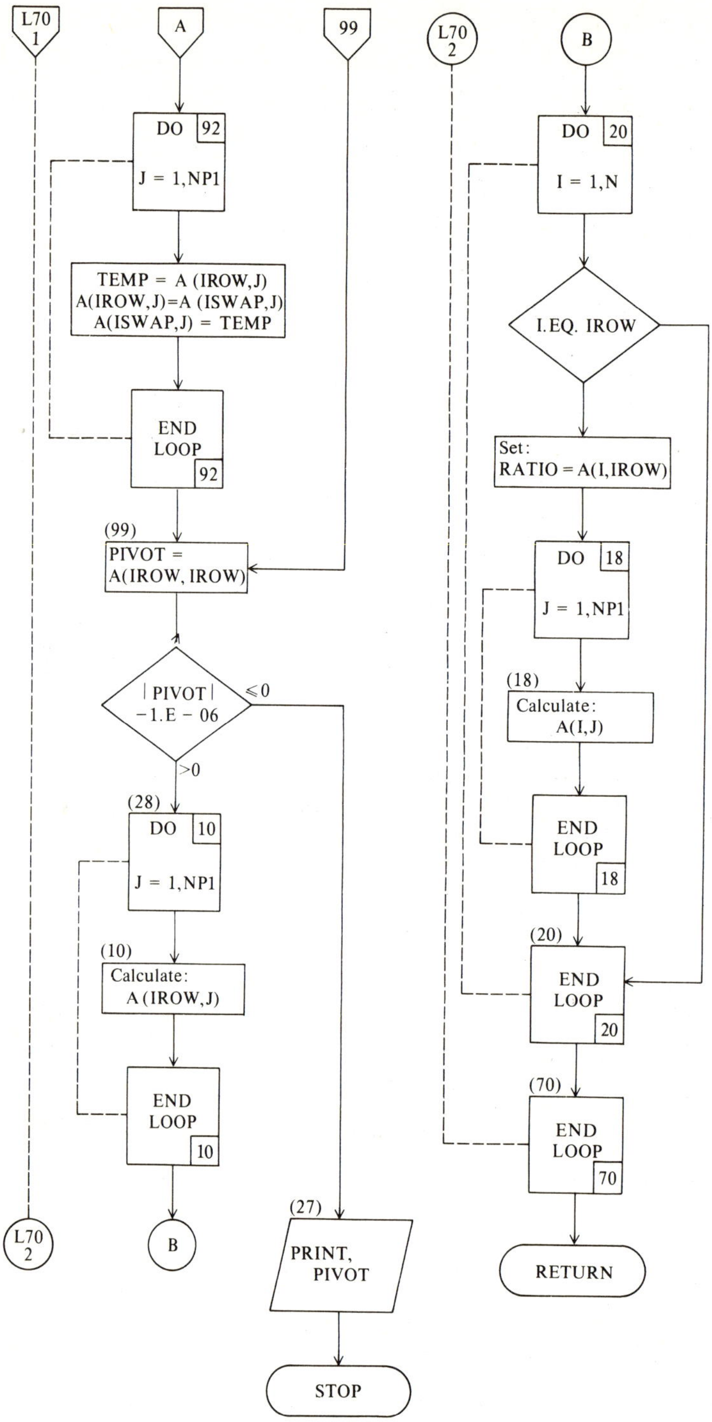

Fig. 12.3.1*b* Flow chart of program SOLVEQ.

```
C     SØLVEQ
      CØMMØN N,A(10,11),NP1
      READ,N
      NP1=N+1
      READ,((A(I,J),J=1,NP1),I=1,N)
      CALL SØLVE
      PRINT,"SØLUTIØN ØF THE SYSTEM ØF EQUATIØNS"
      DØ 11 I=1,N
   11 PRINT 12,I,A(I,NP1)
   12 FØRMAT(1HX,I3,1H=E12.3)
      STØP
      END
      SUBRØUTINE SØLVE
      CØMMØN N,A(10,11),NP1
      K=1
      DØ 70 IRØW=1,N
      K=K+1
      ISWAP=IRØW
      DØ 50 IN=K,N
      IF(ABS(A(ISWAP,IRØW))-ABS(A(IN,IRØW)).GE.0.)GØ TØ 50
      ISWAP=IN
   50 CØNTINUE
      IF(ISWAP.EQ.IRØW) GØ TØ 99
      DØ 92 J=1,NP1
      TEMP=A(IRØW,J)
      A(IRØW,J)=A(ISWAP,J)
   92 A(ISWAP,J)=TEMP
   99 PIVØT=A(IRØW,IRØW)
      IF(ABS(PIVØT)-1.E-06) 27,27,28
   27 PRINT,"PIVØT IS TØØ SMALL. VALUE ØF PIVØT=",PIVØT
      STØP
   28 DØ 10 J=1,NP1
   10 A(IRØW,J)=A(IRØW,J)/PIVØT
      DØ 20 I=1,N
      IF(I.EQ.IRØW) GØ TØ 20
      RATIØ=A(I,IRØW)
      DØ 18 J=1,NP1
   18 A(I,J)=A(I,J)-A(IRØW,J)*RATIØ
   20 CØNTINUE
   70 CØNTINUE
      RETURN
      END
```

Fig. 12.3.2 Program SOLVEQ.

```
INPUT:00020
 ?4

INPUT:00040
 ?2,4,6,2,12,3,5,7,9,0,1,2,3,4,0,4,2,2,1,0

SØLUTIØN ØF THE SYSTEM ØF EQUATIØNS
X  1=   -.100E+01
X  2=   -.466E-09
X  3=   0.300E+01
X  4=   -.200E+01
STØP
```

```
INPUT:00020
 ?3

INPUT:00040
 ?4,4,-8,-20,2,1,1,8,3,-1,1,5

SØLUTIØN ØF THE SYSTEM ØF EQUATIØNS
X  1=   0.100E+01
X  2=   0.200E+01
X  3=   0.400E+01
STØP
```

Fig. 12.3.3 Trial runs for SOLVEQ

reduction procedure are carried out, where N is the number of simultaneous equations and also the maximum value of the index IROW of the outer DO loop.

Two trial runs are shown in Fig. 12.3.3. In these trial runs, the systems of equations discussed in the two examples above have been solved. The results of the first trial run agree with the hand calculation. The results of the second trial run agree with the hand calculation except for the value of x_2. By hand calculation, x_2 was found to be zero, but the computer calculation shows x_2 to have a value of -4.66×10^{-10}. This discrepancy is due to round-off errors that may occur in the computer when decimal numbers are coded.

12.4 METHOD OF LEAST SQUARES: PARABOLIC FIT

In Sec. 12.3 we discussed a program which solves a system of linear equations. As an illustration of the use of that program we will discuss the method of least squares.

There are occasions when it is desirable to represent experimental data approximately by some function $y = f(t)$. One method of determining the function is the method of least squares. Suppose we have n data points (t_i, y_i) and we attempt to represent the data by the straight line $y = a + bt$. In general, the straight line will not pass through all the points, and so we seek a compromise. One criterion for a best fit is that the sum of the squares of the vertical deviations ϵ_i of the plotted points from the straight line be minimized (see Fig. 12.4.1). It is also possible to fit data by higher-order polynomials, for example, by the polynomial $y = a + bt + ct^2$. This is called a least-squares quadratic fit. We will limit the discussion to the case of a quadratic fit and show how the coefficients a, b, and c can be calculated.

As we are seeking a quadratic fit by the least-squares criterion, the vertical deviation ϵ_i at a point (x_i, y_i) is

$$\epsilon_i = y_i - (a + bt_i + ct_i^2) \tag{12.4.1}$$

Squaring both sides of Eq. (12.4.1), we obtain

$$\epsilon_i^2 = y_i^2 - 2ay_i - 2by_it_i - 2cy_it_i^2 + a^2 + 2abt_i + 2act_i^2 + b^2t_i^2 + 2bct_i^3 + c^2t_i^4 \tag{12.4.2}$$

In the next step, sum over the n data points and calculate the partial derivatives with

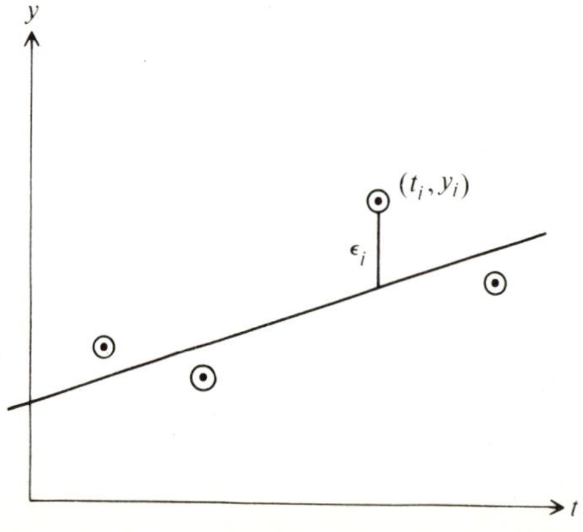

Fig. 12.4.1 Vertical deviation from regression line.

respect to a, b, and c, in turn. Three equations are obtained, and each is set equal to zero to minimize the deviation. After collecting terms, the following equations are obtained:

$$na \quad + b\Sigma t_i \quad + c\Sigma t_i^2 = \Sigma y_i \tag{12.4.3}$$

$$a\Sigma t_i \quad + b\Sigma t_i^2 + c\Sigma t_i^3 = \Sigma y_i t_i \tag{12.4.4}$$

$$a\Sigma t_i^2 + b\Sigma t_i^3 + c\Sigma t_i^4 = \Sigma y_i t_i^2 \tag{12.4.5}$$

The summations indicated above are taken over the range $i = 1, n$. We now have three unknowns and three equations. The equations are in the form $\bar{A}\bar{X} = \bar{B}$, where the matrix $\bar{X}$ is a column matrix consisting of the unknowns a, b, and c, and $\bar{B}$ is the column matrix consisting of the elements Σy_i, $\Sigma y_i t_i$, and $\Sigma y_i t_i^2$.

The program shown in Fig. 12.4.2 calculates the coefficients a, b, and c. The reader will note that the subroutine SOLVE, previously discussed in Sec. 12.3, is used here to solve the system of equations above. Because SOLVE is being used for the specific case of a system of three equations, all row indices are fixed at upper values of 3, and all column indices are fixed at upper values of 4 (see, for example, lines 00330 and 00410). Before subroutine SOLVE can be called, however, the matrix of coefficients A(I,J), upon which SOLVE operates, must be calculated. These calculations are simple. On line 00100, A(1,1) is assigned the value N, the number of data points. Those elements, the calculation of which requires a summation, are computed within the DO loop which begins on line 00110. For example, Σt_i^2, the coefficient of c in Eq. (12.4.3), is calculated on line 00130. Some of the elements of the matrix have

```
00001C       LEAST3
00002C       THIS PRØGRAM CALCULATES THE CØEFFICIENTS FØR A
00003C       LEAST SQUARES PARABØLIC FIT.
00010        DIMENSIØN T(10),Y(10)
00020        CØMMØN A(10,11)
00030        READ,N
00040        PRINT,"ENTER T DATA; THEN ENTER Y DATA"
00050        READ,(T(I),I=1,N)
00060        READ,(Y(I),I=1,N)
00070        DØ 13 I=1,3
00080        DØ 13 J=1,4
00090     13 A(I,J)=0.
00100        A(1,1)=N
00110        DØ 30 I=1,N
00120        A(1,2)=A(1,2)+T(I)
00130        A(1,3)=A(1,3)+T(I)**2
00140        A(1,4)=A(1,4)+Y(I)
00150        A(2,3)=A(2,3)+T(I)**3
00160        A(2,4)=A(2,4)+T(I)*Y(I)
00170        A(3,3)=A(3,3)+T(I)**4
00180     30 A(3,4)=A(3,4)+T(I)**2*Y(I)
00190        A(2,1)=A(1,2)
00200        A(2,2)=A(1,3)
00210        A(3,1)=A(1,3)
00220        A(3,2)=A(2,3)
00230        CALL SØLVE
00240        PRINT,"CØEFFICIENTS FØR PARABØLIC FIT"
00250        DØ 11 I=1,3
00260     11 PRINT 12,I,A(I,4)
00270     12 FØRMAT(1HX,I3,1H=E12.3)
00280        STØP
00290        END
```

Fig. 12.4.2a Program LEAST3.

```
00300       SUBRØUTINE SØLVE
00310       CØMMØN A(10,11)
00320       K=1
00330       DØ 70 IRØW=1,3
00340       K=K+1
00350       ISWAP=IRØW
00360       DØ 50 IN=K,3
00370       IF(ABS(A(ISWAP,IRØW))-ABS(A(IN,IRØW)).GE.0.)GØ TØ 50
00380       ISWAP=IN
00390    50 CØNTINUE
00400       IF(ISWAP.EQ.IRØW) GØ TØ 99
00410       DØ 92 J=1,4
00420       TEMP=A(IRØW,J)
00430       A(IRØW,J)=A(ISWAP,J)
00440    92 A(ISWAP,J)=TEMP
00450    99 PIVØT=A(IRØW,IRØW)
00460       IF(ABS(PIVØT)-1.E-06) 27,27,28
00470    27 PRINT,"PIVØT IS TØØ SMALL. VALUE ØF PIVØT=",PIVØT
00480       STØP
00490    28 DØ 10 J=1,4
00500    10 A(IRØW,J)=A(IRØW,J)/PIVØT
00510       DØ 20 I=1,3
00520       IF(I.EQ.IRØW) GØ TØ 20
00530       RATIØ=A(I,IRØW)
00540       DØ 18 J=1,4
00550    18 A(I,J)=A(I,J)-A(IRØW,J)*RATIØ
00560    20 CØNTINUE
00570    70 CØNTINUE
00580       RETURN
00590       END
```

Fig. 12.4.2*b* Program LEAST3.

identical values, and these are assigned outside the DO loop. For example, the coefficient of b in Eq. (12.4.4) is also Σt_i^2. Therefore on line 00200 the assignment A(2,2) = A(1,3) is made.

A trial run is shown in Fig. 12.4.3. Six sets of data were used. The solution of the simultaneous equations is

$$a = 1.01$$

$$b = 2.71$$

$$c = 7.32$$

Therefore, the equation which represents a least-squares quadratic fit to the data is as follows:

$$y = 1.01 + 2.71t + 7.32t^2$$

```
INPUT:00030
 ?6

ENTER T DATA; THEN ENTER Y DATA
INPUT:00050
 ?.2,.5,1.,1.5,2.,3.

INPUT:00060
 ?1.9,4.,11.5,21.,36.,75.

CØEFFICIENTS FØR PARABØLIC FIT
X  1=    0.101E+01
X  2=    0.271E+01
X  3=    0.732E+01
STØP
```

Fig. 12.4.3 Trial run for LEAST3.

12.5 ROUTH STABILITY CRITERION[1]

The Routh criterion is one of a number of methods that is used to test the stability of a feedback system and has application in the field of control theory. In the Routh method, an array is formed from the characteristic equation of the system. While the roots of the equation are not actually found, the method determines if there are any roots with positive real parts: if there are, the system is unstable.[2] It will be assumed that no terms of the characteristic equation are missing. If there are any missing terms, there are some roots with positive real parts, and the system is unstable.[3]

First we will define the notation used here, and then we will show how the array is calculated. Let us assume a characteristic equation of the form

$$A_n s^m + A_{n-1} s^{m-1} + \cdots + A_2 s + A_1 = 0 \tag{12.5.1}$$

The subscript n is equal to $m + 1$, where m is the order of the equation. By numbering the subscripts from 1 to n, we avoid a zero subscript on the constant term. The coefficients $A_n, A_{n-1}, \ldots, A_1$ are real numbers, and may be negative or positive. If one or more coefficients are negative, the system is known to be unstable. Nevertheless, as discussed below, the Routh criterion may still be used with profit because it indicates not only stability or instability but also the number of roots in the right half-plane, which is useful to the system designer. The first step in the procedure is the arrangement of the coefficients $A_n, A_{n-1}, \ldots, A_1$ in an alternating pattern in two rows. Then, additional coefficients are calculated, row by row, as described below. After calculation, the array appears as follows:

$$\begin{array}{llll}
A_n & A_{n-2} & A_{n-4} & \cdots \\
A_{n-1} & A_{n-3} & A_{n-5} & \cdots \\
\hdashline
B_1 & B_2 & \cdots & \cdots \\
C_1 & C_2 & \cdots & \cdots \\
D_1 & \cdots & \cdots & \cdots \\
E_1 & \cdots & \cdots & \cdots
\end{array}$$

Table 12.5.1 Routh arrays for sixth- and fifth-order equations

Sixth-order equation, N = 7	*Fifth-order equation, N = 6*
$A_7\, A_5\, A_3\, A_1$	$A_6\, A_4\, A_2$
$A_6\, A_4\, A_2$	$A_5\, A_3\, A_1$
$B_1\, B_2\, B_3$	$B_1\, B_2$
$C_1\, C_2$	$C_1\, C_2$
$D_1\, D_2$	D_1
E_1	E_1
F_1	

[1] R. Saucedo and E. E. Schiring, "Introduction to Continuous and Digital Control Systems," Macmillan, New York, 1969, p. 96.

[2] It can also be used to determine if there are roots with zero real parts. However, this case will not be considered here.

[3] D. K. Cheng, "Analysis of Linear Systems," Addison-Wesley, Reading, Mass., 1959, p. 281.

The array will have n rows, and the last two rows will each contain only one element. Arrays for fifth- and sixth-order equations in literal form are shown in Table 12.5.1.

The coefficients below the dashed line are calculated as follows:

$$
\begin{aligned}
B_1 &= \frac{A_{n-1}A_{n-2} - A_nA_{n-3}}{A_{n-1}} \\
B_2 &= \frac{A_{n-1}A_{n-4} - A_nA_{n-5}}{A_{n-1}} \\
B_3 &= \frac{A_{n-1}A_{n-6} - A_nA_{n-7}}{A_{n-1}} \\
&\cdots \quad \cdots
\end{aligned}
\tag{12.5.2}
$$

$$
\begin{aligned}
C_1 &= \frac{B_1A_{n-3} - A_{n-1}B_2}{B_1} \\
C_2 &= \frac{B_1A_{n-5} - A_{n-1}B_3}{B_1} \\
C_3 &= \frac{B_1A_{n-7} - A_{n-1}B_4}{B_1} \\
&\cdots \quad \cdots
\end{aligned}
\tag{12.5.3}
$$

$$
\begin{aligned}
D_1 &= \frac{C_1B_2 - B_1C_2}{C_1} \\
D_2 &= \frac{C_1B_3 - B_1C_3}{C_1} \\
&\cdots \quad \cdots
\end{aligned}
\tag{12.5.4}
$$

$$
\begin{aligned}
E_1 &= \frac{D_1C_2 - C_1D_2}{D_1} \\
&\cdots \quad \cdots
\end{aligned}
\tag{12.5.5}
$$

The Routh criterion states that the system with characteristic Eq. (12.5.1) is stable if and only if all the elements in the first column of the array have the same sign. If the elements in the first column of the array do not have the same sign, the number of changes of sign in that column equals the number of roots with positive real parts.

As a specific example, we will test the characteristic equation

$$s^6 + 2s^5 + 4s^4 + 7s^3 + 5s^2 + 3s + 2 = 0$$

The array is

$$
\begin{array}{cccc}
1 & 4 & 5 & 2 \\
2 & 7 & 3 & \\
\cdots & \cdots & \cdots & \cdots \\
\frac{1}{2} & \frac{7}{2} & 2 & \\
-7 & -5 & & \\
\frac{22}{7} & 2 & & \\
-\frac{12}{22} & & & \\
2 & & &
\end{array}
$$

The elements above the dashed line are the coefficients of the characteristic equation. To illustrate the procedure, we show the calculation of the B elements:

$$B_1 \quad \frac{(2 \times 4) - (1 \times 7)}{2} = \frac{1}{2}$$

$$B_2 = \frac{(2 \times 5) - (1 \times 3)}{2} = \frac{7}{2}$$

$$B_3 = \frac{(2 \times 2) - (1 \times 0)}{2} = 2$$

Let us inspect the first column of the array. There are four sign changes in the first column. Therefore, there are four roots with positive real parts, and the system is unstable.

It should be noted that the procedure discussed above will fail if either of the following cases occur:

1. An element in the first column of a row is zero, but not all the remaining elements of that row are zero.
2. All the elements of a row are zero.

There are several methods for dealing with the first case.[1] The method employed here involves transforming the original characteristic equation by use of the substitution $s = 1/x$. The Routh method, as previously described, is then applied to the transformed equation. An example will clarify the procedure.

Given the characteristic equation

$$s^5 + s^4 + 7s^3 + 4s^2 + 15s + 3 = 0$$

the Routh array is

$$\begin{array}{ccc} 1 & 7 & 15 \\ 1 & 4 & 3 \\ \hline 3 & 12 & \\ 0 & 3 & \end{array}$$

The method fails because $C_1 = 0$. However, it is possible to transform the equation and overcome the difficulty as follows. Let $s = 1/x$. We then obtain

$$\frac{1}{x^5} + \frac{1}{x^4} + \frac{7}{x^3} + \frac{4}{x^2} + \frac{15}{x} + 3 = 0$$

$$3x^5 + 15x^4 + 4x^3 + 7x^2 + x + 1 = 0$$

[1] R. Saucedo and E. E. Schiring, "Introduction to Continuous and Digital Control Systems," Macmillan, New York, 1969, p. 100.

We now apply the Routh method to the transformed equation and obtain

3	4	1
15	7	1
2.6	0.8	
2.385	1.0	
−0.29		
1.0		

There are two sign changes. Therefore, there are two roots with positive real parts, and the system is unstable. We will next give an example of the second case where all the elements in a row are zero.

Let us test the characteristic equation

$$s^5 + 3s^4 - 5s^3 - 15s^2 + 4s + 12 = 0$$

We obtain the following array:

1	−5	4
3	−15	12
0	0	

We see that all the B elements are zero, and we cannot continue the procedure. When all the elements of a row are zero, it is necessary to form an auxiliary equation in order to carry out the procedure. However, we will not treat this special case. In the program to be discussed below, when a *row* of zero elements is encountered, a warning message and the array are printed. When only the first element of a row has a value of zero, but not all the remaining elements of that row are zero, a message is printed, the A elements are transposed (corresponding to the transformation shown above), and the Routh array is calculated.

Before discussing the details of the program, it should be stressed that the success of the program is based upon the detection of zeros. However, a computation which yields a zero answer when calculated by hand does not necessarily yield a zero when calculated by computer. This is so because of round-off errors that occur when decimal numbers are coded in the computer. In order to cope with this problem, we must arbitrarily consider an element to be zero when its magnitude is less than 1×10^{-4}.

A program for carrying out the Routh procedure, and its flow chart, are shown in Figs. 12.5.1 and 12.5.2, respectively. The value of N, the number of coefficients of the characteristic equation, is entered using unformatted input. Then N is tested to determine whether it is even or odd. If N is odd, integer variable K is assigned a value in accordance with statement 50 on line 00040; if N is even, K is assigned a value in accordance with statement 51 on line 00060. The values of the coefficients are entered using unformatted input and implied DO loops. Notice that K is the maximum value of the index of J (line 00080) and determines the correct number of elements in the first and second rows.

```
C     RØUTH1
C     THIS PRØGRAM APPLIES RØUTH'S CRITERIØN TØ
C     CHARACTERISTIC EQUATIØNS. N IS THE NUMBER ØF TERMS
C     IN THE EQUATIØN. THE PRØCEDURE FØR ENTERING THE
C     CØEFFICIENTS IS DISCUSSED IN SECTIØN 12.5
      DIMENSIØN A(10,5),TEMP(10,5)
      READ,N
      IF(N/2*2-N) 50,51,51
   50 K=N/2+1
      GØ TØ 99
   51 K=N/2
   99 INDEX=0
      READ,((A(I,J),J=1,K),I=1,2)
   88 DØ 5 I=3,N
      KØUNT=0
      IF(N/2*2-N) 60,61,61
   60 MT=K-I/2
      GØ TØ 98
   61 MT=K+1-(I+1)/2
   98 DØ 5 J=1,MT
      A(I,J)=(A(I-1,1)*A(I-2,J+1)-A(I-2,1)*A(I-1,J+1))/A(I-1,1)
      IF(ABS(A(I,J))-1.E-04) 16,16,15
   16 KØUNT=KØUNT+1
      IF(KØUNT.EQ.MT) GØ TØ 36
   15 IF(J.EQ.MT.AND.ABS(A(I,1)).LE.1.E-04) GØ TØ 6
    5 CØNTINUE
      GØ TØ 89
    6 INDEX=INDEX+1
      IF(INDEX.GT.1) GØ TØ 71
      PRINT 7,I
    7 FØRMAT(11HZERØ AT RØW,I3,27H. TRANSFØRMATIØN PERFØRMED.)
      DØ 8 I=1,2
      DØ 8 J=1,K
    8 TEMP(I,J)=A(I,J)
      DØ 9 I=1,2
      IF(N/2*2-N) 63,64,64
   63 L=I
      MAX=K-I+1
      GØ TØ 11
   64 L=K-I
      MAX=K
   11 DØ 9 J=1,MAX
      IF(N/2*2-N) 65,66,66
   65 M=K+2-J-I
      GØ TØ 9
   66 M=K-J+1
    9 A(I,J)=TEMP(L,M)
      GØ TØ 88
   36 PRINT,"RØW ØF ZERØS.AUXILIARY EQUATIØN NEEDED."
   89 DØ 90 I=1,N
   90 PRINT 91,(A(I,J),J=1,K)
   91 FØRMAT(5F14.3)
      STØP
   71 PRINT,"INDEX EXCEEDS 1.  TRANSFØRMATIØN FAILED."
      STØP
      END
```

Fig. 12.5.1 Program ROUTH1.

It should be noted that the values of the coefficients are entered in the same order as they are to appear in the first two rows of the array. For example, if the equation

$$s^6 + 4s^5 - 2s^4 - 20s^3 - 11s^2 + 16s + 12 = 0$$

is to be tested, N = 7 and K = 4. The values of the coefficients would be entered as follows:

$$1, -2, -11, 12, 4, -20, 16, 0$$

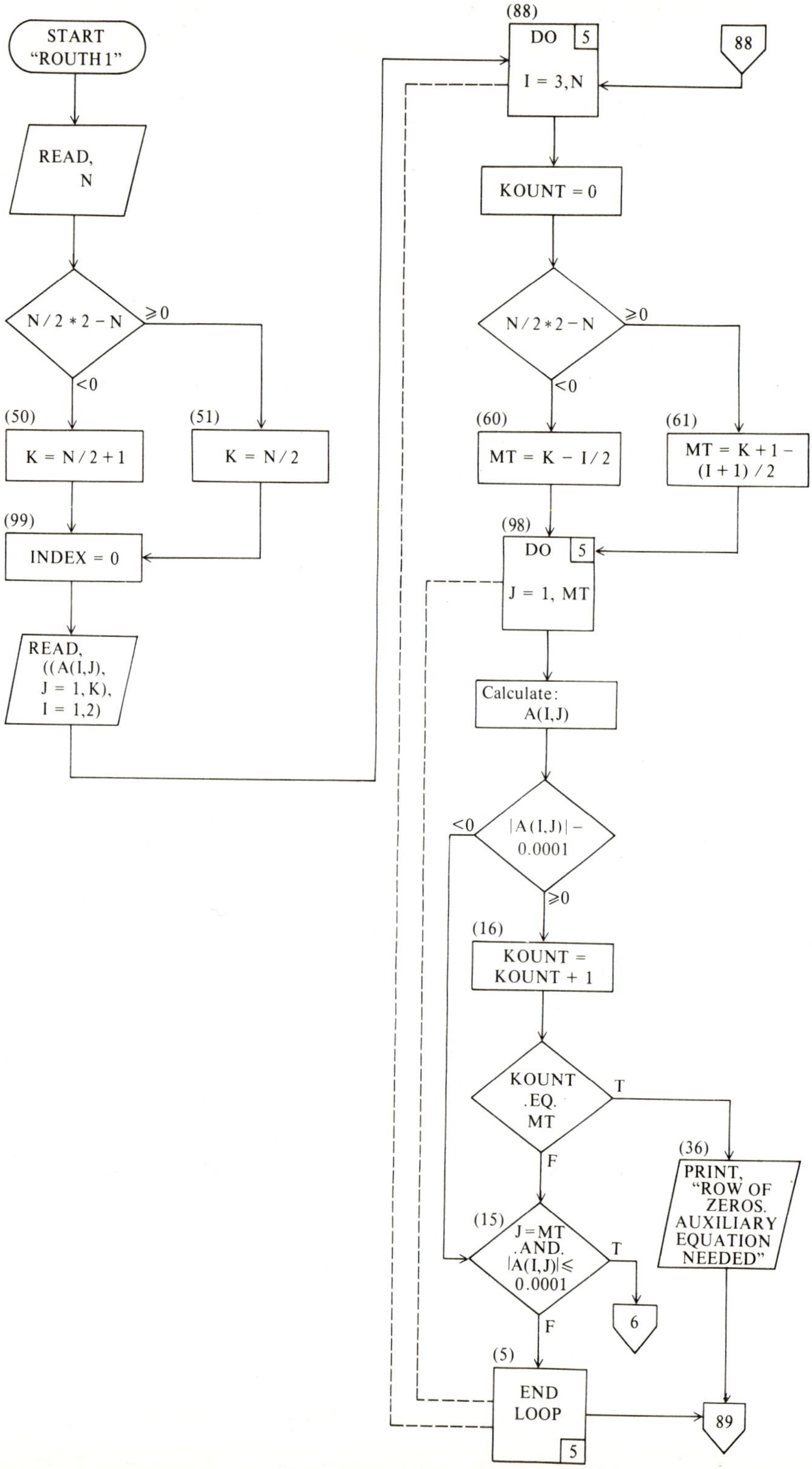
START
"ROUTH1"
READ,
N
N / 2 * 2 − N
≥0
<0
(50)
K = N / 2 + 1
(51)
K = N / 2
(99)
INDEX = 0
READ,
((A(I,J),
J = 1, K),
I = 1,2)
(88)
DO 5
I = 3,N
88
KOUNT = 0
N / 2 * 2 − N
≥0
<0
(60)
MT = K − I / 2
(61)
MT = K + 1 −
(I + 1) / 2
(98)
DO 5
J = 1, MT
Calculate:
A(I,J)
<0
|A(I,J)| −
0.0001
≥0
(16)
KOUNT =
KOUNT + 1
KOUNT
.EQ.
MT
T
F
(36)
PRINT,
"ROW OF
ZEROS.
AUXILIARY
EQUATION
NEEDED"
(15)
J = MT
.AND.
|A(I,J)| ≤
0.0001
T
F
6
(5)
END
LOOP
5
89

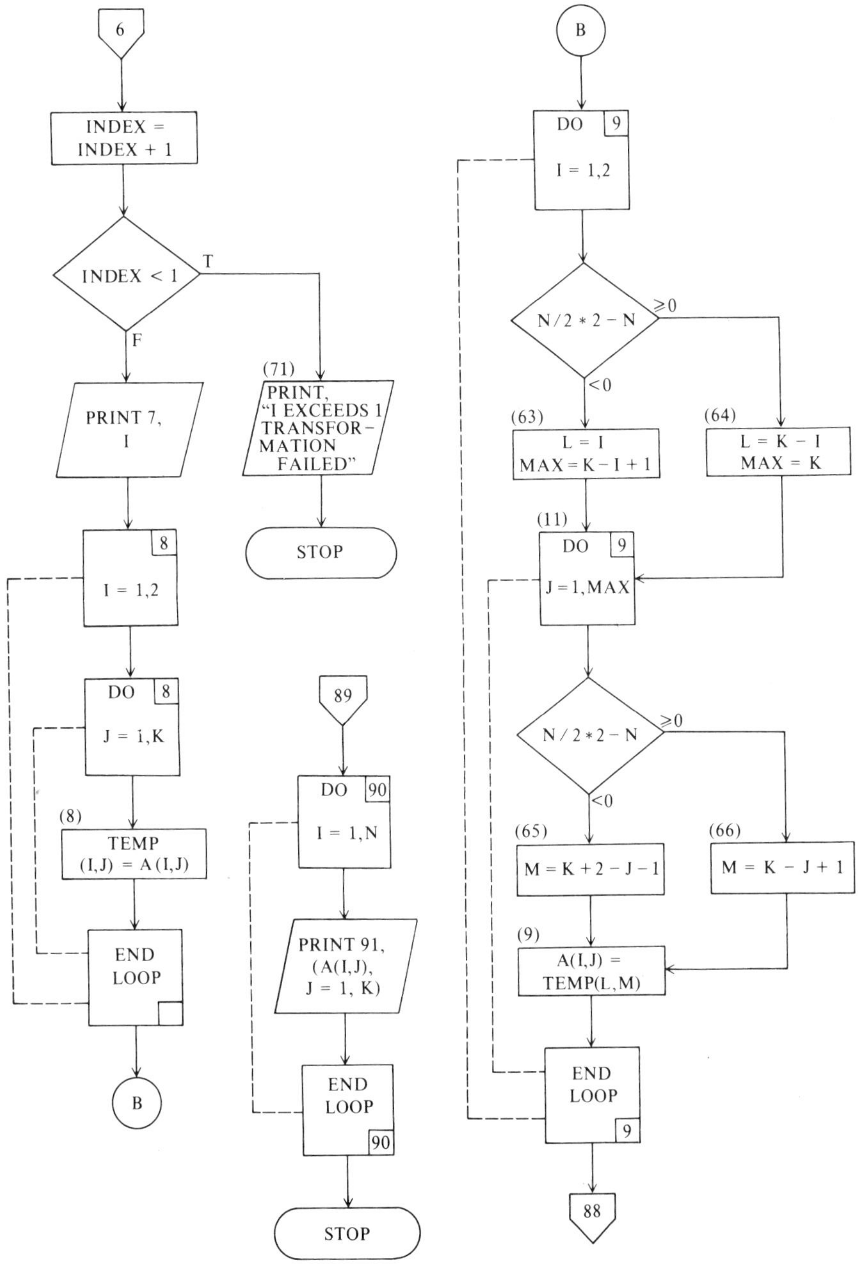

Fig. 12.5.2 Flow chart of program ROUTH1.

These elements would then constitute the following array:

1.	− 2.	−11.	12.	I = 1; J = 1, 2, 3, 4
4.	−20.	16.	0.	I = 2; J = 1, 2, 3, 4

It was necessary to read-in the final zero because, in this example, the nested implied loops were programmed to accept eight values.

The remaining elements of the Routh array are calculated in accordance with the procedure discussed previously in the segment of the program extending from line 00090 to line 00210. There are (N–2) rows of elements to be calculated, and therefore the index I of the outer DO loop (line 00090) varies from 3 to N. The number of elements to be calculated in each row is a function of the number of the row and also depends on whether N is even or odd (see Table 12.5.1). In the program, the number of elements calculated in each row is determined by the maximum value of the index J of the nested DO loop which begins on line 00150. The maximum value of the index J is MT, and MT is assigned values by either statement 60 or statement 61, depending on whether N is odd or even, respectively. The elements of the array are then calculated, row by row.

Immediately after each element is calculated, it is tested to determine if it is zero.[1] If the element is zero, the variable KOUNT is incremented by 1, and the value

```
INPUT:00020
?7

INPUT:00080
?1,4,5,2,2,7,3,0

           1.000          4.000          5.000          2.000
           2.000          7.000          3.000          0.000
           0.500          3.500          2.000          0.000
          -7.000         -5.000          0.000          0.000
           3.143          2.000          0.000          0.000
           -.545          0.000          0.000          0.000
           2.000          0.000          0.000          0.000
STØP


INPUT:00020
?6

INPUT:00080
?1,7,15,1,4,3

ZERØ AT RØW   4. TRANSFØRMATIØN PERFØRMED.
           3.000          4.000          1.000
          15.000          7.000          1.000
           2.600          0.800          0.000
           2.385          1.000          0.000
           -.290          0.000          0.000
           1.000          0.000          0.000
STØP


INPUT:00020
?6

INPUT:00080
?1,-5,4,3,-15,12

RØW ØF ZERØS.AUXILIARY EQUATIØN NEEDED.
           1.000         -5.000          4.000
           3.000        -15.000         12.000
           0.000          0.000          0.000
           0.000          0.000          0.000
           0.000          0.000          0.000
           0.000          0.000          0.000
STØP
```

Fig. 12.5.3 Trial runs for ROUTH1.

[1] An element is considered to be zero if its magnitude is equal to or less than 1×10^{-4}.

of KOUNT is tested by the logical IF statement on line 00190. If KOUNT is equal to MT, there is a row of zeros, and the program branches to statement 36 (line 00440). Then a warning message is printed, the elements of the array that have been determined up to this point are printed, and the program is terminated.

Let us return to the logical IF statement on line 00190 and consider the other possibilities. After all the elements of a particular row have been calculated and it is found that *all* of them are *not* zero, the first element of that row is tested. If the *first* element is *not* zero, then the program branches to statement 89, and the array is printed. If the *first* element of the row *is zero*, then the program branches to statement 6 on line 00230, and the parameter INDEX is incremented by 1. Then a statement is printed indicating in which row the first column zero occurred, and the transformation discussed previously is carried out in the segment of the program extending from line 00270 to line 00420. After the transformation has been completed, the transfer statement on line 00430 causes the program to branch back to statement 88 on line 00090 and the Routh array for the transformed equation is calculated and printed. The purpose of the parameter INDEX is to prevent the program from looping indefinitely if the transformation fails. In this event, the logical IF statement on line 00240 causes a transfer to statement 71. Then a message is printed, and the program halts. Several trial runs are shown in Fig. 12.5.3.

12.6 MATRIX EXPONENTIATION

In certain types of network analysis, it is necessary to raise a matrix to a positive, integer power. In this section we will discuss two programs which perform this operation. An illustration of the use of matrix exponentiation in network analysis will be given in Sec. 12.7.

```
C      MATRX2
C      THIS PRØGRAM RAISES A MATRIX TØ THE M'TH PØWER.
       DIMENSIØN A(10,10),B(10,10),C(10,10)
       READ,N
       READ,((A(I,J),J=1,N),I=1,N)
       READ,M
       KØUNT=1
       DØ 1 I=1,N
       DØ 1 J=1,N
     1 B(I,J)=A(I,J)
     3 DØ 6 I=1,N
       DØ 6 J=1,N
       C(I,J)=0.
       DØ 6 K=1,N
     6 C(I,J)=C(I,J)+A(I,K)*B(K,J)
       KØUNT=KØUNT+1
       IF(KØUNT.EQ.M) GØ TØ 31
       DØ 99 I=1,N
       DØ 99 J=1,N
    99 A(I,J)=C(I,J)
       GØ TØ 3
    31 DØ 23 I=1,N
    23 PRINT 7,(C(I,J),J=1,N)
     7 FØRMAT(6E12.3)
       STØP
       END
```

Fig. 12.6.1 Program MATRX2.

A square matrix $\bar{A}$ can be raised to a positive integer power by successively multiplying it by itself.[1] The program shown in Fig. 12.6.1 carries out this operation. The following data must be entered:

1. The power M to which $\bar{A}$ is to be raised
2. The order N of $\bar{A}$
3. The values of the elements of $\bar{A}$

The DO loop DO 1 assigns the values of the elements of $\bar{A}$ to matrix $\bar{B}$. This step is necessary because if $\bar{A}$ were to be multiplied by itself, the original contents of $\bar{A}$ would be "lost" during multiplication. The trio of nested DO loops which begin on line 00090 multiplies $\bar{A}$ by $\bar{B}$, and stores the result (which is $\bar{A}^2$) in matrix $\bar{C}$. These nested DO loops perform operations identical to the nested DO loops used in the program for matrix multiplication discussed in Sec. 12.1. After the exit from the DO loops, the variable KOUNT, which has been initialized at a value of 1, is incremented by 1. On line 00150, the current value of KOUNT is compared with the value of M, the power to which $\bar{A}$ is to be raised. If KOUNT equals M, the program branches to statement 31, and the answer is printed. If KOUNT is less than M, the program branches to the statement on line 00160. There, the current value of $\bar{C}$ is assigned to $\bar{A}$. Then the program branches back to statement 3 on line 00090, and $\bar{A}$ is again multiplied by $\bar{B}$. (The matrix $\bar{B}$ is always equal to the original value of $\bar{A}$.) In the trial run shown in Fig. 12.6.2, a 3 × 3 matrix is raised to the third power.

Another program which performs matrix exponentiation is shown in Fig. 12.6.3. This program consists of a main program and two subroutines MULMAT and ASIGN. Subroutine ASIGN (X, W, N) assigns the contents of any matrix $\bar{W}$ to matrix $\bar{X}$; subroutine MULMAT (U, V, W, N) "postmultiplies" a matrix $\bar{U}$ by matrix $\bar{V}$ and stores the product in matrix $\bar{W}$.

Let us study the execution of the program after the data have been entered. The statement CALL ASIGN (B, A, N) on line 00040 assigns the contents of $\bar{A}$ to $\bar{B}$. The need for this operation (retention of the original matrix) was explained in connection with the previous program. M1 is then set at a value *1 less than* the power to which the matrix is to be raised (the number of consecutive multiplications to be performed). The DO loop on line 00060 performs these multiplications and terminates when M1 cycles have been executed. Following the multiplication procedure, during each cycle

```
INPUT:00020
 ?3

INPUT:00030
 ?2,-1,1,0,1,2,1,0,1

INPUT:00040
 ?3

   0.110E+02    -.800E+01    0.000E+00
   0.800E+01    -.100E+01    0.800E+01
   0.800E+01    -.400E+01    0.300E+01
STØP
```

Fig. 12.6.2 Trial run for MATRX2.

[1] J. M. Gere and W. Weaver, Jr., "Matrix Algebra for Engineers," Van Nostrand, Princeton, N.J., 1965, p. 25.

```
00001C     MATPØW
00002C     THIS PRØGRAM RAISES A MATRIX TØ THE M'TH PØWER.
00010      DIMENSIØN A(10,10),B(10,10),W(10,10)
00020      READ,M,N
00030      READ,((A(I,J),J=1,N),I=1,N)
00040      CALL ASIGN(B,A,N)
00050      M1=M-1
00060      DØ 1 L=1,M1
00070      CALL MULMAT(A,B,W,N)
00080    1 CALL ASIGN(A,W,N)
00090      DØ 9 I=1,N
00100    9 PRINT 10,(W(I,J),J=1,N)
00110   10 FØRMAT(5E12.3)
00120      STØP
00130      END
00140      SUBRØUTINE MULMAT(U,V,W,N)
00150      DIMENSIØN U(10,10),V(10,10),W(10,10)
00160      DØ 6 I=1,N
00170      DØ 6 J=1,N
00180      W(I,J)=0.
00190      DØ 6 K=1,N
00200    6 W(I,J)=W(I,J)+U(I,K)*V(K,J)
00210      RETURN
00220      END
00230      SUBRØUTINE ASIGN(X,W,N)
00240      DIMENSIØN X(10,10),W(10,10)
00250      DØ 16 I=1,N
00260      DØ 16 J=1,N
00270   16 X(I,J)=W(I,J)
00280      RETURN
00290      END
```

Fig. 12.6.3 Program MATPOW.

```
INPUT:00020
 ?3,3

INPUT:00030
 ?2.,-1.,1.,0.,1.,2.,1.,0.,1.

   0.110E+02   -.800E+01    0.000E+00
   0.800E+01   -.100E+01    0.800E+01
   0.800E+01   -.400E+01    0.300E+01
STØP
```

Fig. 12.6.4 Trial run for MATPOW.

except the last one, the product is transferred from W to A by means of subroutine ASIGN. In the last cycle, the repetitive multiplying process has been completed and the *result* of the exponentiation operation will reside in W. Following the transfer effected by subroutine ASIGN, both A *and* W will contain the answer. In this program, matrix W was printed (PRINT statement on line 00100) although it would have been equally valid to print the contents of A. Figure 12.6.4 shows a trial run.

12.7 *ABCD* PARAMETERS[1]

In Sec. 9.12 we discussed a program which calculates the branch currents and node voltages of a ladder network. In this section, we will view this network from the standpoint of two-port network theory. Many techniques for two-port network description have been developed. Of these, only the *ABCD* parameters will be discussed.

[1] W. R. Le Page and S. Seely, "General Network Analysis," McGraw-Hill, New York, 1959, p. 143.

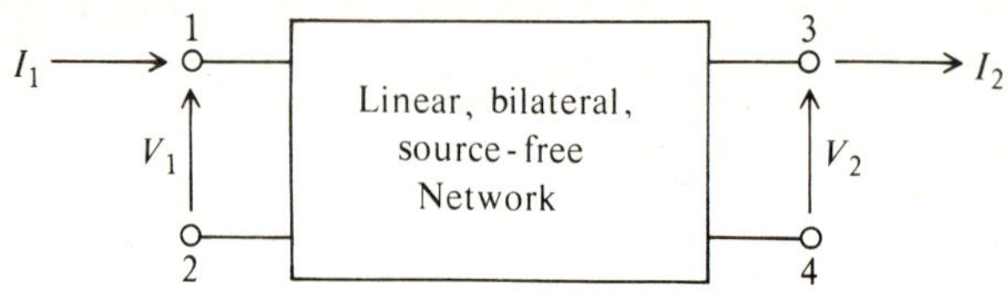

Fig. 12.7.1 A typical two-port network.

The defining equations for *ABCD* parameters are

$$V_1 = AV_2 + BI_2 \tag{12.7.1}$$

$$I_1 = CV_2 + DI_2 \tag{12.7.2}$$

The reference directions of the currents and voltages on which these equations are based are shown in Fig. 12.7.1. The parameters A, B, C, D can be calculated by use of the following equations:

$$A = \left.\frac{V_1}{V_2}\right|_{I_2=0} \quad \text{(Open-circuit reverse voltage gain, dimensionless)} \tag{12.7.3}$$

$$B = \left.\frac{V_1}{I_2}\right|_{V_2=0} \quad \text{(Short-circuit reverse transfer impedance, ohms)} \tag{12.7.4}$$

$$C = \left.\frac{I_1}{V_2}\right|_{I_2=0} \quad \text{(Open-circuit reverse transfer admittance, mhos)} \tag{12.7.5}$$

$$D = \left.\frac{I_1}{I_2}\right|_{V_2=0} \quad \text{(Short-circuit reverse current gain, dimensionless)} \tag{12.7.6}$$

It can also be shown that for a bilateral linear network, which are the conditions assumed here, the following relation must be satisfied:

$$AD - BC = 1 \tag{12.7.7}$$

Equation (12.7.7) serves as a useful check on the accuracy of calculations. At this point it may be helpful to calculate the *ABCD* parameters of two simple networks. Figure 12.7.2 shows a T network (the name derives from the form of the network) to which a voltage source V_1 has been temporarily connected for the purpose of calculating the *ABCD* parameters. To calculate parameters A and C, no connections

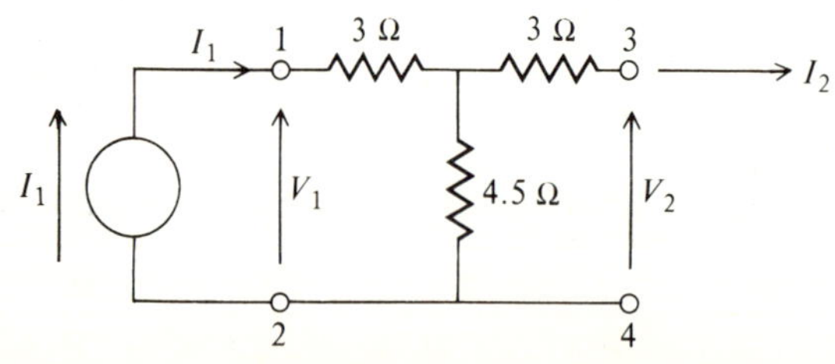

Fig. 12.7.2 A typical T section.

are made to terminals 3 and 4 in order to satisfy the condition $I_2 = 0$ [see Eqs. (12.7.3) and (12.7.5)]. Using Ohm's law, we obtain

$$I_1 = \frac{V_1}{7.5} \tag{12.7.8}$$

$$V_2 = 4.5I_1 \tag{12.7.9}$$

$$V_2 = \frac{4.5}{7.5} V_1 = \frac{3}{5} V_1 \tag{12.7.10}$$

From Eqs. (12.7.3) and (12.7.10), $A = \frac{5}{3}$ and from Eqs. (12.7.5) and (12.7.9), $C = 1/4.5$. Next, we temporarily connect terminals 3 and 4 by a link of zero resistance. Therefore, we satisfy the condition $V_2 = 0$, and can calculate B and D. The 3-ohm resistor at the right and the 4.5-ohm resistor are now in parallel, and equivalent to a resistor of value

$$\frac{3 \times 4.5}{7.5} = \frac{9}{5} \text{ ohm}$$

Using Ohm's law:

$$I_1 = \frac{V_1}{3 + \frac{9}{5}} = \frac{5}{24} V_1 \tag{12.7.11}$$

By the current division theorem:

$$I_2 = \frac{4.5}{7.5} I_1 = \frac{3}{5} I_1 \tag{12.7.12}$$

Now, we replace I_1 and obtain

$$I_2 = \tfrac{3}{5} \cdot \tfrac{5}{24} V_1 = \tfrac{1}{8} V_1 \tag{12.7.13}$$

From Eqs. (12.7.4) and (12.7.13), $B = 8$, and from Eqs. (12.7.6) and (12.7.12), $D = \frac{5}{3}$.

As a second example, consider the shunt section shown in Fig. 12.7.3. Inspecting the network in the open-circuit case, we observe that

$$V_1 = V_2 \tag{12.7.14}$$

$$I_1 = \frac{V_1}{6} = \frac{V_2}{6} \tag{12.7.15}$$

The last step follows from the fact that $V_2 = V_1$. From these equations, it follows that $A = 1$ and $C = 1/6$. Next, we short-circuit terminals 3 and 4 and obtain the B and D parameters. Applying Kirchhoff's Current Law at node 1, we obtain

$$I_2 = I_1 \tag{12.7.16}$$

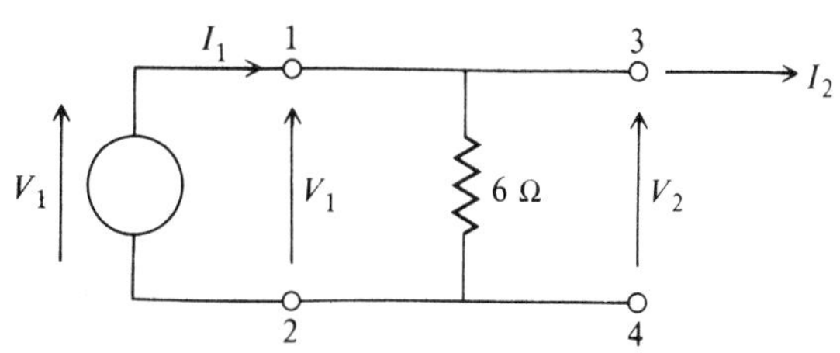

Fig. 12.7.3 A typical shunt section.

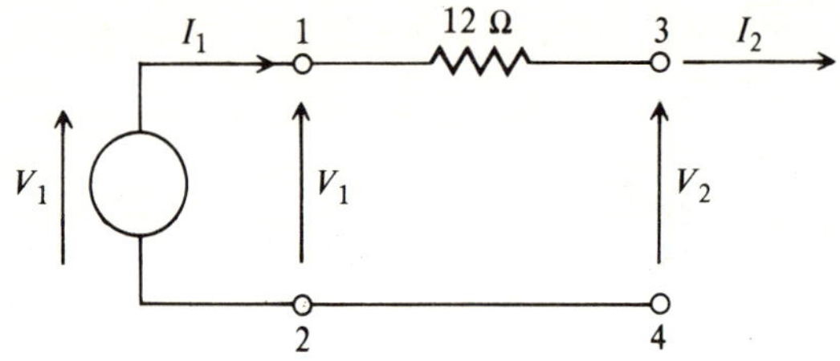

Fig. 12.7.4 A typical series element. The *ABCD* parameters are: $A = 1, B = 12, C = 0, D = 1$.

From Eq. (12.7.16) it follows that $D = 1$. Recalling that

$$B = \left.\frac{V_1}{I_2}\right|_{V_2=0}$$

and as $V_1 = V_2 = 0$, it follows that B=0. The *ABCD* parameters of a series section are shown in Fig. 12.7.4.

ABCD parameters are most useful in matrix form. To introduce this method, let us first write the defining equations, Eqs. (12.7.1) and (12.7.2), in matrix form:

$$\begin{bmatrix} V_1 \\ I_1 \end{bmatrix} = \begin{bmatrix} A & B \\ C & D \end{bmatrix} \begin{bmatrix} V_2 \\ I_2 \end{bmatrix} \tag{12.7.17}$$

Suppose there are 2 two-port networks in cascade as shown in Fig. 12.7.5. The network on the left has parameters *A, B, C,* and *D* and the other network has parameters A', B', C', and D'. Carefully observe the notation used in Fig. 12.7.5 for the voltages and currents. In terms of this notation the matrix equation for the network on the left is

$$\begin{bmatrix} V_1 \\ I_1 \end{bmatrix} = \begin{bmatrix} A & B \\ C & D \end{bmatrix} \begin{bmatrix} V_2' \\ I_2' \end{bmatrix} \tag{12.7.18}$$

The matrix equation for the other network is

$$\begin{bmatrix} V_1' \\ I_1' \end{bmatrix} = \begin{bmatrix} A' & B' \\ C' & D' \end{bmatrix} \begin{bmatrix} V_2 \\ I_2 \end{bmatrix} \tag{12.7.19}$$

Inspecting Fig. 12.7.5, it is evident that

$$V_2' = V_1'$$
$$I_2' = I_1'$$

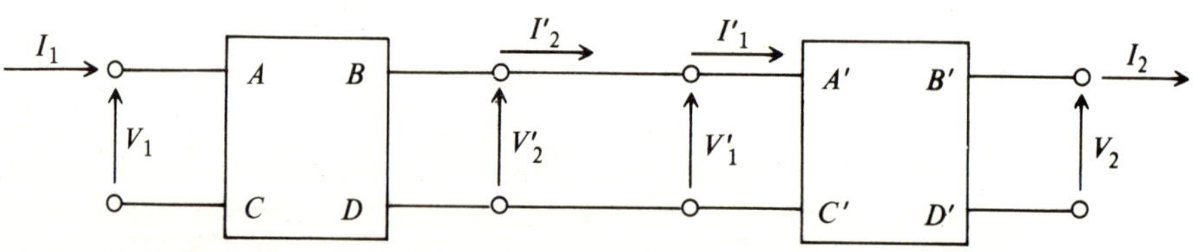

Fig. 12.7.5 Two two-port networks in cascade.

We now replace the matrix $\begin{bmatrix} V_2' \\ I_2' \end{bmatrix}$ by its equivalent in Eq. (12.7.19) and obtain

$$\begin{bmatrix} V_1 \\ I_1 \end{bmatrix} = \begin{bmatrix} A & B \\ C & D \end{bmatrix} \begin{bmatrix} A' & B' \\ C' & D' \end{bmatrix} \begin{bmatrix} V_2 \\ I_2 \end{bmatrix} \tag{12.7.20}$$

After multiplying the two inner matrices, we obtain

$$\begin{bmatrix} V_1 \\ I_1 \end{bmatrix} = \begin{bmatrix} AA' + BC' & AB' + BD' \\ CA' + DC' & CB' + DD' \end{bmatrix} \begin{bmatrix} V_2 \\ I_2 \end{bmatrix} \tag{12.7.21}$$

We now have found the *ABCD* parameters of a *cascade* of two networks. Similarly, by multiplying the n *ABCD* matrices of a cascade consisting of n networks, we can find the *ABCD* matrix of the cascade. This procedure will be illustrated by an example.

The network of Fig. 9.12.5 has been repeated for convenience in Fig. 12.7.6. Also, it has been redrawn as a cascade consisting of 4 two-port sections. (Notice that the 6-ohm resistor has been split into two 3-ohm resistors.) We have already calculated the *ABCD* parameters of the two-port sections. The only task remaining is the multiplication of the matrices. The array of matrices is

$$\begin{bmatrix} 1 & 12 \\ 0 & 1 \end{bmatrix} \begin{bmatrix} 5/3 & 8 \\ 1/4.5 & 5/3 \end{bmatrix} \begin{bmatrix} 5/3 & 8 \\ 1/4.5 & 5/3 \end{bmatrix} \begin{bmatrix} 1 & 0 \\ 1/6 & 1 \end{bmatrix}$$

After the second and third matrices are multiplied, the result is

$$\begin{bmatrix} 1 & 12 \\ 0 & 1 \end{bmatrix} \begin{bmatrix} 41/9 & 80/3 \\ 20/27 & 41/9 \end{bmatrix} \begin{bmatrix} 1 & 0 \\ 1/6 & 1 \end{bmatrix}$$

Two more multiplications yield

$$\begin{bmatrix} 363/27 & 732/9 \\ 20/27 & 41/9 \end{bmatrix} \begin{bmatrix} 1 & 0 \\ 1/6 & 1 \end{bmatrix}$$

$$\begin{bmatrix} 27 & 732/9 \\ 3/2 & 41/9 \end{bmatrix}$$

The reader should note that in each multiplication a matrix is multiplied by a matrix at its right. Thus, for the cascade,

$$A = 27 \qquad B = 732/9 \approx 81.4 \qquad C = 3/2 \qquad D = 41/9 \approx 4.55$$

As a check on these answers, we calculate $AD - BC$:

$$AD - BC = 27 \times \frac{41}{9} - \frac{3}{2} \times \frac{732}{9} = 1$$

We see that Eq. (12.7.7) is satisfied.

As a last calculation, let us relate the values of A and C to the calculations carried out in the first trial run in Sec. 9.12. In that example an 18-V source was connected to the network. Therefore, V_1 was 18 V. The current I_1 delivered by the source was found to be 1 A. The voltage designated as V_6 was found to be $\frac{2}{3}$ V. In the

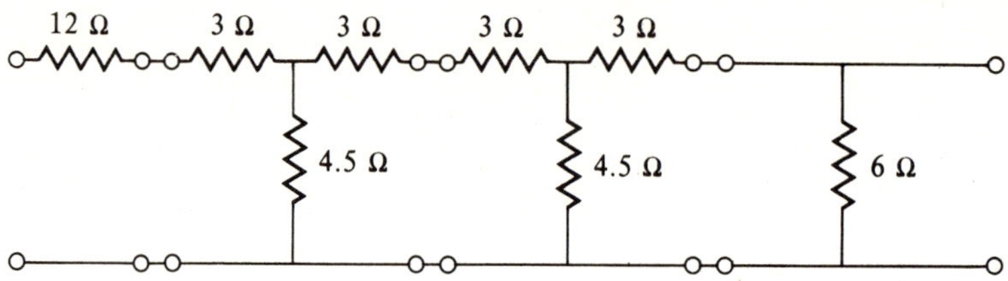

Fig. 12.7.6 Network of Fig. 9.12.5 redrawn as a cascade network.

present context, V_6 is designated as V_2, the voltage across terminals 3 and 4 with terminals 3 and 4 open-circuited. With these data, we can calculate A and C:

$$A = \left.\frac{V_1}{V_2}\right|_{I_2=0} = \frac{18}{\frac{2}{3}} = 27$$

$$C = \left.\frac{I_1}{V_2}\right|_{I_2=0} = \frac{1}{\frac{2}{3}} = 1.5$$

Therefore, we have verified our calculation of A and C. Similarly, the data of the second trial run of Sec. 9.12 enables us to calculate B and D. Again, $V_1 = 18$ V. The current designated I_6 in the trial run is I_2 in the present notation and was found to be 0.2213 A. I_1 was found to be 1.008 A. Therefore:

$$B = \left.\frac{V_1}{I_2}\right|_{V_2=0} = \frac{18}{0.2213} \approx 81.4$$

$$D = \left.\frac{I_1}{I_2}\right|_{V_2=0} = \frac{1.008}{0.2213} \approx 4.55$$

Thus, the previous calculations of B and D have been verified.

The program for calculating the *ABCD* parameters of a cascade of networks if the *ABCD* parameters of the individual sections are known is shown in Fig. 12.7.7. The network is viewed as consisting of three cascaded networks: a section adjacent to the generator; a midsection consisting of M identical sections; and a terminating section. In the program, the *ABCD* parameters of these sections are stored in matrices Z, RESEC, and END, respectively. M is the number of identical sections and RESEC is the repeated section.

The program consists, essentially, of subroutine subprograms MULMAT and ASIGN. MULMAT carries out matrix multiplication in the manner discussed in Sec. 12.6. The product is stored in W. ASIGN assigns the values of the elements of matrix W to matrix X. Let us see how the program calculates the *ABCD* parameters of the network shown in Fig. 12.7.6. For this network, the midsection consists of two identical T networks, and so M = 2. On line 00070, ASIGN is called, and as a result, the values of the elements of matrix RESEC are stored in matrix TEMP. Matrix RESEC is raised to the power M by the action of the DO loop which begins on line 00090. It should be noted that the maximum value of the index of the DO loop is M−1. In the first pass through the DO loop, MULMAT is called (line 00100), and as a result RESEC is squared. The elements of matrix $(\text{RESEC})^2$ are stored in matrix W.

```
00001C       ABCD
00002C       THIS PRØGRAM CALCULATES THE ABCD PARAMETERS ØF A
00003C       CASCADE ØF M IDENTICAL RESISTIVE NETWØRKS TERMINATED
00004C       AT THE GENERATØR AND LØAD ENDS BY ARBITRARY
00005C       RESISTIVE NETWØRKS. SEE FIG.12.7.6.
00010        DIMENSIØN Z(2,2),RESEC(2,2),END(2,2),W(2,2),TEMP(2,2)
00020        READ,((Z(I,J),J=1,2),I=1,2)
00030        READ,M
00040        READ,((RESEC(I,J),J=1,2),I=1,2)
00050        READ,((END(I,J),J=1,2),I=1,2)
00060        IF(M.EQ.1) GØ TØ 19
00070        CALL ASIGN(TEMP,RESEC)
00080        M1=M-1
00090        DØ 1 L=1,M1
00100        CALL MULMAT(RESEC,TEMP,W)
00110      1 CALL ASIGN(RESEC,W)
00120     19 CALL MULMAT(Z,RESEC,W)
00130        CALL ASIGN(TEMP,W)
00140        CALL MULMAT(TEMP,END,W)
00150        PRINT,"ABCD MATRIX ØF THE ØVERALL NETWØRK"
00160        DØ 9 I=1,2
00170      9 PRINT 10,(W(I,J),J=1,2)
00180     10 FØRMAT(2E15.3)
00190        STØP
00200        END
00210        SUBRØUTINE MULMAT(U,V,W)
00220        DIMENSIØN U(2,2),V(2,2),W(2,2)
00230        DØ 6 I=1,2
00240        DØ 6 J=1,2
00250        W(I,J)=0.
00260        DØ 6 K=1,2
00270      6 W(I,J)=W(I,J)+U(I,K)*V(K,J)
00280        RETURN
00290        END
00300        SUBRØUTINE ASIGN(X,W)
00310        DIMENSIØN X(2,2),W(2,2)
00320        DØ 16 I=1,2
00330        DØ 16 J=1,2
00340     16 X(I,J)=W(I,J)
00350        RETURN
00360        END
```

Fig. 12.7.7 Program ABCD.

The last statement of the DO loop (line 00110) assigns the values of the elements of W to RESEC. In the present case, M−1 = 1, and, therefore, there is only one pass through the DO loop. As a result of the "calls" on lines 00120, 00130, and 00140, matrix Z is multiplied by $(\text{RESEC})^2$, the product is assigned to TEMP, and finally TEMP is multiplied by matrix END. The overall matrix product is stored in W, and this is the desired answer. The trial run is shown in Fig. 12.7.8, and the answers are seen to correspond to those previously obtained by hand calculation.

```
INPUT:00020
 ?1,12,0,1

INPUT:00030
 ?2

INPUT:00040
 ?1.667,8,.222,1.667

INPUT:00050
 ?1,0,.1667,1

ABCD MATRIX ØF THE ØVERALL NETWØRK
       0.270E+02       0.813E+02
       0.150E+01       0.455E+01
STØP
```

Fig. 12.7.8 Trial run for ABCD.

PROBLEMS

12.1. In Sec. 12.2, the reduction of the displacement matrix was accomplished indirectly by first storing the value of T(1) in a location named DIV.

Revise the program of Fig. 12.2.2 so the reduction is effected directly, using the current value of T(1).

Suggestion: Remove lines 00500, 00510, 00520, and 00560 from the program and insert the statements that will revise the calculation method.

12-2. Revise the program of Fig. 12.5.1 so that the computer will search the elements in the first column of the array for a *negative* value. If one is found, have the computer print the message SYSTEM UNSTABLE.

12-3. Write a program which will calculate a least-squares fit of order m to a given set of experimental data.

Test the program with $m = 2$ (the case discussed in Sec. 12.4), using the data shown in Fig. 12.4.3.

12-4. Markov processes are important in queuing theory, management science, traffic engineering, and other areas. A sequence of experiments in which the outcome of any trial depends only on the outcome of the immediately preceding trial is called a Markov process.

Calculation of the probabilities of the events of a Markov process is conveniently carried out by matrix multiplication. There are two types of probabilities which may be of interest: conditional probabilities and unconditional probabilities. Let us consider each of these separately.

(*a*) *Calculation of conditional probabilities.* The elements of a square matrix $\bar{P}$, called the *transition matrix*, must be given. Each element denotes a conditional probability. For example, p_{ij} denotes the probability that the process will be in state j at the next step, given that the process is now in state i. To calculate conditional probabilities for later stages of the process, $\bar{P}$ must be raised to an appropriate power. For example, the element $p_{ij}^{(2)}$ of $\bar{P}^2$ (the square of $\bar{P}$) denotes the probability that the process will be in state j *two* steps from now, given that the process is now in state i.

(*b*) *Calculation of unconditional probabilities.* The elements of a row matrix called the *initial probability vector*, denoted $\bar{p}^{(0)}$, and also the elements of the transition matrix $\bar{P}$, discussed above, must be given. If the order of $\bar{P}$ is $m \times m$, the order of $\bar{p}^{(0)}$ must be $1 \times m$. Unconditional probabilities are calculated as follows:

$$\bar{p}^{(n)} = \bar{p}^{(0)}\bar{P}^{n} \tag{1}$$

For example, $\bar{p}^{(2)}$ is calculated by postmultiplying $\bar{p}^{(0)}$ by the second power of $\bar{P}$. Each element of $\bar{p}^{(2)}$ gives the unconditional probability of a particular state. For example, if there are three states $m = 3$, then

$$\bar{p}^{(2)} = [p_1{}^{(2)}, p_2{}^{(2)}, p_3{}^{(2)}]$$

where $p_1{}^{(2)}$ means the probability that the process is in state 1 after *two* steps, $p_2{}^{(2)}$ means the probability that the process is in state 2 after *two* steps, etc.

Write a FORTRAN program that will read-in the values of the elements of matrices $\bar{p}^{(0)}$ and $\bar{P}$ and which will calculate either conditional or unconditional processes as described above. The program should give the user the option of selecting either or both of the probability calculation methods by entering a code number as part of the input data.

Sample data

$$\bar{P} = \begin{bmatrix} 0.8 & 0.2 \\ 0.5 & 0.5 \end{bmatrix}$$

$$\bar{p}^{(0)} = [0.6 \quad 0.4]$$

12-5. In some Markov processes (see Prob. 12-4) it is found that the transition matrix approaches a limit. That is, for n sufficiently large, the difference between each of the elements of $\bar{P}^n$ and the corresponding elements of $\bar{P}^{n+1}$ becomes negligibly small. Write a FORTRAN program which raises a transition matrix to the first, second, third, . . ., nth powers, and calculates the magnitude of the difference between the elements of each power of $\bar{P}$ and the corresponding elements of the next higher power. When this difference becomes less than 1×10^{-4}, the program should print the value of the elements of the power of $\bar{P}$ last calculated, and the value of n.

Sample data

$$\bar{P} = \begin{bmatrix} 0.8 & 0.2 \\ 0.5 & 0.5 \end{bmatrix}$$

12-6. When the dimensions and material of a Belleville washer spring are known, it is relatively simple to analyze the spring to find its load-deflection characteristics and the stresses produced at different loads. This was accomplished in the solution of Prob. 10-9.

However, if the spring must be designed to meet a particular set of load-deflection specifications, the required dimensions cannot be found directly and an iterative method of solution can be employed.

Using the design equations given in Prob. 10-9, write a computer program to design a Belleville spring given the following data:

1. Outside and inside diameter of the washer
2. Maximum deflection at a specified load
3. Physical properties of the material used.

The program should calculate and print the values for the thickness and inside height.

Suggestion: Assume values of thickness and calculate the load required to produce maximum deflection. Compare this load with the specified design load. Continue until a satisfactory design has been obtained. (Note that it may not be possible to find an exact solution when uniform increments of thickness are used; therefore, *bracket* the design load and printout the required dimensions for the calculated loads immediately above and below the specified load.)

Sample data

OD = 2.0 in
ID = 1.25 in
δ = 0.025 in
P = 800 lb
E = 3×10^7 lb/in^2
ν = 0.3
σ = 2×10^5 lb/in^2

12-7. Modify the program discussed in Sec. 12.3 so that it will solve systems of simultaneous homogeneous equations using *complex* coefficients.

Have the program calculate mesh currents I_1 and I_2 in the following network if the impressed voltage is $(10 + j_0)$ V.

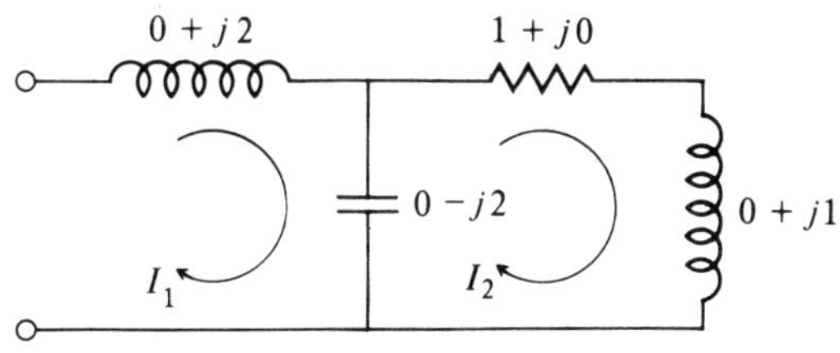

12-8. *Elementary matrices* are modifications of the identity matrix and can be used to perform *elementary transformations* on a square matrix. A common elementary transformation is the interchange of two rows in a matrix. This is accomplished by premultiplying the matrix by an elementary matrix that was obtained by interchanging the corresponding rows in an identity matrix of the same order.

For example, to interchange the second and third rows of a fourth-order square matrix, we premultiply the given matrix by the following modified identity matrix:

$$E = \begin{bmatrix} 1 & 0 & 0 & 0 \\ 0 & 0 & 1 & 0 \\ 0 & 1 & 0 & 0 \\ 0 & 0 & 0 & 1 \end{bmatrix}$$

in which the second and third rows have been interchanged. The result is

$$\begin{bmatrix} 1 & 0 & 0 & 0 \\ 0 & 0 & 1 & 0 \\ 0 & 1 & 0 & 0 \\ 0 & 0 & 0 & 1 \end{bmatrix} \begin{bmatrix} 1 & 2 & 3 & 4 \\ 5 & 6 & 7 & 8 \\ 10 & 20 & 30 & 40 \\ 50 & 60 & 70 & 80 \end{bmatrix} = \begin{bmatrix} 1 & 2 & 3 & 4 \\ 10 & 20 & 30 & 40 \\ 5 & 6 & 7 & 8 \\ 50 & 60 & 70 & 80 \end{bmatrix}$$

Write a subroutine subprogram that can be called by a main program and will return an elementary matrix of order n that can then be used to interchange two rows in a given matrix. The CALL statement should transmit the order of the elementary matrix and indicate which rows in the given matrix are to be interchanged.

Test the subprogram by having a main program execute a call and then print the matrix that is returned.

Sample data

Set	*Order*	*Interchange rows*
1	5	2 and 5
2	3	1 and 2

12-9. In Prob. 12-8 elementary matrices for interchanging rows of a square matrix were described. Interchange of two columns of a square matrix can be effected in a similar manner by postmultiplying the given matrix by an elementary matrix obtained by interchanging the corresponding columns in an identity matrix of the same order.

For example, to exchange the second and fourth columns of a fourth-order square matrix, we perform the following postmultiplication:

$$\begin{bmatrix} 1 & 2 & 3 & 4 \\ 5 & 6 & 7 & 8 \\ 10 & 20 & 30 & 40 \\ 50 & 60 & 70 & 80 \end{bmatrix} \begin{bmatrix} 1 & 0 & 0 & 0 \\ 0 & 0 & 0 & 1 \\ 0 & 0 & 1 & 0 \\ 0 & 1 & 0 & 0 \end{bmatrix} = \begin{bmatrix} 1 & 4 & 3 & 2 \\ 5 & 8 & 6 & 7 \\ 10 & 40 & 30 & 20 \\ 50 & 80 & 60 & 70 \end{bmatrix}$$

Write a subroutine subprogram that will produce an elementary matrix of the proper order for interchanging two specified columns in a square matrix of order n. The test procedure should be similar to that used in Prob. 12-8.

Sample data

Set	*Order*	*Interchange columns*
1	3	1 and 3
2	5	2 and 4

BIBLIOGRAPHY

Dimitry, Donald L., and Thomas H. Mott, Jr.: "Introduction to FORTRAN IV Programming," Holt, New York, 1966.

Golde, Hellmut: "FORTRAN II and IV for Engineers and Scientists," Macmillan, New York, 1966.

Murrill, Paul W., and Cecil L. Smith: "FORTRAN IV Programming for Engineers and Scientists," International Textbook, Scranton, Pa., 1969.

Organick, Elliott I.: "A FORTRAN IV Primer," Addison-Wesley, Reading, Mass., 1966.

Stuart, Fredric: "FORTRAN Programming," Wiley, New York, 1969.

Selected Answers to Problems

CHAPTER 1

1-2 0.00036921

1-4 3.1416

1-6 0.35

1-8 .3216E+02

1-10 .30103E+00

1-12 .186284E+06

1-14 Floating

1-16 Floating

1-18 Fixed

1-20 Fixed

1-22 Floating

1-24 Fixed

1-26 Floating

1-28 +4

1-30 +4

1-32 +1

1-34 (A**2+B**2)/2.

1-36 A*C/D+B/C

1-38 P*XL**3/(3.*E*XI)

1-40 (4.*X**3)/B

1-42 6.*(X**2+Y**3)

1-44 $2x^3 + \dfrac{3y^3 b}{a}$

1-46 $\dfrac{2x^3 + 3y^3}{ab}$

1-48 (A-2.*B)*(A+X)

1-50
1. R*T will be evaluated and stored
2. A will be subtracted from step 1 and stored
3. R-1. will be evaluated and stored
4. Value from step 2 will be divided by the value from step 3

1-52
1. A-B will be evaluated and stored
2. A+B will be evaluated and stored
3. Value from step 1 will be divided by value from step 2 and stored
4. Value from step 3 will be squared and stored
5. Value from step 4 will be divided by 4. and stored

6. 1. will be added to the value from step 5

1-54 1. X/A will be evaluated and stored
2. Exponential of value from step 1 will be evaluated and stored
3. Value of X will be negated and stored
4. Value of step 3 will be divided by A and stored
5. Exponential of value from step 4 will be evaluated and stored
6. Value from step 2 will be added to value from step 5 and stored
7. 0.5 will be multiplied by A and stored
8. Value from step 7 will be multiplied by value from step 6

1-56 A=H*(B1+B2)/2.

1-58 A=P*(1.+XI/XM)**(N*M)

1-60 GR=RHO**2*BETA*G*XL**3*(TS-TF)/VIS**2

1-62 Q=C*A1*A2*SQRT(2.*G*(H1-H2)/(A1**2-A2**2))

1-64 D=(16.*T/(3.1415927*S))**(1./3.)

1-66 H=(V*SIN(ALPHA))**2/(2.*G)

1-68 XN=SIN(THETAI*3.1415927/180.)/SIN(THETAR*3.1415927/180.)

1-70 TM=(T1-TA-T2+TB)/ALOG((T1-TA)/(T2-TB))

CHAPTER 2

2-2 Values of A, B, and C will be read from the teletype according to the unformatted convention

2-4 Values of X, Y, and Z will be read from the teletype according to the unformatted convention

2-6 Values of ITEM and VALUE will be read from the magnetic disk according to the format specified by statement 10

2-8 Values of A and X will be printed by the teletype according to the format specified by statement 65

2-10 Values of A, B, and Z will be printed by the teletype according to the format specified by statement 250

2-12 Values of A and X will be read from paper tape according to the format specified by statement 65

2-14 Values of T and F will be written on magnetic tape according to the format specified by statement 100

2-16 xxx.xxxxxxxxx

2-18 xxx.xxx
xxxxxx

2-20 xxx.xxxxxxx.xxx
xxxx

2-22 bbAb=xxx.xxxbFEET

2-24 bDEPTbbbbbbUNITSb
PRODUCEDbbbbbHOURS

2-26 bb0.xxxxE±xxbbbbbbbbo.
xxxxxxE±xx

2-28 bbbbbxx.xxxxE±xx

2-30 bbbbbbx.xxxxE±xxbbbbbb0.
xxxxxxE±xx

2-32 bbbbbbx.xxxxE±xxbbbbbb0.
0xxxE±xx

2-34 A = +24.00
B = +8.43
C = +0.00421

2-36 A = +2.
B = +400.
C = +7.69
N = +85

2-38 A = +8659000000.
B = +3164.98
N = +40

2-40 AMT = +4060.
MO = +8
NDA = +3
NYR = +70

2-42 A = +86.59
110 FORMAT(F4.2//3X3HTbV)

2-44 Output to teletype: bbbb263.85bb69431.38b-113

2-46 Output to teletype: ROOTb=bbb263.847bbbbbbbb42b ITERATIONS

2-48 Output to printer: bbbbb0.2638E+03bbbbb0.6281E−05

2-50 Output to printer: bbbb263.85bbb0.268E−05

2-52 Output to printer: next page
bbb1.600E+08bb−6.832E+00bbb9.127E+01

2-54 Output to printer: next page
TRIALbNUMBERbb9
bbbbbVELOCITYb=bbbbb91.27bFT/SEC

2-56 Output to teletype: bb2.63847E+02bb6.94314E+04bbbbb−113

2-58 Values stored in locations A and B are transmitted to magnetic disk in binary form

2-60 Output to printer:
bTABLEbOFbRESULTSb
blank line

2-62 Output to teletype:
bb2.63847E+02
blank line
bb6.94314E+04

2-64 Output to printer:
bb912.74bbbb42bb2.6385E+02

CHAPTER 3

```
C     PRØBLEM 3-1
      READ,C,H,Ø,S
      HV=145.*C+620.*(H-Ø/8.)
      PRINT,"HEATING VALUE=",HV
      STØP
      END

INPUT:00010
 ?74.51,5.16,9.19,.87

HEATING VALUE=   13290.9250
STØP
```

```
C     PRØBLEM 3-4
      READ,SØ,VØ,A,T
      S=SØ+(VØ+A*T/2.)*T
      PRINT,"DISPLACEMENT=",S
      STØP
      END

INPUT:00010
 ?390.,8.,-386.,2.

DISPLACEMENT=    -366.0000
STØP
```

```
C     PRØBLEM 3-8
      READ,RHØ,BETA,XL,TS,TF,VIS
      GR=RHØ**2*BETA*32.16*XL**3*(TS-TF)/VIS**2
      PRINT,"GRASHØF NUMBER=",GR
      STØP
      END

INPUT:00010
 ?.0489,.00367,1.,800.,250.,1.669E-05

GRASHØF NUMBER=   5.5725E+08
STØP
```

```
00001C     PRØBLEM 3-13
00010      READ, HP, RPM, S
00011      PRINT, "HØRSEPØWER=", HP
00012      PRINT, "RØTATIØNAL SPEED=", RPM
00013      PRINT, "ALLØWABLE STRESS=", S
00020      T=HP/RPM*63000.
00030      D=(16./3.1415927*T/S)**(1./3.)
00040      PRINT, "DIAMETER=", D
00050      STØP
00060      END

INPUT:00010
 ?600.,85.,6000.

HØRSEPØWER=       600.0000
RØTATIØNAL SPEED=        85.0000
ALLØWABLE STRESS=     6000.0000
DIAMETER=          7.2271
STØP
```

```
00001C     PRØBLEM 3-20
00010      READ,HI,HR,HF,HØ
00020      READ,RC1,RC2,R1,R2
00030      DEL=HI*HØ-HF*HR
00040      RI2=(HI+DEL*RC2)/(1.+HØ*RC2)
00050      AV2=-HF*RC2/(HI+DEL*RC2)
00060      GL1=1./RI2+1./R2+1./R1+1./RC1
00070      RL1=1./GL1
00080      AV1=-HF*RL1/(HI+DEL*RL1)
00090      AV=AV1*AV2
00100      PRINT,AV1,AV2,AV
00110      STØP
00120      END

INPUT:00010
 ?1500.,6.E-4,50.,4.E-5

INPUT:00020
 ?5000.,5000.,100000.,10000.

     -31.5277     -151.5152     4776.9179
STØP
```

CHAPTER 4

```
00001C     PRØBLEM 4-2
00010      READ,A,B,C
00020      IF(C-A)1,2,2
00030    1 TEMP=C
00040      C=A
00050      A=TEMP
00060    2 IF(C-B)3,4,4
00070    3 TEMP=C
00080      C=B
00090      B=TEMP
00100    4 IF(C-(A+B))5,5,6
00110    5 PRINT,"TRIANGLE CANNØT BE FØRMED"
00120      STØP
00130    6 IF(C*C-(A*A+B*B))7,8,7
00140    7 PRINT,"TRIANGLE CAN BE FØRMED"
00150      STØP
00160    8 PRINT,"RIGHT TRIANGLE IS FØRMED"
00170      STØP
00180      END

INPUT:00010
 ?7.,2.,4.

TRIANGLE CAN BE FØRMED
STØP
```

```
C     PRØBLEM 4-4
      READ,A1,B1,C1,A2,B2,C2
      DEL=A1*B2-A2*B1
      IF(DEL)2,1,2
    1 PRINT,"EQUATIØNS ARE DEPENDENT ØR INCØNSISTENT"
      STØP
    2 X=(C1*B2-C2*B1)/DEL
      Y=(A1*C2-A2*C1)/DEL
      PRINT,"X =",X
      PRINT,"Y =",Y
      STØP
      END

INPUT:00010
 ?2.,-3.,5.,2.,5.,13.

X =        4.0000
Y =        1.0000
STØP
```

```
C     PRØBLEM 4-8
      READ,N,T
      GØ TØ(10,20,30,40,50),N
   10 R=53.34
      XK=1.4
      GØ TØ 1
   20 R=35.1
      XK=1.285
      GØ TØ 1
   30 R=386.
      XK=1.66
      GØ TØ 1
   40 R=55.15
      XK=1.4
      GØ TØ 1
   50 R=48.28
      XK=1.395
    1 C=SQRT(32.16*XK*R*(T+460.))
      PRINT,"SØNIC VELØCITY =",C
      STØP
      END

INPUT:00010
 ?3,275.

SØNIC VELØCITY =     3891.7899
STØP
```

```
C     PRØBLEM 4-13
      READ,R,XL,C,ØMEGA
      A=R
      IF(C)51,50,51
   50 B=ØMEGA*XL
      GØ TØ 52
   51 B=ØMEGA*XL-1./(ØMEGA*C)
   52 R=SQRT(A**2+B**2)
      IF(A) 2,3,4
    3 IF(B) 5,6,7
    5 BETA=270.
      GØ TØ 1
    6 PRINT 20
   20 FØRMAT(22HDEGENERATE CASE:A=B=0.)
      STØP
    7 BETA=90.
      GØ TØ 1
    2 IF(B) 8,9,10
    8 BETA=180.+57.296*ATAN(ABS(B/A))
      GØ TØ 1
    9 BETA=180.
      GØ TØ 1
   10 BETA=180.-57.296*ATAN(ABS(B/A))
      GØ TØ 1
    4 IF(B) 11,12,13
   11 BETA=360.-57.296*ATAN(ABS(B/A))
      GØ TØ 1
   12 BETA=0.
      GØ TØ 1
   13 BETA=57.296*ATAN(ABS(B/A))
    1 PRINT 23,R,BETA
   23 FØRMAT(14HRADIUS VECTØR=F8.3,7H ANGLE=F8.3,8H DEGREES/)
      STØP
      END

INPUT:00010
 ?100.,0.01,0.,5000.

RADIUS VECTØR= 111.803 ANGLE=  26.565 DEGREES

STØP
```

CHAPTER 5

```
C     PRØBLEM 5-1
      READ,N1,N2,N3
      PRINT 50
      DØ 1 I=N1,N2,N3
      F=I
      C=5.*(F-32.)/9.
      R=F+459.67
      XK=C+273.15
    1 PRINT 51,F,C,R,XK
      STØP
   50 FØRMAT(6X1HF9X1HC9X1HR9X1HK/)
   51 FØRMAT(4F10.2)
      END

INPUT:00010
 ?450,950,50

      F         C         R         K

   450.00    232.22    909.67    505.37
   500.00    260.00    959.67    533.15
   550.00    287.78   1009.67    560.93
   600.00    315.56   1059.67    588.71
   650.00    343.33   1109.67    616.48
   700.00    371.11   1159.67    644.26
   750.00    398.89   1209.67    672.04
   800.00    426.67   1259.67    699.82
   850.00    454.44   1309.67    727.59
   900.00    482.22   1359.67    755.37
   950.00    510.00   1409.67    783.15
STØP
```

```
C        PRØBLEM 5-7
         READ,N1,N2,N3
         DØ 1 I=N1,N2,N3
         IF(I-90)3,2,3
       2 TAN=.1E+76
         GØ TØ 1
       3 XI=FLØAT(I)*3.1415927/180.
         TAN=SIN(XI)/CØS(XI)
       1 PRINT,I,TAN
         STØP
         END
```

```
INPUT:00010
 ?65,85,5

       65        2.1445
       70        2.7475
       75        3.7321
       80        5.6713
       85       11.4301
STØP
```

```
C        PRØBLEM 5-12
         READ,N,N1,N2,N3
         GØ TØ (10,20,30,40,50),N
      10 R=53.34
         XK=1.4
         GØ TØ 1
      20 R=35.10
         XK=1.285
         GØ TØ 1
      30 R=386.
         XK=1.66
         GØ TØ 1
      40 R=55.15
         XK=1.4
         GØ TØ 1
      50 R=48.28
         XK=1.395
       1 PRINT,"        T              VELØCITY"
         DØ 2 I=N1,N2,N3
         T=FLØAT(I)
         C=SQRT(32.16*XK*R*(T+460.))
       2 PRINT,T,C
         STØP
         END
```

```
INPUT:00010
 ?2,40,135,5

        T           VELØCITY
      40.0000        851.6245
      45.0000        855.8720
      50.0000        860.0986
      55.0000        864.3045
      60.0000        868.4900
      65.0000        872.6554
      70.0000        876.8011
      75.0000        880.9272
      80.0000        885.0341
      85.0000        889.1221
      90.0000        893.1913
      95.0000        897.2421
     100.0000        901.2746
     105.0000        905.2893
     110.0000        909.2861
     115.0000        913.2655
     120.0000        917.2277
     125.0000        921.1727
     130.0000        925.1010
     135.0000        929.0126
STØP
```

```
C       PRØBLEM 5-13
        READ,V,XL,C,EPS
        PRINT,"         W              CURRENT"
        DØ 1 J=4,7
        X=10.**J
        DØ 1 K=1,7,3
        W=X*FLØAT(K)
        REACT=W*XL-1./(W*C)
        IF(ABS(REACT)-EPS) 2,2,3
      2 PRINT,"CURRENT IS INFINITE"
        GØ TØ 1
      3 XI=V/ABS(REACT)
        PRINT,W,XI
      1 CØNTINUE
        STØP
        END

INPUT:00010
 ?10.,.001,1.E-09,1.E-06

        W            CURRENT
   10000.0000         0.0001
   40000.0000         0.0004
   70000.0000         0.0007
  100000.0000         0.0010
  400000.0000         0.0048
  700000.0000         0.0137
CURRENT IS INFINITE
   4.0000E+06         0.0027
   7.0000E+06         0.0015
   1.0000E+07         0.0010
   4.0000E+07         0.0003
   7.0000E+07         0.0001
STØP
```

CHAPTER 6

```
C       PRØBLEM 6-3
        X=1.
        XØLD=0.
        PRINT," NUMBER          RATIØ"
        PRINT 7,X
      5 X=X+XØLD
        XØLD=X-XØLD
        RAT=X/XØLD
        PRINT 6, X,RAT
        IF(ABS(RAT-1.618)-0.0001)4,4,5
      4 STØP
      6 FØRMAT(F8.0,F15.7)
      7 FØRMAT(F8.0)
        END

 NUMBER       RATIØ
      1.
      1.      1.0000000
      2.      2.0000000
      3.      1.5000000
      5.      1.6666667
      8.      1.6000000
     13.      1.6250000
     21.      1.6153846
     34.      1.6190476
     55.      1.6176471
     89.      1.6181818
    144.      1.6179775
STØP
```

```
C       PRØBLEM 6-6
        PRØD=1.
        SUM=1.1
        DØ 20 I=1,1000
        Y=I
        PRØD=PRØD*SUM
        GMEAN=PRØD**(1./Y)
        IF(GMEAN-1.9)20,30,30
     20 SUM=SUM+0.3
        PRINT,"THE TEST FAILED"
        STØP
     30 PRINT,"THE GEØMETRIC MEAN ØF",I,"  TERMS=",GMEAN
        PRINT,
        STØP
        END

THE GEØMETRIC MEAN ØF          7   TERMS=         1.9043

STØP
```

```
C       PRØBLEM 6-10
        C=3.E08
        Q=1.6E-19
        RMASS=9.11E-31
        RAT=Q/RMASS
        DØ 20 I=3000,8000,10
        V=I
        U=SQRT(2.*RAT*V)
        UPRYM=C*SQRT(1.-1./(1.+U**2/(2.*C**2))**2)
        IF(0.99-UPRYM/U) 20,30,30
     20 CØNTINUE
        PRINT,"DID NØT CØNVERGE"
        STØP
     30 PRINT,"THE VØLTAGE AT WHICH U'/U = 0.99 IS",V
        PRINT,
        STØP
        END

THE VØLTAGE AT WHICH U'/U = 0.99 IS     6930.0000

STØP
```

```
C       PRØBLEM 6-14
        READ,P
        DØ 10 I=100,1000
        X=FLØAT(I)/100.
        IF(X-1.)1,1,2
      1 TØTAL=7.+18.8*X
        PCTCØ2=300.*X/TØTAL
        PCTCØ=300.*(1.-X)/TØTAL
        PCTØ2=0.
        GØ TØ 3
      2 TØTAL=2.+23.8*X
        PCTCØ2=300./TØTAL
        PCTCØ=0.
        PCTØ2=500.*(X-1.)/TØTAL
      3 PCTN2=1880.*X/TØTAL
        PCTH2Ø=400./TØTAL
        IF(P-PCTCØ2)10,10,20
     10 CØNTINUE
     20 PRINT 100,I,PCTCØ2
        STØP
    100 FØRMAT(4HWITHI5,28H  PERCENT ØF THEØRETICAL AIR/
&        23HTHE CØ2 CØNTENT WILL BEF7.2,9H  PERCENT/)
        END

INPUT:00010
 ?5.

WITH  244  PERCENT ØF THEØRETICAL AIR
THE CØ2 CØNTENT WILL BE    4.99   PERCENT

STØP
```

CHAPTER 7

```
C      PRØBLEM 7-1
       DIMENSIØN X(10)
       READ,N,(X(I),I=1,N)
       SUMSQ=0.
       DØ 3 I=1,N
     3 SUMSQ=SUMSQ+X(I)**2
       RMS=SQRT(SUMSQ)
       PRINT 5
     5 FØRMAT(2H N,8X3HRMS/)
       PRINT 6,N,RMS
     6 FØRMAT(I3,4XF8.2)
       STØP
       END

INPUT:00020
 ?3,2.,.9,.5

 N          RMS

  3          2.25
STØP
```

```
C      PRØBLEM 7-3
       DIMENSIØN A(10,10)
       READ,NRØ,MCØL
       DØ 1 I=1,NRØ
     1 READ,(A(I,J),J=1,MCØL)
       BIG=A(1,1)
       DØ 10 I=1,NRØ
       DØ 10 J=1,MCØL
       IF(BIG-A(I,J)) 11,10,10
    11 BIG=A(I,J)
       K=I
       L=J
    10 CØNTINUE
       PRINT 7,K,L,BIG
     7 FØRMAT(2HA(I3,1H,I3,2H)=F8.2/)
       STØP
       END

INPUT:00020
 ?3,3

INPUT:00040
 ?16.,8.,-9.

INPUT:00040
 ?22.,96.,48.

INPUT:00040
 ?96.,51.,8.

A(  2,  2)=    96.00

STØP
```

```
00001C      PRØBLEM 7-6
00010       DIMENSIØN A(10,10)
00020       READ,NR,NC
00030       DØ 10 I=1,NR
00040    10 READ,(A(I,J),J=1,NC)
00050       DØ 1 I=1,NR
00060       DØ 1 J=I,NC
00070       TEMP=A(J,I)
00080       A(J,I)=A(I,J)
00090     1 A(I,J)=TEMP
00100       DØ 20 I=1,NR
00110    20 PRINT 7,(A(I,J),J=1,NC)
00120     7 FØRMAT(6E11.3)
00130       STØP
00140       END

INPUT:00020
 ?3,3

INPUT:00040
 ?5.,9.,4.

INPUT:00040
 ?2.,7.,3.

INPUT:00040
 ?8.,1.,6.

  0.500E+01  0.200E+01  0.800E+01
  0.900E+01  0.700E+01  0.100E+01
  0.400E+01  0.300E+01  0.600E+01
STØP
```

```
00001C      PRØBLEM 7-7
00010       DIMENSIØN A(10,20)
00020       READ,N,((A(I,J),J=1,N),I=1,N)
00030       K=2*N
00040       DØ 1 I=1,N
00050       L=N+I
00060       DØ 1 J=L,K
00070       IF(J-N-I)3,2,3
00080     2 A(I,J)=1.
00090       GØ TØ 1
00100     3 A(I,J)=0.
00110     1 CØNTINUE
00120       DØ 4 I=1,N
00130     4 PRINT 5,(A(I,J),J=1,K)
00140     5 FØRMAT(10F6.1)
00150       STØP
00160       END

INPUT:00020
 ?3,5.,9.,4.,2.,7.,3.,8.,1.,6.

   5.0   9.0   4.0   1.0   0.0   0.0
   2.0   7.0   3.0   0.0   1.0   0.0
   8.0   1.0   6.0   0.0   0.0   1.0
STØP
```

```
C     PRØBLEM 7-11
      DIMENSIØN Y(24),A(10),B(10)
      READ,NDATA,NHARM
      READ,(Y(I),I=1,NDATA)
      TAØ=0.
      DØ 10 I=1,NDATA
   10 TAØ=TAØ+Y(I)
      XN=NDATA
      XH=NDATA/2
      AØ=TAØ/XN
      PI=3.14159
      PRINT 98,AØ
   98 FØRMAT("AØ=",F8.2)
      DØ 20 N=1,NHARM
      XN=N
      TEMPA=0.
      TEMPB=0.
      DØ 2 J=1,NDATA
      XJ=J
      TEMPA=TEMPA+Y(J)*CØS(XN*XJ*PI/XH)
    2 TEMPB=TEMPB+Y(J)*SIN(XN*XJ*PI/XH)
      A(N)=TEMPA/XH
      B(N)=TEMPB/XH
   20 PRINT 99,N,A(N),N,B(N)
   99 FØRMAT(2HA(I3,2H)=E12.3,4X2HB(I3,2H)=E12.3)
      STØP
      END

INPUT:00020
 ?24,4

INPUT:00030
 ?2.59,5.,7.07,8.66,9.66,10.,9.66,8.66,7.07,5.,2.59,13*0.


AØ=      3.16
A(  1)=    0.663E-05      B(  1)=    0.500E+01
A(  2)=    -.216E+01      B(  2)=    0.573E-05
A(  3)=    0.291E-10      B(  3)=    -.332E-05
A(  4)=    -.463E+00      B(  4)=    0.245E-05
STØP
```

CHAPTER 8

```
C     PRØBLEM 8-2
      CØMPLEX Y,Z
      READ,Z
      Y=Z*CEXP((0.,1.)*Z)
      PRINT,Y
      STØP
      END
INPUT:00020
 ?4.,-3.

     -98.1175       -21.4168
STØP
```

```
C     PRØBLEM 8-3
      CØMPLEX A1,B1,C1,A2,B2,C2,DEL,X,Y
      READ,A1,B1,C1,A2,B2,C2
      DEL=A1*B2-A2*B1
      IF(CABS(DEL))2,1,2
    1 PRINT,"EQUATIØNS ARE DEPENDENT ØR INCØNSISTENT"
      STØP
    2 X=(C1*B2-C2*B1)/DEL
      Y=(A1*C2-A2*C1)/DEL
      PRINT,"X =",X
      PRINT,"Y =",Y
      STØP
      END
INPUT:00020
 ?2.,1,5.,-2.,8.,3.,2.,0.,4.,-2.,8.,-1.

X =         4.2500        0.0000
Y =         0.0000        -.2500
STØP
```

```
C     PRØBLEM 8-5
      CØMPLEX ZTR,Z1,Z2,Z3,YA,YB,YC,YD,YE,YF
      READ,YA,YB
      READ,YC,YD
      READ,YE,YF
      Z1=1./(YA+YB)
      Z2=1./(YC+YD)
      Z3=1./(YE+YF)
      ZTR=Z1*Z3/(Z1+Z2+Z3)
      PRINT 1,ZTR
    1 FØRMAT("TRANSFER IMPEDANCE=",2F15.6//)
      STØP
      END

INPUT:00020
 ?1.E-04,0.,.2E-04,0.

INPUT:00030
 ?0.,.188E-04,0.,0.

INPUT:00040
 ?.2E-04,0.,0.,0.

TRANSFER IMPEDANCE=     3900.050948     3556.277460


STØP
```

CHAPTER 9

```
C     PRØBLEM 9-1
      LØGICAL A,B,Y
      DIMENSIØN A(4),B(4),Y(4)
      DATA A/F,F,T,T/
      DATA B/F,T,F,T/
      DØ 1 I=1,4
    1 Y(I)=(A(I).AND..NØT.B(I)).ØR.(B(I).AND..NØT.A(I))
      PRINT,"    A    B    Y"
      DØ 2 I=1,4
    2 PRINT 3,A(I),B(I),Y(I)
    3 FØRMAT(4L5)
      STØP
      END

    A    B    Y
    F    F    F
    F    T    T
    T    F    T
    T    T    F
STØP
```

```
00001C     PRØBLEM 9-3
00010      CØMPLEX Z,ZE,I,V,ZTØT
00020      DIMENSIØN Z(29),ZE(29),I(29),V(20)
00030      READ,V(1)
00040      READ,N,(Z(J),J=1,N)
00050      DØ 2 J=1,N
00060    2 ZE(J)=Z(J)
00070      MAX=N-2
00080      DØ 4 J=2,MAX,2
00090      K=N-J+1
00100      L=N-J+2
00110      M=N-J
00120    4 ZE(M)=ZE(M)*(ZE(K)+ZE(L))/(ZE(M)+ZE(K)+ZE(L))
00130      ZTØT=Z(1)+ZE(2)
00140      I(1)=V(1)/ZTØT
00150      V(2)=V(1)-I(1)*Z(1)
00160      I(2)=V(2)/Z(2)
00170      DØ 6 J=4,N,2
00180      I(J-1)=I(J-3)-I(J-2)
00190      V(J)=V(J-2)-I(J-1)*Z(J-1)
00195      IF(J.EQ.N) GØ TØ 10
00200    6 I(J)=V(J)/Z(J)
00210   11 PRINT 7
00220      PRINT,(I(J),J=1,N)
00240      PRINT 8
00250      PRINT,(V(J),J=2,N,2)
00260      STØP
00270   10 IF(CABS(Z(N)).GT.0.) GØ TØ 6
00280      I(N)=I(N-1)
00290      V(N)=(0.,0.)
00300      GØ TØ 11
00310    7 FØRMAT(15HBRANCH CURRENTS)
00320    8 FØRMAT(13HNØDE VØLTAGES)
00330      END

INPUT:00030
 ?4.,0.

INPUT:00040
 ?4,0.,2.,0.,-2.,1.,0.,0.,1.

BRANCH CURRENTS
      1.0000        -1.0000         1.0000         1.0000         0.0000
     -2.0000         0.0000        -2.0000
NØDE VØLTAGES
      2.0000        -2.0000         2.0000         0.0000
STØP
```

CHAPTER 10

```
C       PRØBLEM 10-1
        ANG(A,B)=57.296*ATAN(ABS(B/A))
     30 READ,A,B
        R=SQRT(A**2+B**2)
        IF(A) 2,3,4
      3 IF(B) 5,6,7
      5 BETA=270.
        GØ TØ 1
      6 PRINT 20
     20 FØRMAT(22HDEGENERATE CASE:A=B=0.)
        GØ TØ 30
      7 BETA=90.
        GØ TØ 1
      2 IF(B) 8,9,10
      8 BETA=180.+ANG(A,B)
        GØ TØ 1
      9 BETA=180.
        GØ TØ 1
     10 BETA=180.-ANG(A,B)
        GØ TØ 1
      4 IF(B) 11,12,13
     11 BETA=360.-ANG(A,B)
        GØ TØ 1
     12 BETA=0.
        GØ TØ 1
     13 BETA=ANG(A,B)
      1 PRINT 23,R,BETA
     23 FØRMAT(14HRADIUS VECTØR=F8.3,7H ANGLE=F8.3,8H DEGREES/)
        GØ TØ 30
        END

INPUT:00020
 ?-5.,2.

RADIUS VECTØR=   5.385 ANGLE= 158.199 DEGREES

INPUT:00020
 ?4.,-3.

RADIUS VECTØR=   5.000 ANGLE= 323.130 DEGREES

INPUT:00020
 ?STØP
```

```
00001C      PRØBLEM 10-5
00010       REAL LAMBDA
00020       READ,LAMBDA,P
00030       SERIES=1.
00040       DØ 4 N=1,1000
00050       SERIES=SERIES+LAMBDA**N/FACT(N)
00060       IF(SERIES-P*EXP(LAMBDA)) 4,5,5
00070     4 CØNTINUE
00080       PRINT,"TEST FAILED"
00090       STØP
00100     5 PRINT 27,N
00110    27 FØRMAT("N= ",I5//)
00120    99 STØP
00130       END
00140       FUNCTIØN FACT(J)
00150       IF(J-1)10,10,11
00160    10 FACT=1.
00170       RETURN
00180    11 X=J
00190       FACT=1.
00200     8 FACT=FACT*X
00210       X=X-1.
00220       IF(X.EQ.0.) RETURN
00230       GØ TØ 8
00240       END

INPUT:00020
 ?1.,.995

N=       4


STØP
```

```
00001C      PRØBLEM 10-6
00010       CØMMØN A(10),B(10),N,DØT
00020       READ,N
00030       READ,(A(I),I=1,N)
00040       READ,(B(I),I=1,N)
00050       CALL SCAPRØ
00060       IF(DØT.LE.1.E-07) GØ TØ 10
00070       PRINT 7,DØT
00080     7 FØRMAT("DØT PRØDUCT=",E15.8//)
00090       STØP
00100    10 PRINT,"A AND B ARE ØRTHØGØNAL"
00110       STØP
00120       END
00130       SUBRØUTINE SCAPRØ
00140       CØMMØN A(10),B(10),N,DØT
00150       DØT=0.
00160       DØ 5 I=1,N
00170     5 DØT=DØT+A(I)*B(I)
00180       RETURN
00190       END

INPUT:00020
 ?5

INPUT:00030
 ?1,3,1,0,-6

INPUT:00040
 ?4,1,4,2,-3

DØT PRØDUCT= 0.29000000E+02


STØP
```

CHAPTER 11

```
C     PRØBLEM 11-1
      DIMENSIØN B(5)
      DATA B/3HAIR,3HCØ2,3HHE ,3HN2 ,3HO2 /
      READ 60,GAS,N1,N2,N3
   60 FØRMAT(A3,3I3)
      DØ 100 N=1,5
      IF(GAS.EQ.B(N))GØ TØ 70
  100 CØNTINUE
   70 GØ TØ(10,20,30,40,50),N
   10 R=53.34
      XK=1.4
      GØ TØ 1
   20 R=35.10
      XK=1.285
      GØ TØ 1
   30 R=386.
      XK=1.66
      GØ TØ 1
   40 R=55.15
      XK=1.4
      GØ TØ 1
   50 R=48.28
      XK=1.395
    1 PRINT,"        T              VELØCITY"
      DØ 2 I=N1,N2,N3
      T=FLØAT(I)
      C=SQRT(32.16*XK*R*(T+460.))
    2 PRINT,T,C
      STØP
      END
```

```
INPUT:00030
 ?CØ2  40135  5

        T              VELØCITY
      40.0000          851.6245
      45.0000          855.8720
      50.0000          860.0986
      55.0000          864.3045
      60.0000          868.4900
      65.0000          872.6554
      70.0000          876.8011
      75.0000          880.9272
      80.0000          885.0341
      85.0000          889.1221
      90.0000          893.1913
      95.0000          897.2421
     100.0000          901.2746
     105.0000          905.2893
     110.0000          909.2861
     115.0000          913.2655
     120.0000          917.2277
     125.0000          921.1727
     130.0000          925.1010
     135.0000          929.0126
STØP
```

```
C       PRØBLEM 11-3
C       NØTE: THE FIRST TWØ CØLUMNS ØF THE ARRAY ARE USED
C        TØ STØRE THE MATERIAL NAME.  THE REMAINING TEN
C        CØLUMNS ARE USED FØR THE MATERIAL PRØPERTIES.
C        "K" IS THE NUMBER ØF THE CØLUMN TØ BE SØRTED.
        CØMMØN A(50,12)
        READ,M,N,K
        READ 10
     10 FØRMAT(12X50H12345678901234567890123456789012345678901234567890
&1234567890)
        IF(N.LE.7)GØ TØ 25
        READ 20
     20 FØRMAT(12X50H12345678901234567890123456789012345678901234567890
&1234567890)
     25 READ 30,((A(I,1),A(I,2)),I=1,M)
     30 FØRMAT(10A6)
        DØ 1 J=3,N
      1 READ,(A(I,J),I=1,M)
        MLAST=M-1
        DØ 2 KØUNT=1,M
        DØ 2 I=1,MLAST
        IF(A(I,K).LE.A(I+1,K))GØ TØ 2
        CALL SWAP(I,N)
      2 CØNTINUE
        PRINT 10
        DØ 3 I=1,M
      3 PRINT 40,(A(I,J),J=1,N)
     40 FØRMAT(2A6,5F10.3)
        IF(N.LE.7)STØP
        PRINT,
        PRINT 20
        DØ 4 I=1,M
      4 PRINT 30,A(I,1),A(I,2),(A(I,J),J=8,N)
        STØP
        END
        SUBRØUTINE SWAP(I,N)
        CØMMØN A(50,12)
        DØ 1 J=1,N
        TEMP=A(I,J)
        A(I,J)=A(I+1,J)
      1 A(I+1,J)=TEMP
        RETURN
        END
```

```
INPUT:00020
 ?13,6,3

INPUT:00030
 ?               SP.WT.       E          G       TAU

INPUT:00100
 ?GRAY C.I.    STL (1020)  STL (4130)  BRØNZE      STL (STNLS)

 ?MØNEL        INCØNEL     MG ALLØY    AL ALLØY    TI ALLØY

 ?PHENØL       PØLYSTYRENE NYLØN

INPUT:00130
 ?.26,.283,.283,.32,.273,.319,.307,.066,.1,.163,.048,.038,.041

INPUT:00130
 ?13.,29.8,29.8,16.,29.8,26.,31.,6.5,10.6,16.,.71,.5,.4

INPUT:00130
 ?5.5,11.5,11.5,6.,11.5,9.5,11.,2.4,4.,6.,.28,.18,.14

INPUT:00130
 ?3000.,6000.,9000.,5000.,6500.,3700.,4100.,4000.,4000.,16000.,

 ?900.,800.,1200.

                SP.WT.       E          G       TAU
PØLYSTYRENE      0.038     0.500      0.180    800.000
NYLØN            0.041     0.400      0.140   1200.000
PHENØL           0.048     0.710      0.280    900.000
MG ALLØY         0.066     6.500      2.400   4000.000
AL ALLØY         0.100    10.600      4.000   4000.000
TI ALLØY         0.163    16.000      6.000  16000.000
GRAY C.I.        0.260    13.000      5.500   3000.000
STL (STNLS)      0.273    29.800     11.500   6500.000
STL (1020)       0.283    29.800     11.500   6000.000
STL (4130)       0.283    29.800     11.500   9000.000
INCØNEL          0.307    31.000     11.000   4100.000
MØNEL            0.319    26.000      9.500   3700.000
BRØNZE           0.320    16.000      6.000   5000.000
STØP
```

CHAPTER 12

```
C     PRØBLEM 12-3
      DIMENSIØN T(10),Y(10)
      CØMMØN A(10,11),MAX2,MAX3
      READ,M,N
      READ,(T(I),I=1,N)
      READ,(Y(I),I=1,N)
      MAX1=2*M
      MAX2=M+1
      MAX3=M+2
      DØ 13 I=1,MAX2
      DØ 13 J=1,MAX3
   13 A(I,J)=0.
      DØ 30 K=1,MAX2
      DØ 30 J=1,MAX2
      KK=K+J-2
      DØ 30 I=1,N
   30 A(K,J)=A(K,J)+T(I)**KK
      A(1,1)=N
      DØ 31 J=1,MAX2
      KJ=J-1
      DØ 31 I=1,N
   31 A(J,MAX3)=A(J,MAX3)+Y(I)*T(I)**KJ
      CALL SØLVE
      PRINT 1,M
    1 FØRMAT("CØEFFICIENTS FØR FIT ØF ØRDER",I3)
      DØ 11 I=1,MAX2
   11 PRINT 12,I,A(I,MAX3)
   12 FØRMAT(1HX,I3,1H=E12.3)
      STØP
      END
      SUBRØUTINE SØLVE
      CØMMØN A(10,11),MAX2,MAX3
      K=1
      DØ 70 IRØW=1,3
      K=K+1
      ISWAP=IRØW
      DØ 50 IN=K,3
      IF(ABS(A(ISWAP,IRØW))-ABS(A(IN,IRØW)).GE.0.)GØ TØ 50
      ISWAP=IN
   50 CØNTINUE
      IF(ISWAP.EQ.IRØW) GØ TØ 99
      DØ 92 J=1,4
      TEMP=A(IRØW,J)
      A(IRØW,J)=A(ISWAP,J)
   92 A(ISWAP,J)=TEMP
   99 PIVØT=A(IRØW,IRØW)
      IF(ABS(PIVØT)-1.E-06) 27,27,28
   27 PRINT,"PIVØT IS TØØ SMALL. VALUE ØF PIVØT=",PIVØT
      STØP
   28 DØ 10 J=1,4
   10 A(IRØW,J)=A(IRØW,J)/PIVØT
```

```
      DØ 20 I=1,MAX2
      IF(I.EQ.IRØW) GØ TØ 20
      RATIØ=A(I,IRØW)
      DØ 18 J=1,4
   18 A(I,J)=A(I,J)-A(IRØW,J)*RATIØ
   20 CØNTINUE
   70 CØNTINUE
      RETURN
      END
INPUT:00030
 ?2,6

INPUT:00040
 ?.2,.5,1.,1.5,2.,3.

INPUT:00050
 ?1.9,4.,11.5,21.,36.,75.

CØEFFICIENTS FØR FIT ØF ØRDER  2
X  1=   0.101E+01
X  2=   0.271E+01
X  3=   0.732E+01
STØP
```

```
C     PRØBLEM 12-5
      DIMENSIØN A(10,10),B(10,10),C(10,10)
      READ,N,((A(I,J),J=1,N),I=1,N)
      KØUNT=1
      DØ 1 I=1,N
      DØ 1 J=1,N
    1 B(I,J)=A(I,J)
    3 INDEX=0
      DØ 6 I=1,N
      DØ 6 J=1,N
      C(I,J)=0.
      DØ 6 K=1,N
    6 C(I,J)=C(I,J)+A(I,K)*B(K,J)
      KØUNT=KØUNT+1
      IF(KØUNT.EQ.50) GØ TØ 31
      DØ 8 I=1,N
      DØ 8 J=1,N
      IF(ABS(A(I,J)-C(I,J)).GT.1.E-04) INDEX=1
    8 CØNTINUE
      IF(INDEX.EQ.0) GØ TØ 31
      DØ 99 I=1,N
      DØ 99 J=1,N
   99 A(I,J)=C(I,J)
      GØ TØ 3
   31 DØ 23 I=1,N
   23 PRINT 7,(C(I,J),J=1,N)
    7 FØRMAT(6E12.4)
      STØP
      END

INPUT:00020
 ?2,.8,.2,.5,.5

  0.7143E+00  0.2857E+00
  0.7143E+00  0.2857E+00
STØP
```

```
C    PRØBLEM 12-7
     CØMMØN N,A(20,21),NP1
     CØMPLEX A
     READ,N
     NP1=N+1
     READ,((A(I,J),J=1,NP1),I=1,N)
     CALL SØLVE
     PRINT,"SØLUTIØN ØF THE SYSTEM ØF EQUATIØNS"
     DØ 11 I=1,N
  11 PRINT 12,I,A(I,NP1)
  12 FØRMAT(1HXI3,1H=2E12.3)
     STØP
     END
     SUBRØUTINE SØLVE
     CØMMØN N,A(20,21),NP1
     CØMPLEX PIVØT,RATIØ,TEMP,A
     K=1
     DØ 70 IRØW=1,N
     K=K+1
     ISWAP=IRØW
     DØ 50 IN=K,N
     IF(CABS(A(ISWAP,IRØW))-CABS(A(IN,IRØW)).GE.0.)GØ TØ 50
     ISWAP=IN
  50 CØNTINUE
     IF(ISWAP.EQ.IRØW) GØ TØ 99
     DØ 92 J=1,NP1
     TEMP=A(IRØW,J)
     A(IRØW,J)=A(ISWAP,J)
  92 A(ISWAP,J)=TEMP
  99 PIVØT=A(IRØW,IRØW)
     IF(CABS(PIVØT)-1.E-06) 27,27,28
  27 PRINT,"PIVØT IS TØØ SMALL. VALUE ØF PIVØT=",PIVØT
     STØP
  28 DØ 10 J=1,NP1
  10 A(IRØW,J)=A(IRØW,J)/PIVØT
     DØ 20 I=1,N
     IF(I.EQ.IRØW) GØ TØ 20
     RATIØ=A(I,IRØW)
     DØ 18 J=1,NP1
  18 A(I,J)=A(I,J)-A(IRØW,J)*RATIØ
  20 CØNTINUE
  70 CØNTINUE
     RETURN
     END
INPUT:00030
 ?2

INPUT:00050
 ?0.,0.,0.,2.,10.,0.,0.,2.,1.,-1.,0.,0.

SØLUTIØN ØF THE SYSTEM ØF EQUATIØNS
X  1=   0.250E+01   -.250E+01
X  2=   0.000E+00   -.500E+01
STØP
```

Index